MICROWAVE THEORY

AND

APPLICATIONS

View
Sept 81
Aug I

PRENTICE-HALL, INC.
Englewood Cliffs, New Jersey

MICROWAVE THEORY

AND

APPLICATIONS

STEPHEN F. ADAM

Microwave Division, Hewlett-Packard Company
Instructor, Foothill College
Los Altos Hills, California

HEWLETT  PACKARD COMPANY

Palo Alto, California

13-581488-X

Library of Congress Catalogue Card No.: 76-82902

Current printing (last digit):
10 9 8

Printed in the United States of America

PRENTICE-HALL INTERNATIONAL, INC., *London*
PRENTICE-HALL OF AUSTRALIA, PTY. LTD., *Sydney*
PRENTICE-HALL OF CANADA, LTD., *Toronto*
PRENTICE-HALL OF INDIA PRIVATE LTD., *New Delhi*
PRENTICE-HALL OF JAPAN, INC., *Tokyo*

PREFACE

A thorough understanding of the physical aspects of microwave electronics can be gained through a rigorous mathematical analysis but such a study is impractical or impossible for many people working in the field. This book has been written to supply that thorough understanding by examining transmission theory and applications with a minimum of mathematics. It may be helpful to engineering students who up to now have been given much field theory but who have not gained an understanding of the physical phenomena. Product designers and others working in associated service groups developing microwave equipment may find that this treatment will give them a better comprehension of the principles underlying their tasks. Laboratory and production technicians who desire more knowledge about their work can learn from this material because it relies on physical rather than mathematical explanations.

The book was developed by the author during seven years of teaching experience at Foothill College in Los Altos Hills, California. Originally, Theodore Moreno's book, *Microwave Transmission Design Data,* was used to give the basic understanding of transmission line theory and physical phenomena required for further study of advanced techniques. Students in the evening courses included MSEEs with no microwave experience, mechanical engineers involved in microwave design work, and electronic technicians. The approach taken has fulfilled the diverse needs of this group. The book is based on this teaching experience.

Transmission line theory is covered first; then the principles of microwave measurements are discussed in Chapter 3 to enable the reader to bridge the gap between theory and practice and to understand the last three chapters. Chapter 4 deals with laboratory sources of microwave signal generation, covering the problems faced in the early stages of microwave technology as well as present state-of-the-art techniques. Chapter 5 explains signal analysis. It covers power measurements, how to obtain frequency information about a signal, and, equally important, how to use spectrum analysis to determine the purity and the modulation of signals. Chapter 6 explains network analysis, the technique whereby devices connected between signal sources and indicators are analyzed. Reflection and transmission measurements are dealt with from the old customary approach all the way to modern, fully characterized, even computer-controlled, automatic network analysis using scattering *s* parameters. Network analysis is covered not only in the frequency domain but through time domain reflectometry as well.

The questions that conclude various sections throughout the text are designed to stress significant points and should be answered. They will help the student assess his understanding of the material.

The laboratory manual accompanying this text book will provide sound laboratory experience and should be used in conjunction with the text. The experiments are designed to complement and emphasize the material covered in reading and class discussions.

ACKNOWLEDGMENTS

Many contributions are necessary to complete a book covering so wide a subject as this one. Much information was taken from existing publications —books, professional journals, company documentation—and references to them appear throughout the text. The following publishers gave permission to use copyrighted illustrative material: The Institute of Electrical and Electronics Engineers, Inc., McGraw-Hill Book Company, Microwave Journal, Lenkurt Electric Co., Inc., Microwave Development Laboratories, Inc., and Sperry Rand Corporation Gyroscope Division.

The material was reviewed by specialists within Hewlett-Packard Company; many helpful suggestions were given by Richard Anderson (network analysis), Rod Carlson (signal analysis), Harley Halverson and Del Hanson (signal generation). Bob Mangold also contributed to the work in its early development stages.

The work was supported by the Hewlett-Packard Microwave Division and the entire microwave engineering staff deserves credit for help and contributions to this effort. Preliminary production was handled by Virginia Reynolds, editor, and Mee Chow, who did the artwork. Actual production was coordinated by Ed Lauffenburger, with the brief assistance of Bruce Hanson. Without the effort of all these people, and many others who have not been mentioned, the book would not have been possible.

Palo Alto, California   Stephen Adam

CONTENTS

1 INTRODUCTION

In the early stages of the development of modern radio techniques, only the low end of the frequency spectrum was utilized because, not only was there a limited number of users of these techniques, but the number of applications was limited as well. As members of the scientific and technical community became more cognizant of the field of radio and as applications of radio waves became more extensive, the lower range of the RF spectrum became overpopulated, resulting in confusion and conflict in the areas of transmission and reception. To clear up this confusion, international conventions had to be formed to allocate the usable frequency spectrum among the multitude of national and international research, industrial, commercial, and government services employing it.

As techniques and components became more sophisticated, a wider bandwidth was absolutely essential for full utilization of these improvements; the obvious direction for expansion was into the higher reaches of the frequency spectrum. This was done, but the state of the art kept advancing at an ever-increasing rate. As a result, by the start of World War II the bandwidth requirements of sophisticated communication and radar equipment were in the microwave regions.

The increase in frequency meant a decrease in wavelength, as well as an increase in problems because the geometry of the components used in equipment became comparable to the wavelengths used. Transmission line theory provides the solution to these problems in microwave technology.

The low-frequency boundary of the microwave range has not been set officially, but it is usually considered to be the point at which lower-frequency techniques and lumped-circuit elements cannot be used efficiently—generally about 1 gigahertz. The high-frequency end of the range is considered to be that point at which radio waves overlap with light waves, that is, where optical techniques must be used to manipulate signal transmission. Transmission line theory, distributed line analysis, and electromagnetic field theory are used to analyze circuits operating in this range.

Microwaves are widely used in modern technology. One major field is the television industry. Transcontinental transmission of program material is in the microwave frequency range, using complex transmitter, relay, and receiver networks throughout the country. Local stations convert the signals to lower frequencies for transmission to individual sets. On a local level, microwaves are used from the studio to transmitter locations.

Microwaves are used in national and local security applications, such as early-warning radar, missile guidance systems, and Doppler radars, to detect and control the speed of vehicles.

Microwave communication networks and relay stations are used commercially for routine multichannel communication transmissions, both long distance and local. Air and sea navigation is much more reliable now that microwave technology has been applied to that area. And, on a consumer level, microwave ovens are becoming more common. These are just some of the examples of the everyday uses of microwaves. The field is a relatively new one, but it is very progressive and fast growing.

2
TRANSMISSION LINE THEORY

2.1 TRANSMISSION LINES

2.1.1 GENERAL

As the uses of electromagnetic spectra increase, telecommunication bandwidth requirements increase, and equipment must be designed for higher and higher frequencies. As the frequency increases, the values of the components used in networks keep decreasing. As one approaches ultrahigh frequencies, for example, the values of inductors and capacitors become so small that the ordinary techniques used at lower frequencies are not usable any more. Even in the megahertz region, "lead dressing" is standard procedure to alleviate parasitic effects due to stray fields. The use of "lumped-circuit" elements at very high frequencies is impractical, so it is necessary to examine other approaches for carrying high frequencies from one place to another and to find some other ways of making resonant circuit components. If the frequency of operation gets so high that the wavelength on a device or transmission system is generally comparable with its size, transmission line theory should be applied to analyze the behavior of that device or transmission system.

High-frequency signals in such devices are usually sinusoidal or can be analyzed as though they are composed of sinusoidal waves that are traveling or propagating through a device from the input to output or outputs. If one considers a voltage wave traveling through a device from input to output, it is simple to see that, if the wavelength of the applied signal is comparable with the device dimensions at different points on the device, the instantaneous values of the voltage will vary due to its sinusoidal nature. This fact leads one to conclude that some parameters other than voltage or current would be needed to define such devices, since these values vary along the length of propagation. Transmission line theory is devised to handle such problems.

A transmission line is essentially a system of material boundaries forming a continuous path from one place to another and capable of directing the transmission of electromagnetic energy along this path.[1] When analyzing transmission lines, it is generally assumed that the cross-sectional geometry is constant, forming a uniform transmission line. The transmission line will be considered to be uniform if there is no change in the geometry of the line along the path of propagation. If there *is* a change in the geometry at any point, there will be a "discontinuity" in the line.

Traveling Waves

When a sine wave is applied to an infinitely long transmission line, the wave will propagate along the line. Figure 2.1-1 shows this wave at three

---

[1] Moreno, T., *Microwave Transmission Design Data* (New York: Dover Publications, Inc., 1958; © 1948, Sperry Gyroscope Company), p. 1.

successive instants in time. (Note that the crest of the wave progresses down the transmission line.) The voltage wave on a uniform, lossless transmission line is always accompanied by a current wave of similar shape, and, regardless of their shape, the two waves will be propagated without any change in magnitude or shape. These waves have definite electrical characteristics. The length of the wave λ is defined as the distance between successive points which have the same electrical phase. This wavelength depends upon the frequency

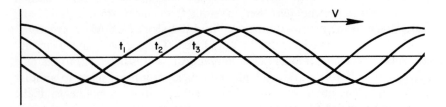

Fig. 2.1-1. A traveling wave.

of variation of the wave and the dielectric constant (a physical characteristic of the medium) of the medium through which the wave is traveling. In free space a wave will travel with a velocity of approximately 300,000,000 meters per second. However, in a medium other than free space, the velocity will be reduced by the factor $1/\sqrt{\epsilon_r}$, where ϵ_r is the relative dielectric constant of the medium. (For practical purposes the relative dielectric constant of air can be considered to be unity.)

The following formula shows the relationship between the various factors which determine wavelength:

$$\lambda = \frac{1}{\sqrt{\epsilon_r}} \cdot \frac{v}{f}$$

where v = velocity of propagation in free space,
$\quad f$ = frequency of oscillation,
$\quad \epsilon_r$ = relative dielectric constant of the medium the wave is traveling in.

Wavelength can also be defined as the distance in which the phase changes by 2π radians, where 2π radians = 360°. (See Fig. 2.1-2.)

Using Equivalent Circuits for Transmission Line Analysis

To understand fully the behavior of signal propagation on a transmission line, it is not enough to understand the voltages between conductors and the currents carried in the conductors. If a signal is applied to an infinitely long, uniform transmission line, electromagnetic waves will be carried

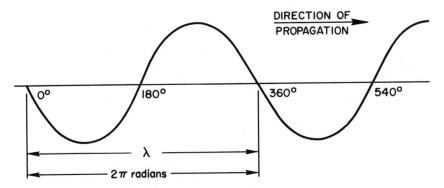

Fig. 2.1-2. Wavelength and phase change.

down its path. Voltage exists between the conductors, and current flows through them. Electric and magnetic fields are formed between and around the conductors respectively, and their behavior and field configurations are also very important.

A small section of such a transmission line can be analyzed by using lumped-circuit analogy. For example, a unit-length-long piece of a parallel wire transmission line and its equivalent circuit are shown in Fig. 2.1-3. The circuit contains a series inductance. An inductor is defined as a current-carrying conductor forming a magnetic field around itself that delays the voltage. Since a piece of wire does establish a magnetic field around itself, according to the Biot-Savart law, it does have inductance. Since the conductor can have finite resistance, a series resistor will define it adequately. The two conductors are a finite distance apart, and they form some parallel capacitance. A dielectric medium keeping the two conductors a constant distance apart can have dielectric losses, so parallel conductance would describe this effect sufficiently. In the Giorgi (MKSA) system, inductance is measured in henries/unit length, capacitance in farads/unit length, resistance in ohms/unit length, and conductance in mhos/unit length. It can be imag-

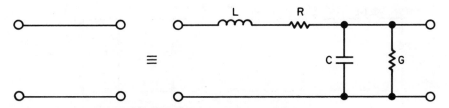

Fig. 2.1-3. Unit-length piece of parallel wire and its equivalent lumped-circuit model.

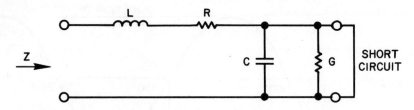

Fig. 2.1-4. Impedance of the equivalent circuit.

ined that a transmission line is built of an infinite number of infinitely short lengths of this type of "two-port" network "cascaded" one after the other (connected in tandem).

Impedance/Admittance Relationships

In transmission systems, impedance relationships take a leading role in defining propagation characteristics. It is desirable to analyze the series and parallel elements of the equivalent circuit separately. Kirchhoff's law allows us to add impedances in series and admittances in parallel configuration. The impedance of the equivalent circuit can be measured as shown in Fig. 2.1-4 with the output short-circuited. The parallel circuit components are shorted; only the series components are measured.

Impedance can be expressed as

$$Z = R + j\omega L$$

Admittance information can be gained by measuring from the other end when the front end is open-circuited. Figure 2.1-5 shows the admittance measurement setup. Since the series elements are left open, only the parallel components will be measured. Admittance is expressed as

$$Y = G + j\omega C$$

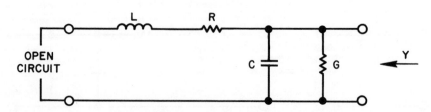

Fig. 2.1-5. Admittance of the equivalent circuit.

Transmission Line Parameters

The four components of the equivalent circuit of a uniform transmission line are divided into series and parallel groups defining the impedance and the admittance of the transmission line, respectively. Two parameters can be derived using the impedance and admittance expressions. It is convenient to define the propagation constant as

$$\gamma = \sqrt{ZY} = \sqrt{(R + j\omega L)(G + j\omega C)}$$

Since the square root of the product of two complex numbers is also complex, the propagation constant is generally expressed as

$$\gamma = \alpha + j\beta$$

where α is the attenuation constant in nepers/unit length (if the circuit components are given in the MKSA system, as described above) and β is the phase constant in terms of radians/unit length. By definition, the other parameter is the characteristic impedance:

$$Z_0 = \sqrt{\frac{Z}{Y}} = \sqrt{\frac{R + j\omega L}{G + j\omega C}} \text{ ohms}$$

If R and G are negligible in size, that is, if there is no absortive loss on the transmission line, then

$$Z_0 = \sqrt{\frac{L}{C}} \text{ ohms}$$

The characteristic impedance becomes a real number that is independent of frequency changes (if $R = G = 0$, $j\omega$ will cancel).

The reciprocal of the characteristic impedance is defined as the characteristic admittance:

$$Y_0 = \frac{1}{Z_0} \text{ mho}$$

2.1.2 WAVES ON TRANSMISSION LINES

Incident and Reflected Waves

Voltage applied to a transmission line can be written in exponential form:

$$V_1 = V_{1p}\epsilon^{j\omega t}$$

where p stands for peak value. Current resulting from the voltage applied to the transmission line can then be written:

$$I_1 = I_{1p}\epsilon^{j\omega t}$$

These voltages and currents are periodical waves. If that voltage is applied to a transmission line, a voltage wave will proceed along that line. The voltage wave may be written in exponential terms:

$$V = V_1 \epsilon^{\gamma l}$$

The associated current wave flowing in the line is

$$I = I_1 \epsilon^{\gamma l}$$

If the transmission line is not infinitely long, it is terminated with an impedance Z_l, as shown in Fig. 2.1-6. Since that load impedance is not equal to the

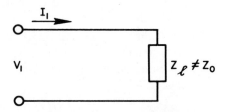

Fig. 2.1-6. Transmission line terminated with an impedance not equal to characteristic impedance.

characteristic impedance, not all the energy propagated down the transmission line will be absorbed, and part of the signal is reflected because it is mismatched. This signal is traveling in the opposite direction from the "incident" signal. The voltage and current waves are:

$$V = V_1 \epsilon^{\gamma l} + V_2 \epsilon^{-\gamma l}$$

$$I = I_1 \epsilon^{\gamma l} - I_2 \epsilon^{-\gamma l}$$

V_1 and V_2 are periodical voltage waves; I_1 and I_2 are periodical current waves. The voltage across the load impedance will become

$$V_L = V_1 + V_2$$

The current flowing through the load is

$$I_L = I_1 - I_2 = \frac{V_1}{Z_0} - \frac{V_2}{Z_0}$$

Then the load impedance is

$$Z_L = \frac{V_L}{I_L}$$

Two wave trains are traveling opposite to each other: the incident wave and the reflected wave. Since both are really traveling on the same line, which has a characteristic impedance of Z_0, the equation can be written as

$$\frac{V_1}{I_1} = \frac{V_2}{I_2} = Z_0$$

The following equation can be derived from the preceding equations:

$$\frac{V_2}{V_1} = \frac{Z_L - Z_0}{Z_L + Z_0}$$

This equation shows that the *relative* amplitudes and phases of both waves are determined by the terminating impedance only. The *absolute* magnitudes of the waves are dependent of the impedance of the source.

Transmission Modes

Associated electric and magnetic fields are formed as voltage and current waves travel down a transmission line. Since these fields are the result of voltage and current waves, which are periodical, the electric and magnetic waves also vary in a periodic manner. As propagation frequency increases, an appreciable portion of the wavelength of that propagating signal becomes comparable to the cross-sectional geometry of the transmission line, more than one kind of electromagnetic field configurations can be imagined. As the frequency increases, more and more different types of propagation modes can exist on a certain transmission line. If propagating frequency increases to infinity, an infinite number of propagation modes can exist. These modes are the so-called high-order modes of propagation. The principal mode is the one which can carry energy at all frequencies. Higher-order modes are those that propagate only above a definite frequency range. The point at which these frequencies start to propagate is called the cutoff frequency for that particular mode.

The following analogy is presented to show how these high-order modes are established. Figure 2.1-7 shows rice being blown down the inside of a blowgun. As the figure clearly shows, the rice can fit in the blowgun in only

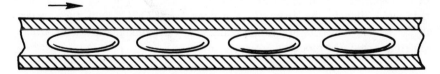

Fig. 2.1-7. Rice propagating down a blowgun.

one way. If rice is continuously blown through the tube with a constant velocity, a certain propagation will exist. This is analogous to a single mode propagation. If one either increases the inside diameter of the blowgun or decreases the particle size of the rice, the rice could propagate down the tube

Fig. 2.1-8. Rice propagating down a blowgun when the particle size of rice is small compared to the cross-sectional geometry of the tube.

in different modes, as shown in Fig. 2.1-8. The particle size of the rice is small compared to the cross-sectional geometry of the tube. Consequently, the rice will not be required to move down in the tube in a predetermined way. It can tumble around and move all over inside the tube, slowing down the propagation of each particle and at the same time increasing it to a certain extent as rotational velocity may be added to the motion. This is analogous to some high-order modes of propagation on transmission lines.

One can imagine a number of patterns the rice particles can assume while propagating down that particular transmission line. If the rice size is decreased again, one can imagine more and more types of patterns that are analogous again to some even higher-order modes. This clearly shows that certain patterns can occur only when a definite size change occurs either in the transmission line or in the propagating frequency (wavelength).

The transmission line in Fig. 2.1-9 shows the principal mode of propagation and the electric and magnetic field configuration of the pattern on the parallel wire. Since there is a difference in potential between the wires, an electric field is established in between them. The solid lines in the figure show the electric field configuration. Since current flows in the conductors, magnetic fields are established around them. At any point in space, the electric and the magnetic field lines are perpendicular to each other. The figure also clearly shows that these fields are all transverse to the direction of propagation. In other words, no components of either electric or magnetic field align in the direction of propagation. That is why these waves, in the principal mode, are called transverse electromagnetic waves, abbreviated as the TEM mode of propagation.

Now, if the propagating frequency increases so much that the length of the wave traveling down the transmission line is comparable in size to the cross-sectional geometry of that transmission line, higher-order modes can propagate. These higher modes will have at least one of their field components showing in the direction of propagation. Depending on which component shows in the direction of propagation, it will be called the H- or the E-wave of propagation. The H-wave is that in which at least one component of the magnetic field shows in the direction of propagation. This

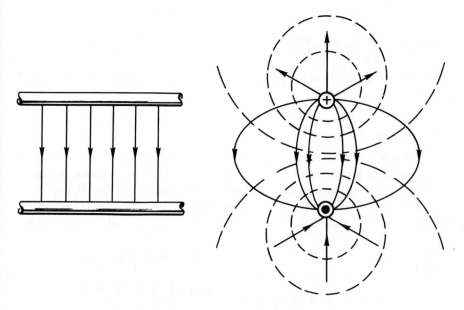

Fig. 2.1-9. Electric and magnetic field configuration of
the parallel wire transmission line.

mode is called the transverse electric mode, abbreviated as the TE mode.
The E-wave is that in which the electric field will have at least one component showing in the direction of propagation. This is called the transverse
magnetic, or TM, mode. The number of these modes on a transmission
line is infinite. Although each TE and TM mode can be infinite in number,
at a certain frequency there can be only a finite number of higher-order
modes propagating. The number of these modes is dependent entirely upon
the geometry of the transmission line.

As was mentioned in the blowgun analogy, the velocity of propagation
of the TE and TM modes is different from that of the TEM mode. In fact, two
kinds of velocities can be imagined: group velocity and phase velocity. Group
velocity means the velocity of the entire group moving down in the transmission line, and phase velocity includes all rotations and turns of the individual
moving particles. Generally, in the TEM or principal mode, phase and group
velocities are identical to each other. In a standard transmission line with no
dielectric material around it, they would move with exactly the velocity of
light,

$$c = 300,000 \text{ km/s}$$

In higher-order modes, group velocity and phase velocity are related to each
other by the following equation:

$$c = \sqrt{v_g v_p}$$

where v_g = group velocity,
v_p = phase velocity.

The geometric means of the phase velocity and the group velocity are equal to the velocity of light.

Discontinuities

When a sudden change in geometry occurs on a transmission line and when uniform transmission lines exist before and after the plane of that discontinuity, the problem can be handled as two uniform transmission lines joined together. The only question is what happens at the plane or near the plane of the discontinuity. Figure 2.1-10 shows a discontinuity formed at the

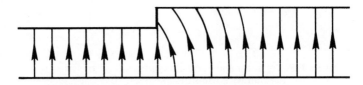

Fig. 2.1-10. Discontinuity on a transmission line.

plane where two uniform transmission lines are joined. The electric field between the conductors is drawn. As is apparent, the electric field lines are bent in the region near the discontinuity; but after some distance, they straighten out again. When either electric or magnetic field components are aligned in the direction of propagation, higher-order modes are launched. Although it is assumed that higher-order modes cannot be propagated on this particular transmission line, this does not mean that they cannot be launched. Since they cannot be propagated, they will decrease their effectiveness exponentially, and they will be attenuated accordingly. Discontinuities in the transmission line will effectively launch certain higher-order modes, and energy will be stored when they do. It is known from lumped-circuit theory that energy storage will occur where either capacitance or inductance is present. Discontinuities can be understood as reactive components on a transmission line.

Another effect can be observed from Fig. 2.1-10. The electric field distortion occurs only at the right (widest) side of the discontinuity; no field distortion occurs immediately left of the plane of discontinuity. Whether the discontinuity bends the electric or the magnetic field determines the equivalent circuit. If there are more discontinuities, there are more steps one after the other on the transmission line; if they are close enough to each other, they might interact (see Fig. 2.1-11). As can be seen, interference will occur when

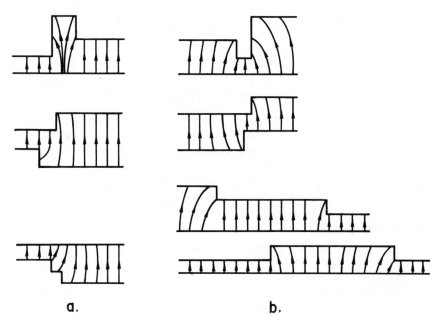

Fig. 2.1-11. Multiple discontinuities (a) interfering
(b) not interfering with each other.

the field lines due to one discontinuity are not straightened before another
discontinuity occurs. Some discontinuities (as shown in Fig. 2.1-11b) are
close to each other, but, as they do not distort the field in a common direction,
they do not interfere with each other. If they interfere, a third effect will
occur. This modifies their single and simple effect by a mutual coupled effect.
This field distortion is very similar to the effect in capacitors due to fringing
field effects. In fact, the very same term is used for these field distortions
in the literature.

2.1.3 PROPAGATION CHARACTERISTICS

Attenuation Constant, Phase Constant

As an electromagnetic wave is propagated down a transmission line, it
is continuously attenuated by the lossy elements of the line. The propagation
constant

$$\gamma = \sqrt{(R + j\omega L)(G + j\omega C)} = \alpha + j\beta$$

clearly shows that it is composed of the attenuation and phase constants. (α

is negative and is not shown since it is not gain but attenuation.) In previous paragraphs the incident wave was shown to be

$$V_1 = \epsilon^{\gamma l},$$

which can be further expressed as

$$V_1 = \epsilon^{(\alpha + j\beta) l} = \epsilon^{\alpha l} \times \epsilon^{j\beta l}.$$

The first part of this equation shows that the voltage gets attenuated exponentially as the wave travels down the line. It was also seen that the attenuation constant α will be expressed in terms of nepers per unit length if it is calculated in the MKSA system. To convert nepers to the more commonly used decibel per unit length, multiply by 8.69: 1 neper = 8.69 dB.

Sources of Attenuation

Attenuation can be contributed by many factors, such as the following:

1. Conductor losses (skin effect)
2. Dielectric losses
3. Hysteresis losses
4. Mismatch losses
5. Losses due to radiation

The first three—conductor, dielectric, and hysteresis losses—are absorptive losses by nature; they dissipate energy. Mismatch loss and losses due to radiation reflect and guide the energy away from the transmission line, respectively.

Conductor Losses (Skin Effect)

Conductor losses are caused by the series resistance of the conducting medium. Existence of the skin effect is known from lower-frequency techniques. As frequency increases, skin effect becomes more critical. In other words, in higher frequencies in the transmission line, the current is restricted to travel in only the surface layer of the conductor. The penetration of the current flow is defined by the skin depth (δ). The skin depth is the thickness of the layer where the current density drops to $1/\epsilon$ the value on the surface. Skin depth can be calculated from the following formula:

$$\delta = \frac{1}{2\pi} \sqrt{\frac{\rho}{f \mu_r}} \ (\text{cm})$$

where ρ is the specific resistivity of the conductor in terms of Ω cm, μ_r is the relative permeability of the conductor (this is considered only when the material is ferromagnetic and has a relative permeability different from nonferrous material; in other words, this value is equal to 1), and f is the operating frequency in terms of gigahertz. Table 2.1-1 gives the specific resistivity of a

Table 2.1-1. Specific Resistivity of a Few Metals

Metal	Specific Resistivity (ohm-cm) $\times 10^{-6}$
Aluminum	2.83
Brass	6.4–8.4
Copper	1.724
Gold	2.44
Lead	22.0
Palladium	11.0
Phosphor Bronze	10.5
Platinum	10.0
Rhodium	5.0
Silver	1.629
Zinc	6.1

few metals. Since skin depth at microwave frequencies is a very small fraction of the conductor thickness (portions of thousandths of an inch), it is common practice to plate the surfaces of conductors. Plating with a highly conductive metal a couple of thousandths of an inch thick completely masks off the effect of the base material since all the current will be flowing in the plating.

Caution should be exercised when calculating skin depth and effective losses with the specific resistivity of plated metals. Solid metals and plated surfaces have different resistivities since plated metals are not so dense as solid metals. They are porous in nature. If high conductivity is desired, plated surfaces are burnished to pack the plating more densely. It is a good practice not to plate only to the thickness of the skin depth but to take at least three to five times the thickness which has been calculated. The surface finish of the plated material also comes into consideration. Because current penetration into the surface is so shallow, the currents have to follow all surface imperfections, consequently increasing the path of the current and thus increasing the actual resistance of the conductor. When very low losses are desired, microwave transmission line components are usually highly polished so that all the machining marks are essentially polished out. Even microscopic scratches crossing the current flow can appreciably increase the equivalent resistance of the conductor. However, if the machining operation is planned so that the machining marks will be in line with the current flow, the effect of these marks on the resistance will be negligible.

Dielectric Losses

It was shown in the first paragraph that propagation velocity is slowed down if a dielectric insulator is placed around and between the conductors.

It is also known that the dielectric material in a capacitor increases the effective capacitance between said conductors; however, most dielectric material has losses associated with this space-saving effort (longer electrical length in the same physical length). These losses can be taken into account if the dielectric constant is handled as a complex value, as in the formula

$$\epsilon = \epsilon' - j\epsilon''$$

where ϵ' is the real part of the dielectric constant and ϵ'' is the imaginary part of the dielectric constant. Losses of the dielectric material are usually expressed by the loss tangent. Using the complex expression of the dielectric constant, the loss tangent can be defined as

$$\tan \delta = \frac{\epsilon''}{\epsilon'}$$

Since the loss tangent of commonly used dielectric material is very small, it is approximately equal to the power factor of a capacitor. It is true that the power factor is defined by $\cos \Theta$, where

$$\Theta = 90° - \delta$$

Since very small angles, tangents and sines, are approximately equal, power factor and loss tangent can also be taken as equal.

Hysteresis Losses

For most practical purposes hysteresis losses are included in the skin-effect formula. If material with permeability differing from nonmagnetic materials is used, plating is usually applied, and the effect of hysteresis losses is negligible or entirely alleviated. Hysteresis will be discussed more thoroughly in later chapters in which ferrite materials are covered.

MISMATCH LOSSES and LOSSES DUE TO RADIATION are not absorptive losses. Mismatch loss occurs when a discontinuity appears on a transmission line or when a termination of the transmission line is not equal to the characteristic impedance. Consequently, not all the power available at that point on the transmission line is transmitted or propagated; part of it is reflected. So, as far as the transmission line and the terminating load are concerned, not all the energy is delivered into the load. The losses from the available signal and the dissipated signal are due to mismatch. Discontinuities or openings on transmission lines at higher frequencies are more serious than at lower frequencies. The microwave energy, the electromagnetic wave, will radiate out of the line and cause losses. For all practical purposes, this type of loss is not intentional, and cracks should be looked for and repaired. At high microwave frequencies, some cables have to be double- and triple-shielded to alleviate losses due to radiation.

The attenuation of a transmission line, as defined by the propagation constant, includes all the attenuations described above. Usually the different attenuations due to different sources of losses are calculated and handled separately, and finally they are totaled, giving the total attenuation.

2.1.4 REFLECTIONS ON TRANSMISSION LINES

Reflection Coefficient

If a signal is applied to a uniform, practically lossless, transmission line, and if that transmission line is terminated with an impedance not equal to the characteristic impedance of the line, that impedance will not be able to absorb all the energy. Part of the signal will be reflected. Figure 2.1-12 shows a

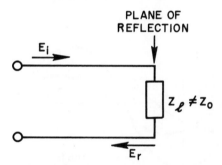

Fig. 2.1-12. A transmission line terminated with impedance not equal to characteristic impedance will reflect part of the incident signal.

transmission line not terminated in its characteristic impedance. E_i is the incident signal traveling toward the termination. E_r is the reflected signal traveling in the opposite direction. The ratio of these two voltages, the reflected signal over the incident signal, gives the so-called reflection coefficient.

$$\frac{E_r}{E_i} = \Gamma$$

The reflection coefficient, Γ, is a vector, since it has magnitude as well as phase information. Both the incident waves and the reflected waves are traveling on the same transmission line but in opposite directions. Their relative phases are dependent on the terminating impedance and only on the terminating impedance and the distance from the termination to the point of measure-

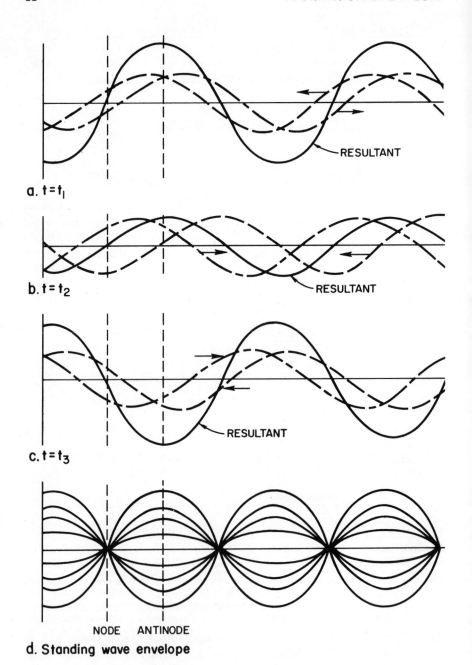

a. $t = t_1$

b. $t = t_2$

c. $t = t_3$

NODE   ANTINODE

d. Standing wave envelope

Fig. 2.1-13. Formation of standing waves.

ment. The magnitude of the reflected voltage depends on how much the terminating impedance is mismatched. That is why the reflection coefficient serves as a figure of merit for the termination at the end of any particular transmission line. The absolute value of the reflection coefficient,

$$\left|\frac{E_r}{E_i}\right| = |\Gamma| = \rho$$

is usually given with the Greek letter ρ.

Standing Waves

If two waves of the same amplitude and frequency are traveling on a transmission line in opposite directions, they will alternately add to and subtract from each other. The result is known as a standing wave. See Fig. 2.1-13 for a diagram showing how two traveling waves combine to form standing waves. (Note that the maximum and zero voltage points do not shift with respect to time. This is the difference between traveling waves and standing waves.) The zero crossings are called nodes, and the positions of maximum amplitude are called antinodes. The following discussion will describe the manner in which standing waves are produced, detected, and suppressed, and it will also discuss their effect on the transmitted signal.

Waves having the same length but not necessarily the same magnitude will form an interference pattern. This is called the standing-wave pattern. The bottom line of Fig. 2.1-13 shows the standing-wave pattern as it is formed on the transmission line. In practice, this pattern has to be detected to enable one to plot it, and only the envelope will be shown as it is drawn in Fig. 2.1-14. It is worthwhile to mention that these standing waves were built of

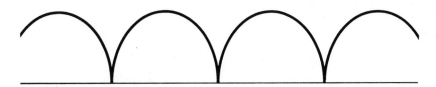

Fig. 2.1-14. Detected standing-wave pattern of total reflection.

total reflection. Figure 2.1-15 shows a standing wave built of a nontotal reflection. The reflection coefficient will be less than 1. The peak value of the standing-wave pattern will be called $E_{\max}$. The smallest value will be called $E_{\min}$. The voltage standing-wave ratio is defined as the ratio of these two values.

$$\text{VSWR} = \sigma = \frac{E_{\max}}{E_{\min}}$$

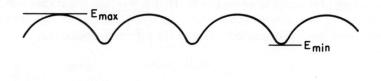

Fig. 2.1-15. Standing-wave pattern of a load not
forming a total reflection.

$E_{\max}$ is really the sum of the absolute values of the incident and reflected
voltages,

$$E_{\max} = |E_i| + |E_r|$$

and $E_{\min}$ is the difference of the absolute value of the incident voltage and the
absolute value of the reflected voltage:

$$E_{\min} = |E_i| - |E_r|$$

Substituting these equations into the standing-wave ratio equation,

$$\text{VSWR} = \sigma = \frac{E_{\max}}{E_{\min}} = \frac{|E_i| + |E_r|}{|E_i| - |E_r|}$$

Dividing with $|E_i|$,

$$\sigma = \frac{1 + \left|\dfrac{E_r}{E_i}\right|}{1 - \left|\dfrac{E_r}{E_i}\right|}$$

gives the expression for the voltage standing-wave ratio in terms of the reflection coefficient,

$$\sigma = \frac{1 + \rho}{1 - \rho}$$

This equation may be rewritten to solve for the absolute value of the reflection
coefficient in terms of the voltage standing-wave ratio:

$$\rho = \frac{\sigma - 1}{\sigma + 1}$$

As the standing-wave pattern was built of voltage waves, it can be imagined
that it can be built of power waves also, and then the power standing-wave
ratio can be defined. The power standing-wave ratio is really equal to the
square of the voltage standing-wave ratio:

$$\text{PSWR} = \sigma^2$$

If the voltage standing-wave ratio must be expressed in terms of decibles, as when large standing-wave ratios are involved, then

$$\text{SWR (dB)} = 20 \log \sigma$$

One might want to know how far down the reflected wave is from the incident wave in terms of decibels. This information is called the return loss:

$$\text{return loss (dB)} = -20 \log \rho$$

The negative sign on the return loss merely shows that it is a loss and not a gain.

Mismatch Loss

One more fact can be given about the reflection of a termination, that is, the so-called mismatch loss. This describes how many decibels less than the incident voltage available are being absorbed by that termination. If the termination happens to be an antenna, that information can really be very valuable. It would tell how many decibels less power are being radiated from the antenna due to the mismatch than are available at the port of the antenna. It can be calculated from the following equation.

$$\text{mismatch loss (dB)} = -10 \log (1 - \rho^2)$$

2.1.5 IMPEDANCE AND ADMITTANCE TRANSMISSION LINE RELATIONSHIPS

The impedance of a termination is generally a complex value,

$$Z = R + jX$$

where X can be either positive or negative. If X is positive, it is called the inductive reactance; if it is negative, it is called capacitive reactance. If one desires to plot an impedance on a chart, the Cartesian (rectangular) coordinate system would be an excellent one to start with. Figure 2.1-16 shows an impedance plotted in a rectangular coordinate system. From the figure it can be clearly seen that the absolute value of the impedance can be calculated by the Pythagorean formula,

$$|Z| = \sqrt{R_1^2 + X_1^2}$$

and the angle of that impedance is given as

$$\tan \phi_1 = \frac{X_1}{R_1}$$

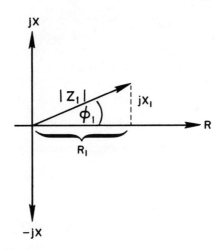

Fig. 2.1-16. Impedance plot in a rectangular
coordinate system.

Reactance Charts

Historically, reactance charts were used to present a graphic approach
to wave relations on transmission lines. As can be seen from the figure, the
reactance lines are on the vertical axis: inductive reactance is in the positive
direction and capacitive reactance is in the negative direction. Resistance is
plotted along the horizontal axis. Note that the left half of the chart representing negative resistances is not used in this case. Other charts (discussed
in Chap. 6) can be expanded into negative resistances.

Figure 2.1-17 shows a reactance chart. On the chart, the values of the
scales are *normalized* so that a characteristic impedance of 1 occurs at the
point $(1, 0)$. The reactance chart method is considered more graphic and
easier to understand than the equation method, in which impedance at a particular point on the transmission line is determined from the following
equation:

$$Z_X = Z_0 \frac{(Z_L + Z_0)e^{\gamma X} + (Z_L - Z_0)e^{-\gamma X}}{(Z_L + Z_0)e^{\gamma X} - (Z_L - Z_0)e^{-\gamma X}}$$

where γ is the propagation constant, Z_L is the impedance of the load, X is the
distance along the line from the load, and Z_0 is the characteristic impedance
of the line. However, there are serious limitations to this reactance chart:
(1) not all values of impedance out to infinity are presented on the chart;
(2) it is difficult to interpolate between constant standing-wave ratio circles
since the pattern is an ever-expanding circle array; (3) these angle indications
are not radial.

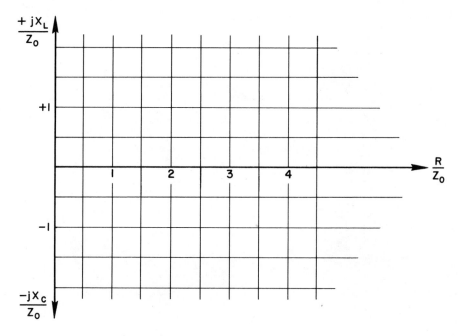

Fig. 2.1-17. Reactance chart.

The Smith Chart

In the late 1930's, Philip H. Smith devised a chart that avoided all the limitations mentioned above by a conformal transformation of that reactance chart. This method posed the infinity points of the jX_L and jX_C axis around to a common point. At infinity on the resistance axis, the resultant graph shows positive and negative infinities of the reactances also emerging at that point. This is known as the Smith Chart (Fig. 2.1-18). The values of the axis are normalized by dividing the impedance of interest by the characteristic impedance. For example, if the impedance of a 50Ω transmission line was found to be 100Ω at a particular point, the normalized impedance at that point would be 2. In the case of complex impedances, the normalized impedance is determined by the following equation:

$$\frac{Z}{Z_0} = \frac{R}{Z_0} + j\frac{X}{Z_0}$$

Points along the center line on the Smith Chart represent pure resistance; points around the outer edge of the chart represent pure reactance with no resistance. The circles tangent to the right edge are circles of constant resistance. The lines tangent to the resistance horizontal axis at infinity represent lines of constant reactance. Positive values are inductive reactances, and

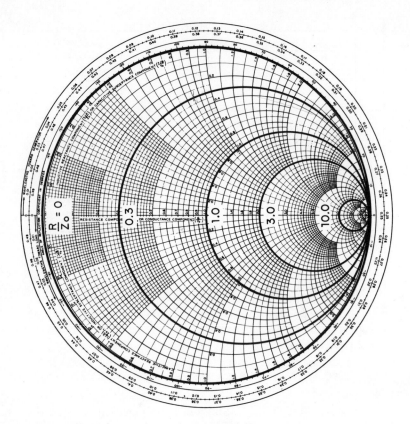

Fig. 2.1-18. Smith Chart—circles of constant resistance.

negative values are capacitive reactances. The Smith Chart is one of the most useful devices available to those working with transmission lines. It can be used for the graphic solution of a large variety of transmission line problems. Fundamentally a special kind of impedance coordinate system, the Smith Chart is almost indispensable in the solution of impedance problems when data are obtained through measurements made on a transmission line with special impedance measuring devices, which will be discussed later in another paragraph.

Normalized Impedance

Referring to the previous discussion on transmission lines and standing-wave ratio, remember the manner in which voltage and current vary along a transmission line. Standing waves are not present on the line if it is terminated by its characteristic impedance. Under this circumstance both voltage and current are constant throughout the length of the line. This is the situation in which all the power flowing down the line is absorbed by the load. If the load does not equal the characteristic impedance, there is a mismatch, with some of the energy being reflected from the load. Likewise, reflection may occur from discontinuities at any point on the line. Where there is a condition of reflected power, the forward and reflected waves combine to form both voltage and current standing waves on the transmission line.

The voltage standing-wave ratio (commonly called VSWR) is the ratio of the minimum and maximum values of the voltage standing wave. The VSWR can range from unity, the condition of no reflection, to infinity, in which complete reflection occurs at the load, as in the case of an open-circuit or a short-circuit termination. Refer again to the Smith Chart in Fig. 2.1-18. The values of resistance and reactance shown on this chart are all on the basis of normalized values. The circles (which are labeled 0.3, 1.0, 3.0, etc.) having their centers on the straight line across the chart are circles of constant resistance which have been normalized. Now it may be seen that the line of zero reactance is the straight line. Positive and negative reactance lines curve away from this line perpendicular at any point with the constant resistance lines and represent normalized reactance. For example, if the transmission line were terminated in its characteristic impedance, the point of impedance would be represented by the exact center of the chart with a resistance of 1.0 and no reactive component. On the chart of Fig. 2.1-19 these lines are marked $+0.3$, $+1.0$, and $+3.0$ on the positive side and -0.3, -1.0, and -3.0 on the negative side.

Using the Smith Chart

An impedance represented by the value of $5 + j25\Omega$ would be plotted on the Smith Chart as follows. The $+j$ indicates a reactive component above the 0 reactance line; $-j$ is below that line. Normalized for a 50Ω line, this becomes $0.1 + j0.5\Omega$. This point is plotted at the position A illustrated in Fig. 2.1-19. Now consider the scales on the periphery of the chart. They are calibrated both in wavelength and in degrees. Note that one complete revolution of the chart is equivalent to one-half a wavelength, or 180°. Now the impedance at any point along the transmission line is normally considered to be that impedance which would be measured if the line were cut at that point and measurements were made looking into the line section which is connected to the load. The impedance of the generator has no effect on this value. If the transmission line is terminated with the value of its characteristic impedance, then, no matter where the line is cut, the observer will measure the characteristic impedance. However, if the line is not terminated in its characteristic impedance, as it is cut at various points, it will be found that the impedance will vary cyclically. However, the impedance of the load will be the

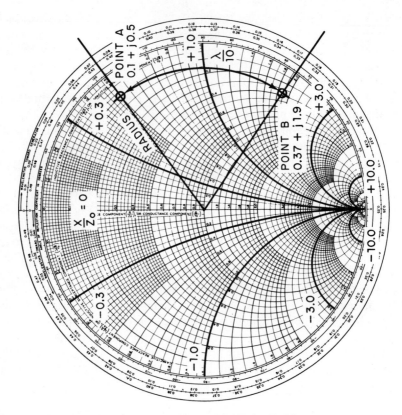

Fig. 2.1-19. Normalized impedance plot on the Smith
Chart.

same at each half-wave point, assuming that there are no losses on the line.
Thus, as complete revolutions of the Smith Chart are traversed, the beginning
point will be passed on each revolution, representing the cyclical variations
of standing waves on a transmission line.

Assume that point A of Figure 2.1-19 represents a terminating impedance on the transmission line and that the impedance one-tenth of a wavelength from the load is desired. Using the center of the chart $1.0 + j0$ as a
pivot point, swing an arc equal to the radius from the center to the point A
in the direction toward the generator by an amount equivalent to one-tenth
wavelength, as determined by the outer calibration circle. Thus the new point
B is arrived at, and the normalized impedance viewed at this point is
$0.37 + j1.9$. For this specific example, remember that the load remained constant and that conditions along the transmission line were being considered.
Therefore the arc which was drawn is one of constant voltage standing-wave

ratio. All circles drawn from the center of the Smith Chart are circles of constant VSWR.

The following problems are examples of how the Smith Chart can be used to solve transmission line problems.

PROBLEM 1. It is given that the normalized impedance at a point on a transmission line is 0.1 + j0.5. The frequency of the source is 1,000 MHz. What impedance can be measured on the transmission line 3.8 cm toward the source? What is the admittance at this point? What is the VSWR on the line?

SOLVING THE PROBLEM. This is a typical problem that demonstrates the transformer properties of a transmission line. To find the impedance 3.8 cm away, one must know the starting point on the chart, the length of transmission line in wavelengths, and the direction of travel. The frequency of 1,000 MHz corresponds to a free-space wavelength of 30 cm. Then the length of the section of the line in wavelength is 3.8/30 = 0.126 wavelength. The direction of travel is toward the generator. A circle is drawn (see Fig. 2.1-20) using as a pivot the 1 + j0 point at the center of the chart. Using the radius from this point to point A, the arc is drawn toward the generator a distance of 0.126 wavelength. The new point gives a normalized impedance of approximately 0.8 + j3. The admittance of this point is obtained by going around the circle exactly one-quarter wavelength from point B. In other words, locate the point exactly on the opposite side of the chart, and the admittance becomes approximately .09 − j0.32 (see point C on the chart). This is, of course, the normalized admittance value of that point. This characteristic of the chart is extremely useful in design work with parallel impedances because the admittance values are far more convenient for parallel circuit considerations. The VSWR may now be found by continuing the arc until it intercepts the 0 reactance line. Continue the circle to intercept the 0 reactance line on the right side from the center of the chart to point D in Fig. 2.1-20. Here we can read the constant resistance value of 11.8, which is the VSWR.

PROBLEM 2. Find the impedance of the load on a transmission line when this load is located an unknown distance away from a slotted-line section. A slotted line is a piece of transmission line with a longitudinal slot in it that allows insertion of a probing device to measure the field strength along the path of the slotted line, causing the least amount of disturbance to the fields in the transmission line. The assumption is that the slotted line is inserted into the line some distance from the load. The operating frequency in this example is 3 GHz. $\Rightarrow \lambda = \frac{3 \times 10^{10}}{3 \times 10^9} = 10\,cm.$

SOLVING THE PROBLEM. To find the impedance of the load, it is necessary to know the VSWR as well as the distance to the load. The procedure is as follows.

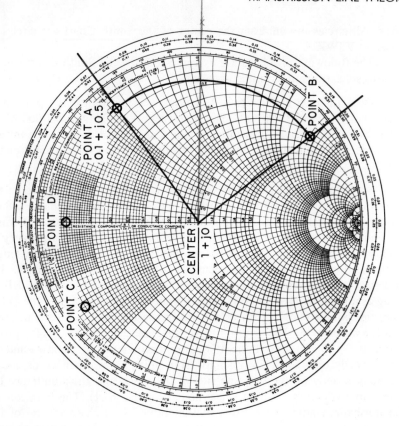

Fig. 2.1-20. Problem 1.

1. Place the load to be measured on the end of the slotted line.
2. Locate a minimum on the slotted line (point *B* in Fig. 2.1–21).
3. Measure the VSWR.
4. Replace the load with a short at the plane of the load.
5. Locate the minimums, which are nulls now because of the short, around point *B* (points *A* and *A'*).
6. Plot the impedance on the Smith Chart.

The VSWR has to be drawn before plotting the impedance. The constant VSWR circle drawn in Fig. 2.1-22 assumes a measured VSWR of 3.3:1. The minimum shift in terms of wavelengths can be clearly seen by taking the distance from *B* to *A* and dividing it by the wavelength.

To plot impedance on the chart, one has to know in which direction to measure the shift. Since a short circuit was used for reference, wavelength

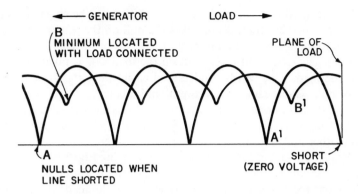

Fig. 2.1-21. Standing-wave pattern of load and short
circuit, measuring impedance.

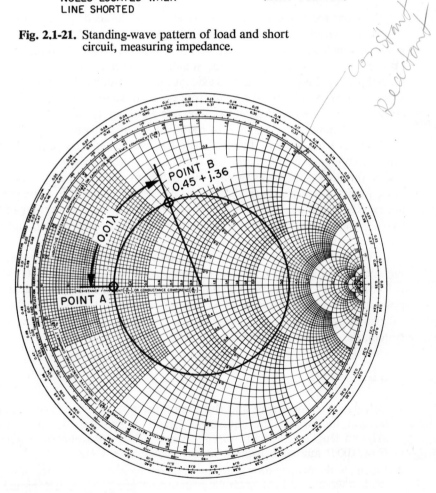

Fig. 2.1-22. Problem 2.

must be measured from the short-circuit position on the chart. It is a good practice to remember that, if one takes the null on the generator side of the minimum to calculate the minimum shift in terms of wavelength, it is plotted toward the generator; but if the null on the load side is taken, it is plotted toward the load. This can be understood easily since the distance between adjacent nulls is one-half a wavelength, and a complete circle on the Smith Chart also describes one-half a wavelength. Consequently, if the null on the generator side were used, it would give exactly the same locus on the chart as if the other null were used.

We can understand this procedure by studying Fig. 2.1-21. It shows an assumed case where the standing-wave indicator is about four half-wavelengths from the plane of the load. If we short the load, the short-circuit impedance repeats itself every half-wavelength back toward the generator to point A, which was located in step 4. Then, when the load was replaced, the positions of the minimums and the standing-wave pattern changed so that point B in step 2 was located. It is evident that the distance between A and B is the same at each half-wave point from the load, so that the measurements could have been made at repetitive intervals all along the line.

The distance from point B to the load is four half-wavelengths minus the distance from A to B. Each of these half-wavelengths corresponds to a full circumference of the Smith Chart. Thus the measurements at points A and B were just the same as though they had been made at points A' and B'. Since points A and A' always fall on the line of zero reactance, these points are exactly determined and fall on the vertical line through the center of the chart. Now the distance from B' to the load is a half-wavelength minus the distance from A' to B'.

We started on the Smith Chart at a point A on the line of zero reactance and moved along the VSWR circle 0.1 wavelength (wavelength at 3,000 MHz is 10 cm) toward the generator to point B. The normalized impedance at this point is approximately $0.45 + j0.63$. The impedance of the load can then be calculated by denormalizing. Assuming a 50-ohm characteristic impedance system, the impedance will be $22.5 + j31.5$.

QUESTIONS

1. Calculate the impedance and the admittance of a unit-length-long transmission line where the equivalent circuit components are $L = 0.002$ μH/unit length, $C = 0.012$ pF/unit length, $R = 0.015$ ohm/unit length, $G = 0.0001$ mho/unit length. The frequency is 1.6 GHz.

2. Calculate the propagation constant and characteristic impedance of a network where the equivalent circuit components are $L = 2 \times 10^{-9}$ H/meter, $C = 1.5 \times 10^{-14}$ F/meter, $R = 0.01$ ohm/meter, $G = 10^{-5}$ mho/meter. The frequency is 3.2 GHz.

3. What is the ratio of reflected voltage over incident voltage when the characteristic impedance of the transmission line is 200 ohms and the terminating resistance has a value of 270 ohms?

4. What is the skin depth on a transmission line made of silver-plated copper, 0.005 in. thick, at 8 GHz?

5. What is the complex value of the dielectric constant of an insulator if the relative dielectric constant is given as $\epsilon_r = 2.56$ and the dielectric loss is $\tan \delta = 0.0015$?

6. What is the VSWR and the equivalent return loss of a load if the absolute value of the reflection coefficient is 0.20?

   $- 20 \log \rho$

7. Calculate the mismatch loss on an antenna having 2:1 VSWR.

   $\rho = \dfrac{VSWR - 1}{VSWR + 1} = \dfrac{1}{3}$

8. What will the VSWR be on a $Z_0 = 50$ ohm transmission line if it is terminated with an impedance of $Z_L = 27.5 + 20j$?

9. In problem 1, what is the value of impedance on the transmission line 0.193 wavelength away from the plane of load in the direction toward the generator?

   $1 - \rho^2 = \dfrac{8}{9} \Rightarrow$ mismatch loss $= -10 \log \dfrac{8}{9} = 0.51$

2.2 COAXIAL TRANSMISSION LINES

The preceding discussions are based on parallel-wire transmissions; another name for this is Lecher wire. The Lecher wire has very serious limitations as far as radiation and losses are concerned. Coaxial transmission line is far superior to Lecher wire and is preferred in higher frequencies. A coaxial transmission line consists of a center conductor with another conductor around it. Since it is a two-wire system, it carries the TEM wave of propagation, the principal mode of transmission. The operation of coaxial lines is limited to the principal mode; higher-order modes aren't wanted. Therefore the useful range of coaxial transmission is restricted to the principal mode, which is under the first higher-order mode cutoff frequency.

2.2.1 DEFINING EQUIVALENT CIRCUIT COMPONENTS

To analyze the coaxial transmission line, the equivalence of parameters has to be determined: the series inductance in terms of henries per unit length, the series resistance in terms of ohms per unit length, the parallel capacitance in terms of farads per unit length, and the parallel conductance in terms of mhos per unit length. Centimeters will be used for unit length in this discussion. The following equations describe these parameters.

$$L = 0.4605 \, \mu_r \left(\log \frac{b}{a} \right) \times 10^{-8} \text{ henry/cm}$$

This equation neglects current penetration into the conductor. μ_r stands for the relative permeability factor. In

$$C = \frac{0.241\,\epsilon_r}{\log \dfrac{b}{a}} \times 10^{-12}\,\text{farad/cm}$$

$\epsilon_{\text{relative}}$ stands for the relative dielectric constant compared to vacuum or air. In

$$R = \frac{\rho}{2\pi\delta}\left(\frac{1}{b}+\frac{1}{a}\right) = \sqrt{\frac{f\mu_r\rho}{10^9}}\left(\frac{1}{b}+\frac{1}{a}\right)\,\Omega/\text{cm}$$

δ stands for skin depth in centimeters, f is the frequency in hertz, ρ is the resistivity in terms of ohms/cm. It can be seen that the resistance is proportional to the square root of the frequency. This is due to the skin depth. If only copper conductors are considered, this can be expressed in the following equation.

$$R = 4.14 \times 10^{-8}\sqrt{f}\left(\frac{1}{b}+\frac{1}{a}\right)\,\Omega/\text{cm}$$

where a and b are not diameters but are the radii in terms of centimeters in these equations. The following equation shows the effective line resistance if the inner and outer conductors are made of different metals.

$$R = \sqrt{\frac{f}{10^9}}\left(\frac{\sqrt{\rho_a\mu_a}}{a}+\frac{\sqrt{\rho_b\mu_b}}{b}\right)\,\Omega/\text{cm}$$

where ρ_a and μ_a are the specific resistance and the relative permeability factor of the conductor, respectively, and ρ_b and μ_b are the same for the outer conductor. As μ is a permeability factor, it provides information on the hysteresis losses if ferromagnetic materials are used.

If a coaxial transmission line is vacuum insulated or air insulated, there is practically no parallel loss conductance involved. But in most of the cables used for coaxial transmission lines the dielectric losses involved correspond to conductance loss. The following equation gives the equivalent parallel conductance equivalent circuit for coaxial transmission lines.

$$G = \omega C \tan \delta\,\text{mho/cm},$$

where C is the capacitance, $\tan \delta$ stands for the dielectric loss tangent, and ω is equal to $2\pi f$ (f is the operating frequency). These equations provide the equivalent circuit parameters in terms of geometrical dimensions.

Characteristic Impedance of Coaxial Transmission Lines

It is well known from ordinary transmission line theory that

$$Z_0 = \sqrt{\frac{R + j\omega L}{G + j\omega C}}$$

If losses are small, and they can be neglected; then the equation becomes

$$Z_0 = \sqrt{\frac{L}{C}}$$

Substituting the values of L and C from the above equation gives

$$Z_0 = 138 \sqrt{\frac{\mu_r}{\epsilon_r}} \log \frac{b}{a} = 60 \sqrt{\frac{\mu_r}{\epsilon_r}} \ln \frac{b}{a} \, \Omega$$

where μ_r is relative permeability and ϵ_r is the relative dielectric constant. Furthermore, b stands for the outer conductor's inner radius and a stands for the inner conductor's radius (Fig. 2.2-1). For example, if one has an air-insulated coaxial transmission line made of nonmagnetic material and having an inner conductor radius of 0.05 inch and an outer conductor inner radius of 0.15 inch, the characteristic impedance will be

$$Z_0 = 138 \sqrt{\frac{\mu_r}{\epsilon_r}} \log \frac{b}{a} = 138 \log \frac{0.15}{0.05} = 138 \times 0.477 = 66\Omega$$

where μ_r and ϵ_r are equal to 1, as described above. In case the losses are excessive and the simplified formula cannot be used, the complex formula for characteristic impedance must be used and obviously will result in a complex value for the characteristic impedance. This complex value will become dependent upon frequency variation.

2.2.2 PROPAGATION CHARACTERISTICS OF COAXIAL TRANSMISSION LINES

In all the transmission line theory the propagation constant has been defined as

$$\gamma = \sqrt{(R + j\omega L)(G + j\omega C)} = \alpha + j\beta$$

where α is the attenuation constant in terms of nepers per unit length and β is the phase constant in terms of radians per unit length. The phase constant β is related to wavelength. It is the information about electrical length when

$$\beta = \frac{2\pi}{\lambda_1}$$

where λ_1 is the wavelength on the transmission system. The velocity of propagation on the transmission line is given by

$$v = \frac{\omega}{\beta}$$

If losses are small, this formula will reduce to

$$v = \frac{c}{\sqrt{\mu_r \epsilon_r}}$$

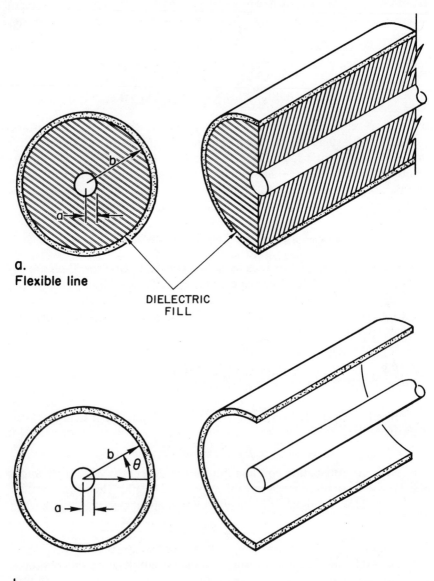

a.
Flexible line

DIELECTRIC
FILL

b.
Rigid line

Fig. 2.2-1. Coaxial line construction (a) flexible line
(b) rigid line.

where c is the velocity of propagation in a vacuum. From this it can be seen whether μ_r and ϵ_r are equal to 1, which means that the coaxial transmission line is vacuum insulated and the conductors are nonmagnetic. Then

$$v = c$$

From this it can be said that the wavelength

$$\lambda_1 = \frac{\lambda}{\sqrt{\mu_r \epsilon_r}}$$

This clearly proves that on an air-filled coaxial transmission line, the wavelength will measure exactly the same as in free space or on a Lecher wire system. Consequently, wavelength measurements can be performed directly on air-filled coaxial transmission lines.

Attenuation

The real part of a propagation constant provides information on the attenuation in terms of nepers per unit length. If losses are small, they can be neglected in most cases. Furthermore, where the characteristic impedance can be assumed to be a real value not having imaginary components independent of frequency, the attenuation due to *conductor losses* can be calculated by

$$\alpha_C = \frac{R}{2Z_0}$$

and the attenuation due to *dielectric losses* by

$$\alpha_D = \frac{G}{2Y_0}$$

The total attenuation for the above equation is

$$\alpha = \alpha_C + \alpha_D = \frac{R}{2Z_0} + \frac{G}{2Y_0} \text{ nepers/unit length}$$

The formulas for the equivalent circuit components will have to be calculated to get R and G values. The characteristic impedance also has to be calculated to use these expressions.

Using the geometrical parameters and the parameters of materials comprising the coaxial line in question, more direct expressions can be given for gaining information on attenuation.

ATTENUATION DUE TO CONDUCTOR LOSSES.[2]

$$\alpha_C = 13.6 \frac{\delta \mu_r}{\lambda} \frac{1}{b} \left(1 + \frac{b}{a}\right) \frac{\sqrt{\epsilon_r}}{\ln \frac{b}{a}} \text{ dB/unit length}$$

[2] Moreno, T., *Microwave Transmission Design Data* (New York: Dover Publications, Inc., 1958; © 1948, Sperry Gyroscope Company).

where the wavelength (λ) and the inner (a) and outer (b) conductor radii are given in terms of unit length; δ is the skin depth in the same unit length.

For copper conductors the expression is

$$\alpha_C = 2.98 \times 10^{-9}\sqrt{f}\,\frac{1}{b}\left(1 + \frac{b}{a}\right)\frac{\sqrt{\epsilon_r}}{\ln\dfrac{b}{a}} \text{ dB/cm}$$

where b and a are given in centimeters and f stands for frequency in terms of hertz. If the inner and outer conductors are made of different material, or if they are plated with different metals, the expression can be written

$$\alpha_C = \frac{13.6}{\lambda}\left(\frac{\delta_a\mu_a}{a} + \frac{\delta_b\mu_b}{b}\right)\frac{\sqrt{\epsilon_r}}{\ln\dfrac{b}{a}} \text{ dB/unit length}$$

The subscripts refer to the particular conductor.

If one would like to make the least lossy coaxial transmission line (as far as conductor losses are concerned), for a given inner or outer conductor dimension an optimum ratio of radii can be derived for an air-insulated line. This ratio is

$$\frac{b}{a} = 3.6$$

which gives a characteristic impedance of 77Ω.

ATTENUATION DUE TO DIELECTRIC LOSSES depends upon the properties of the dielectric material being used between the conductors. As in ordinary transmission line theory, the dielectric constant of a dielectric material is expressed as a complex quantity,

$$\epsilon = \epsilon' - j\epsilon''$$

where the real part of the expression is the relative dielectric constant ϵ_r and the imaginary part contains the information about the shunt losses of that material. The loss tangent is the customary way in which manufacturers of dielectric materials provide information about dielectric losses. The loss tangent is

$$\tan \delta = \frac{\epsilon''}{\epsilon'}$$

Dielectric losses of a coaxial line can be calculated by

$$\alpha_D = 27.3 \frac{\sqrt{\epsilon_r}}{\lambda} \tan \delta \text{ dB/unit length}$$

where the wavelength λ is given in terms of unit length.

ATTENUATION DUE TO BOTH CONDUCTOR AND DIELECTRIC LOSSES is found by simply adding to get the total attenuation:

$$\alpha_{total} = \alpha_C + \alpha_D$$

It is noteworthy to realize that the conductor losses increase only with the square root of frequency, but the dielectric losses increase linearly with frequency. Since dielectric losses at low frequencies are usually much smaller than conductor losses, they are often neglected. A word of caution is due. Dielectric losses will become increasingly important at higher frequencies since they are linearly proportional to frequency, whereas conductor losses increase with the square root of frequency.

Some other interesting relations and optimum values can be established in air-insulated coaxial lines. For a predetermined outer or inner conductor dimension, an optimum voltage breakdown b/a ratio can be derived using maximum voltage gradient calculations. Maximum *voltage* between conductors can be maintained when

$$\frac{b}{a} = \epsilon \cong 2.718$$

where the characteristic impedance becomes 60Ω.

Maximum power can be carried on a coaxial transmission line having 30Ω characteristic impedance, which relates to a b/a ratio of 1.65.

2.2.3 FIELD CONFIGURATIONS ON COAXIAL TRANSMISSION LINES

Since coaxial transmission lines have two conductors, according to transmission line theory they are capable of carrying the principal TEM mode. TEM (transverse electromagnetic mode) means that neither the electric nor the magnetic field has components in the direction of propagation. Transmission line theory states that both electric and magnetic fields can only be perpendicular at any place to each other. Consequently there is only one way the electric and magnetic fields can exist in a coaxial structure. This is shown in Fig. 2.2-2.

As the electric field is being established only between two conductors in the principal mode, it has to be radial. The magnetic field is placed around the center conductor between it and the outer conductor. On the longitudinal cross-sectional part in Fig. 2.2-2, the electric field intensity distribution is periodical according to the wavelength. There are high-intensity and low-intensity planes periodically changing places as the traveling wave goes down the transmission line. If total reflection occurs on a line, it is obvious that, where electric field intensity is high, the magnetic field intensity will be low.

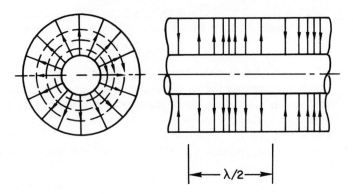

Fig. 2.2-2. Field distribution for the principal mode
in a coaxial line.

Where the electric field intensity is low, magnetic field intensity is consequently high.

2.2.4 HIGHER-ORDER MODES

As the propagation frequency increases, the wavelength decreases. After a definite limit where the cross-sectional dimensions of the coaxial line are comparable with the wavelength, the boundary conditions of higher-order modes are established. In other words, high-order modes will be able to propagate beside the principal mode. In any transmission line any number of modes—in fact an infinite number of modes—can exist, but at a certain frequency, as was described before, only a definite number of frequencies can exist. Below the first higher-order mode cutoff frequency only the principal mode can be propagated; but above that frequency the first high-order mode will be able to exist, and it will carry energy. For the lowest cutoff frequency higher-order mode the cutoff wavelength is approximately equal to the length of the circle drawn between the inner and outer conductors. Figure 2.2-3 shows this condition. The approximate cutoff wavelength for the first high-order mode is shown mathematically as follows:[3]

$$\lambda_c = \left(\frac{b-a}{2} + a\right) 2\pi = \pi(a+b)$$

If the coaxial line is filled with dielectric material, then this equation must be multiplied by the square root of the relative dielectric constant. Mathematically,

$$\lambda_c = \pi(a+b)\sqrt{\epsilon_r}$$

[3] Moreno, T., *Microwave Transmission Design Data.*

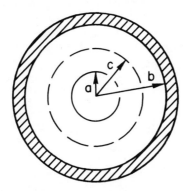

Fig. 2.2-3. Cross section of a coaxial line showing the midradius and its circle determining the cutoff wavelength of the TE₁₁ mode.

This formula will provide the information to an accuracy of about 8%. The higher-order modes are denoted depending upon their H-waves or E-waves by TE and TM, respectively. Two indices after this notation are given depending upon their field configuration. These are quite similar to the ones which will be discussed later for cylindrical waveguides.

Figure 2.2-4 shows a number of field configurations with the approximate equations for calculating them. These are the first few higher-order modes. This clearly shows that the design of a coaxial system for very high frequencies must take into consideration the geometry of the line so that it will always operate below the first higher-order mode. It was shown before that, as one uses larger-diameter coaxial lines, the losses will become smaller. However, a consideration that would limit increasing the size of the particular coaxial line in question is the first higher-order mode's cutoff frequency. As an example, the cutoff frequency of the 7 mm air line shown in Fig. 2.2-5 is calculated below.

$$\lambda_{c_{TE_{11}}} \approx \pi(0.06 + 0.138) = \pi \times 0.198 = 0.6217 \text{ in.} = 1.579 \text{ cm}$$

The cutoff frequency is

$$f_c \cong \frac{c}{\lambda_c} \cong 19 \text{ GHz}$$

This is why 7 mm, 50Ω air lines are usually specified to operate up to 18 GHz. It is a good practice not to approach the cutoff frequency too closely. Mechanical tolerances may allow the cutoff frequency to move up and down according to the magnitude of these tolerances. Furthermore, discontinuities on the transmission lines excite the higher-order modes, although it is true they do not carry energy and they attenuate exponentially. Close to cutoff, this exponential decay becomes quite slow, and discontinuities otherwise far

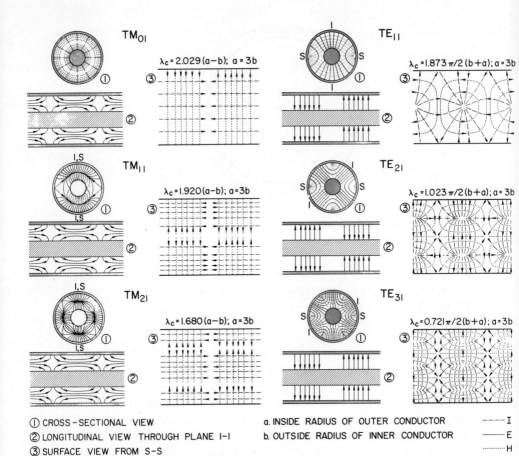

① CROSS-SECTIONAL VIEW
② LONGITUDINAL VIEW THROUGH PLANE I-I
③ SURFACE VIEW FROM S-S

a. INSIDE RADIUS OF OUTER CONDUCTOR   ---- I
b. OUTSIDE RADIUS OF INNER CONDUCTOR   ——— E
                                             .... H

Fig. 2.2-4. Higher-order modes in coaxial lines.

apart from one another can interfere under these conditions, causing some unwanted difficulties. Attenuation of any of the higher-order modes near cutoff is

$$\alpha = \frac{54.6}{\lambda_c} \sqrt{1 - \left(\frac{\lambda_c}{\lambda}\right)^2} \text{ dB/cm}$$

where the wavelengths are given in centimeters.

2.2.5 OBSTACLES IN COAXIAL TRANSMISSION LINES

It has been shown that attenuation due to dielectric losses is increasingly important at high frequencies. For low-loss, precision transmission lines, air

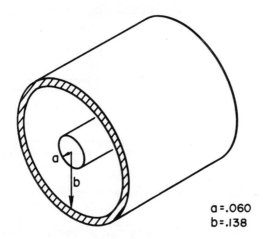

a =.060
b =.138

Fig. 2.2-5. 7-mm coaxial air line dimensions.

lines are used. Since the conductors in the center of the line have to be held concentric, holding structures are necessary. Several approaches are used in coaxial air-line transmission lines.

Low DIELECTRIC CONSTANT, LOW-LOSS, PLASTIC OR CERAMIC BEADS were developed to accommodate very wide frequency band coverage. These beads certainly represent discontinuities in the transmission line, but this effect is negligible because the reactive component at the junction where the bead starts is matched with the opposite sign of the same magnitude of reactive component. Where the discontinuity occurs, energy is stored and some losses will occur. In very high-power transmission lines, this shows up as excessive heat, and plastic beads are not customarily used because they would melt or deform.

QUARTER-WAVELENGTH STUBS ARE OFTEN USED where narrow bandwidth or single-frequency transmission systems are involved. A quarter-wavelength stub is another transmission line which is connected in parallel with the main transmission line, as shown in Fig. 2.2-6. As can be seen, the stub is short-circuited at the far end and is connected to the center conductor at the junction. Since the stub is a quarter-wavelength long, it will transform that short circuit into an open circuit a quarter-wavelength away (a transmission line has transformer properties and a short circuit will look like an open circuit a quarter-wavelength away). As was mentioned above, this can only be used at a single frequency or with a very narrow bandwidth since a quarter-wavelength will not provide the required reactance at a wider range of frequencies and will cause high reflections. Consequently, in a narrow band application, a quarter-wavelength stub effectively will not show up at all as a discontinuity,

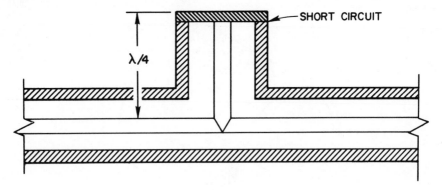

Fig. 2.2-6. Cross section of a quarter-wavelength stub.

and, at the same time, it will give a very rigid support to the center conductor to which it is connected.

There are several ways in which beads can be placed into transmission lines. One approach is shown in Fig. 2.2-7, where a quarter-wavelength-

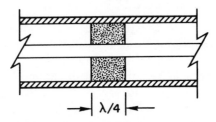

Fig. 2.2-7. Quarter-wavelength-long bead.

long dielectric bead holds the center conductor in the outer conductor. In this structure the quarter-wavelength-long dielectric-filled coaxial transmission line will have a different characteristic impedance. At the two ends of that bead the reflections will cancel, since the discontinuities are of the same magnitude but a quarter-wavelength apart. Transmission lines have transformer properties; assuming the discontinuity is capacitive, a quarter-wavelength away another capacitance will cancel the reflection, since it is transformed into an inductance at that plane. The structures shown in Fig. 2.2-8 keep the characteristic impedance constant by undercutting the center conductor, the outer conductor, or both. Although there are still discontinuities at the input and the output of the bead, these cancel because they are a quarter-wavelength long, and the characteristic impedance remains constant. The undercut was needed due to the dielectric constant of the material

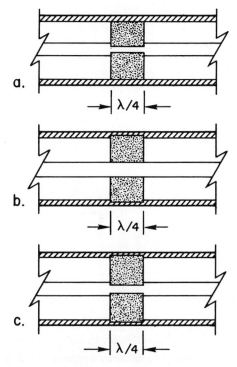

Fig. 2.2-8. Quarter-wavelength-long beads (a) center
conductor undercut (b) outer conductor
undercut (c) both conductors undercut.

used to make the bead in order to maintain the same characteristic impedance
as in the air-filled coaxial line. Quarter-wavelength-long beads are also narrow band or single-frequency type structures, since, if the frequency changes,
cancellation of the discontinuities will not occur. Because these structures are
transforming small reflections and not a short circuit, the mismatch will not
be so large as that of a quarter-wavelength stub. Narrow beads can be used
in pairs if they are spaced in quarter-wavelength distances from each other,
as shown in Fig. 2.2-9. These structures, of course, are also narrow band
devices.

   Stubs and beads can be designed to accommodate broader band
ranges. Figure 2.2-10 shows a broadband stub support for a coaxial transmission line. Here the characteristic impedance is changed a quarter-wavelength in each direction from the stub support in order to have a double
cancellation effect that achieves wider bandwidth characteristics for this support. In coaxial transmission lines in lower powers where beads are used, the
best approach to broadbanding is to place the matching reactance into the

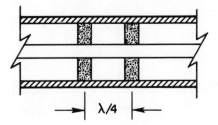

Fig. 2.2-9. Narrow beads spaced a quarter-wavelength apart.

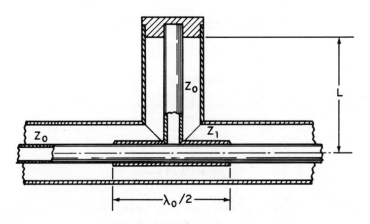

Fig. 2.2-10. Broadband stub support for coaxial transmission line.

same plane where the discontinuity occurs. The discontinuity due to the bead is usually a step capacitance. To cancel the effect of that capacitance, series inductance has to be established in the same plane. This will provide very wide band impedance-matching. Such a design is shown in Fig. 2.2-11. The

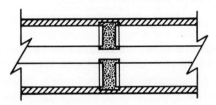

Fig. 2.2-11. Broadband matched narrow dielectric bead.

undercut on the bead acts as a series inductance. The equivalent circuit of this matching network is shown in Fig. 2.2-12. This structure has a very wide band response, and it is used in most modern bead designs.

Inductance due to undercut

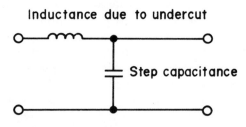

Step capacitance

Fig. 2.2-12. Equivalent circuit for half of the undercut bead.

SLEEVES. In more complex coaxial transmission networks it is necessary to eliminate reflections by the use of matching networks. The simplest type of matching network is a so-called quarter-wave *sleeve*. This reacts as an impedance-matching network in a coaxial system. There are two basic types of quarter-wave sleeves: one that is made by increasing the diameter of the center conductor, the other one made by decreasing the outer conductor's diameter. Figure 2.2-13 shows both these quarter-wavelength sleeves. These

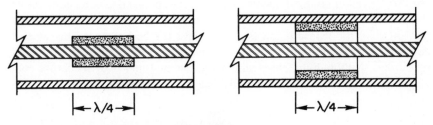

Fig. 2.2-13. Quarter-wavelength sleeves.

sleeves react as transformers providing the necessary impedance match. The diameters of the sleeve sections depend on the magnitude of the standing-wave ratio which is being matched. To gain the least frequency sensitivity, the sleeve should be placed as close as possible to the discontinuity being matched, because a shorter line length between them results in less phase difference. Multiple quarter-wavelength sleeves can be arranged to gain wider bandwidths by using more complicated mathematical approaches. Effectively, a sleeve changes or adjusts the characteristic impedance of the line, and this can be achieved by using good conductive material or by using dielectric material. Of course, the effective dielectric constant must be known to enable one to calculate the necessary length of the quarter-wavelength section, since the dielectric material will adjust the effective electrical length by the square root of the dielectric constant of that material. These are called the dielectric slug tuners. Some matching networks are made adjustable by moving two of these slugs or sleeves against and away from each other to establish certain kinds of mismatches a distance away from the discontinuity to be tuned out in order to get as good a match as possible. If one takes a stub such as was described above and moves the shorting plane, he is able to place a reactance to the main transmission line. Using two or more of these so-called adjustable moving stubs at certain distances from each other, a very effective tuning transformer can be achieved. These are the so-called double- or triple- or multistub tuners which effectively place certain reactances at definite planes on the transmission line in order to tune out all the existing mismatches, effectively making the line really perfectly matched. This technique is still used quite a bit in single-frequency measurement techniques. Generally, a double-stub tuner within a certain bandwidth would be able to tune out any amount of reflection; another tuner with a different spacing of the two stubs is needed if the rated frequency range is passed over.

SUSCEPTANCE TRANSFORMERS. Other types of so-called susceptance transformers are also used quite often for impedance-matching purposes. One of these is the slide screw tuner, in which a probe is inserted into a coaxial line through a longitudinal opening made on the coaxial line outer conductor; this provides a parallel capacitance. This so-called screw is then moved along the length of the transmission line to achieve the best match. This type of device is easier to use than the double-stub or triple-stub tuner for tuning out reflections. Although matching with a slide screw tuner is by its nature convergent, the mismatch can be reduced in steps by alternately varying the probe

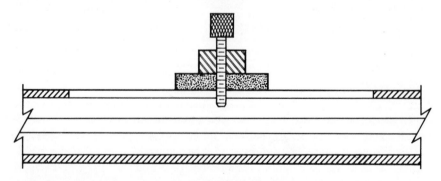

Fig. 2.2-14. Cross section of a simple coaxial slide screw tuner.

penetration and moving along the line axially while measuring the reflection. Figure 2.2-14 shows a cross section of a simple slide screw tuner. Simple slide screw tuners or a double-, triple- or multistub tuner can provide a nearly perfect match at any single frequency but must be retuned for any other frequency since they are only single-frequency, not broadband, tuning devices. This is quite easily understood by noticing that the discontinuity is being tuned out not at the plane of the discontinuity but at a certain distance away, using the transformer properties of the transmission line.

QUESTIONS

1. Give the inductance, capacitance, resistance, and conductance of a coaxial transmission line having an outer conductor inner diameter of 0.300 in. and an inner conductor diameter of 0.125 in. when the dielectric material between conductors has a relative dielectric constant of 2.3 with a loss tangent of 0.0008, the length of the line is 15 in., and the frequency of operation is 3 GHz. The conductors are made of copper.

2. What is the characteristic impedance of a coaxial transmission line having 0.4375 in. diameter outer conductor and 0.100 in. diameter inner conductor with a dielectric fill $\epsilon_r = 2.56$?

3. Give the attenuation due to both conductor and dielectric losses on a coaxial transmission line having an outer conductor inner diameter of 0.375 in. and an inner conductor diameter of 0.092 in. and the relative dielectric constant of the insulator between conductors $\epsilon_r = 2.32$, with the dielectric loss tan $\delta = 0.001$. The outer conductor is made of copper and the inner conductor of phosphor bronze. The length of the line is 38 in., and the frequency of operation is 4 GHz.

4. What is the cutoff frequency of the first higher-order mode on a coaxial transmission line having an outer conductor inner diameter of 0.312 in. and an inner conductor diameter of 0.120 in. and a dielectric constant of the insulator between conductors of $\epsilon_r = 2.1$?

2.3 WAVEGUIDE TRANSMISSION LINES

2.3.1 SINGLE-MODE PROPAGATION

The reader has already seen in this chapter that there are three possible modes which can exist on any arbitrary transmission line. One of them is the TEM, or principal mode, which needs at least two conductors. If the operating frequency is high enough and the boundary conditions for higher-order modes to exist are satisfied, TE and TM types of higher-order modes can also exist. Their number is infinite, and they can exist in either one- or two-conductor lines if the frequency is high enough. Certain transmission lines will carry only a certain number of higher-order modes in a certain frequency range. Generally, all higher-order modes have different cutoff frequencies, and these modes will be able to propagate just above their cutoff frequencies. Waveguides generally are defined as transmission lines that cannot carry the principal mode of transmission. Only the higher-order modes can exist on them. These are one-conductor type transmission lines. Waveguide can be an arbitrarily shaped hollow pipe with or without conductive, metallic boundaries. As mentioned above, there is an infinite number of modes that are able to propagate in a particular transmission line, but there is only a finite number of modes in a certain frequency range that can propagate in a particular transmission line. All those have different field configurations. Their field patterns are discussed later in this section. All modes are excited in different ways.

All modes are solutions of Maxwell's equations that will fit the described boundary conditions. Basically the high-order modes can be divided into two families: the TE and the TM modes. The TE, or transverse electric, modes are those in which the magnetic waves have at least one component in the direction of propagation, but all electric field components are transverse to the axis. Because of this, they are sometimes called the H-waves. The TM, or transverse magnetic, modes are such that some component of the electric field will be in the direction of propagation. All the magnetic field components are transverse to the axis. TM modes are also called E-waves. The identification of different modes is done with two indices after their designation, for instance TE_{mn}. They are dependent upon the field configuration.

The lowest possible high-order mode in a waveguide is called the dominant mode. It is quite important to differentiate from all other modes, since this is the only type of mode in waveguide which can exist and can propagate without any interference from other high-order modes.

The useful bandwidth in a waveguide is called the range of frequencies where single-mode propagation exists—in other words, the bandwidth where only the dominant mode can propagate. This region of frequencies lasts only to the cutoff frequency of the second higher-order mode. Even at this frequency or very close to it (the second higher-order mode's cutoff frequency), it is not a good practice to increase the effective bandwidth, since machining tolerances of the waveguide would possibly allow the higher-order modes to come in a little bit over or about that particular frequency. Furthermore, if discontinuities occur on the line somewhere and the particular second higher-mode would be excited by the discontinuity, close to cutoff beyond cutoff attenuation will not be so rapid and the whole thing could cause some other difficulties which would not be encountered if a little more room had been given below this cutoff frequency. That is why a little space is generally left from the second higher-order mode's cutoff frequency to the top of the useful bandwidth of the single-mode propagating dominant mode. The bottom of the bandwidth close to the first higher-order modes (the dominant modes) cutoff frequency is also not too useful since the losses in a waveguide are quite excessive very close to cutoff. The useful bandwidth of the dominant mode starts usually about 20% higher in frequency from cutoff. It will also be seen later in this chapter that the cross-sectional geometry of the waveguide transmission line will have quite an important effect on where the second higher-order mode comes in, since the cutoff frequencies of the higher-order modes are entirely dependent upon the geometry of the waveguide. As was mentioned before, the phase velocity and the group velocity of the propagation are identical in the principal mode and, in an air-filled transmission line, are exactly equivalent to the velocity of light.

$$v_{\text{TEM}} = c \approx 3 \times 10^8 \text{ m/s} \qquad (2.3\text{-}1)$$

In waveguides where only higher-order modes can exist, by the same token, for any high-order modes, the group velocities and phase velocities differ from each other. In fact, in an air- or vacuum-insulated waveguide, the geometric means of the phase velocity and group velocity are equal to the velocity of light.

$$\sqrt{v_p v_g} = c \qquad (2.3\text{-}2)$$

which means that the phase velocity v_p could be higher than the velocity of light, in which case the group velocity would be slower than the velocity of light. If one makes a waveguide slotted line, he would find out that, for a particular frequency, the equivalent waveguide wavelength would be longer than the so-called free-space wavelength because the effective velocity measured would be the phase velocity. The waveguide wavelength relates to the cutoff frequency and to the free-space wavelength. The relationship is

$$\lambda_g = \frac{\lambda}{\sqrt{1 - \left(\frac{\lambda}{\lambda_c}\right)^2}} \qquad (2.3\text{-}3)$$

where λ_g stands for the guide wavelength, λ is the free-space wavelength, and λ_c is the cutoff wavelength. It can be seen that, if the free-space wavelength approaches the cutoff wavelength, the guide wavelength gets very long; in fact, when the former reaches cutoff wavelength, the latter will become infinitely long. Furthermore, if one goes higher and higher in frequency (which would mean that the frequency would go farther and farther away from the cutoff frequency), one would find that the waveguide wavelength gets closer and closer to the free-space wavelength. When the frequency becomes very high and tends to go to infinity, then the waveguide wavelength and free-space wavelength become the same. Of course, it is worth mentioning at this point that many higher-order modes are already coming in and can propagate.

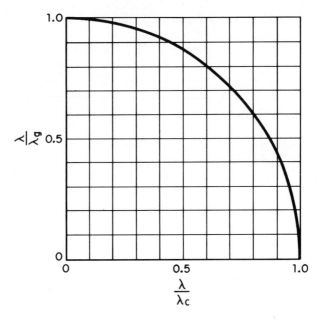

Fig. 2.3-1. Wavelength relationships in any air-filled
waveguide for any mode of propagation.

The relationship shown in Eq. (2.3-3) can be plotted in terms of λ/λ_c versus λ/λ_g in Fig. 2.3-1.

If one takes a waveguide completely filled with very-low-loss dielectric material where the dielectric constant is ϵ_r, the waveguide wavelength expression will be modified:

$$\lambda_g = \frac{\lambda}{\sqrt{\epsilon_r - \left(\dfrac{\lambda}{\lambda_{co}}\right)^2}} \qquad (2.3\text{-}4)$$

where λ_{co} is the cutoff wavelength of the empty waveguide, not the dielectric-filled guides.

2.3.2 RECTANGULAR WAVEGUIDE

In the previous section useful bandwidth was defined as the case when only a single mode can propagate. This means that only the dominant mode can propagate. It was found that one of the widest single-mode-propagating waveguide cross sections is not the cylindrical waveguide cross section but the rectangular waveguide cross section, in which the second higher-order mode would come quite higher in frequency than in other structures. Figure 2.3-2 shows such a rectangular waveguide in a rectangular coordinate system

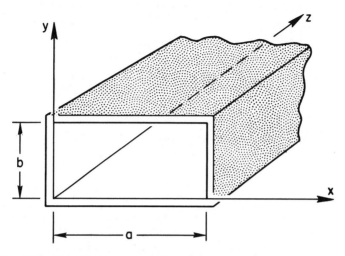

Fig. 2.3-2. Rectangular waveguide.

where a is the wider inside dimension and b is the narrower inside dimension. The rectangular waveguide, being a waveguide and a one-conductor propagating transmission line, can propagate only high-order modes. The modes of transmission can consequently be only in TE and TM modes. Subscripts are used to describe the electric and magnetic field configurations. The general symbols of TE_{mn} or TM_{mn} are used to describe transverse electric or transverse magnetic waves. The subscript m indicates the number of half-wave variations of the electric field along the wide dimension of the waveguide, and n indicates the number of half-wave variations of electric or magnetic field in the narrow dimension of the guide. The TE_{10} mode, which has the longest operating wavelength, is designated as the dominant mode. This is the mode for the lowest frequency which will propagate in the waveguide. If a dimension (the wide dimension on the waveguide) is less than $\frac{1}{2}$ wavelength, no propagation will occur. Therefore waveguide acts as a high-pass filter. For a rectangular piece of waveguide, the cutoff frequency can be found from

$$f_c = \frac{1}{2} c \sqrt{\left(\frac{m}{a}\right)^2 + \left(\frac{n}{b}\right)^2} \qquad (2.3\text{-}5)$$

where c is the velocity of propagation of light, m and n are the subscripts of the particular TE or TM mode, and a and b are the wide and narrow inside dimensions of the rectangular guide, respectively. The cutoff wavelength is then

$$\lambda_c = \frac{2}{\sqrt{\left(\frac{m}{a}\right)^2 + \left(\frac{n}{b}\right)^2}} \qquad (2.3\text{-}6)$$

The cutoff for the dominant mode can be calculated easily. The dominant mode being the TE_{10}, substituting 1 and 0 in place of m and n, respectively,

$$\lambda_{c_{TE_{10}}} = \frac{2}{\sqrt{\left(\frac{1}{a}\right)^2 + \left(\frac{0}{b}\right)^2}} \qquad (2.3\text{-}7)$$

yields

$$\lambda_{c_{TE_{10}}} = 2a. \qquad (2.3\text{-}8)$$

The field configuration in a rectangular waveguide propagating the commonly used TE_{10} mode is shown in Fig. 2.3-3. Rectangular waveguides were origi-

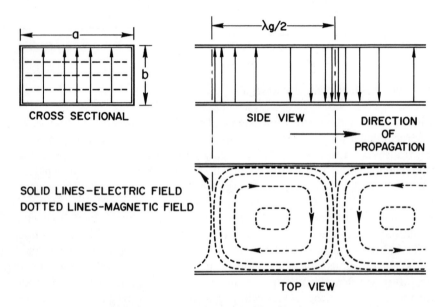

Fig. 2.3-3. Fields in a rectangular guide (dominant TE_{10} mode).

nally chosen so that the dominant mode would exist over a certain frequency range. This frequency determines the a dimension. The b dimension is chosen on the basis of the following criteria: (1) the attenuation loss is greater as the b dimension is made smaller; and (2) the b dimension determines the voltage breakdown characteristics and therefore determines the maximum power-handling capacity. For larger power capacity, a and b should be made as large as possible. In practice, b is made about half of the a dimension. It is desirable in all types of transmission systems to select sizes so that only one mode of propagation is possible. In other words, the physical size of the guide is related to the frequency band under consideration. Because of the possibility of higher-order modes, it has become common to operate waveguides over an approximately 50% frequency band. By properly selecting this frequency range, it is possible to operate far enough from cutoff so that the guide parameters do not vary too rapidly and to avoid also the frequency region where other modes are possible. Table 2.3-1 lists some of the characteristics of various sizes of commonly used waveguides. Figure 2.3-4 shows the field configurations of some of the other higher-order modes.

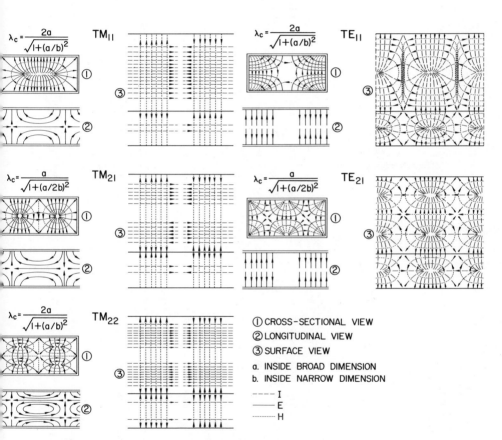

Fig. 2.3-4. Modes in rectangular waveguide.

2.3.3 WAVEGUIDE IMPEDANCE

The characteristic-wave-impedance is analogous to the characteristic impedance of two-wire and coaxial lines. The characteristic-wave-impedance represents the ratio of the electric and magnetic field, since the electric field

Table 2.3-1. Reference table of rigid rectangular waveguide data and fittings.†

EIA WG WR ( )	MDL Band	Recommended Operating Range for TE₁₀ Mode — Frequency (kmc/sec)	Recommended Operating Range — Wavelength (cm)	Cut-off for TE₁₀ Mode — Frequency (kmc/sec)	Cut-off — Wavelength (cm)	Range in 2λ/λc	Range in λg/λ	Theoretical cw power rating lowest to highest frequency (mw)	Theoretical attenuation lowest to highest frequency (db/100 ft.)	EIA WG WR ( )	JAN WG RG ( )	Material Alloy	JAN FLANGE UG ( )/U Choke	JAN FLANGE UG ( )/U Cover	Inside	Tol. (±)	Outside	Tol. (±)	Wall Thickness (nom.)	Tol. (±)
2300		0.32-0.49	93.68-61.18	0.256	116.84	1.60-1.05	1.68-1.17	153.0-212.0	.051-.031	2300		Alum.			23.000-11.500	0.020	23.376-11.876	.020	0.188	.020
2100		0.35-0.53	85.65-56.56	0.281	106.68	1.62-1.06	1.68-1.18	120.0-173.0	.054-.034	2100		Alum.			21.000-10.500	0.020	21.376-10.876	.020	0.188	.020
1800		0.41-0.625	73.11-47.96	0.328	91.44	1.60-1.05	1.67-1.18	93.4-131.9	.056-.038	1800	201	Alum.			18.000-9.000	0.020	18.250-9.250	.020	0.125	.020
1500		0.49-0.75	61.18-39.97	0.393	76.20	1.61-1.05	1.62-1.17	67.6-93.3	.069-.050	1500	202	Alum.			15.000-7.500	0.015	15.250-7.750	.015	0.125	.015
1150		0.64-0.96	46.84-31.23	0.513	58.42	1.60-1.07	1.82-1.18	35.0-53.8	.128-.075	1150	203	Alum.			11.500-5.750	0.015	11.750-6.000	.015	0.125	.015
975		0.75-1.12	39.95-26.76	0.605	49.53	1.61-1.08	1.70-1.19	27.0-38.5	.137-.095	975	204	Alum.			9.750-4.875	0.010	10.000-5.125	.010	0.125	.010
770		0.96-1.45	31.23-20.67	0.766	39.12	1.60-1.06	1.66-1.18	17.2-24.1	.201-.136	770	205	Alum.			7.700-3.850	0.010	7.950-4.100	.010	0.125	.010
650	L	1.12-1.70	26.76-17.63	0.908	33.02	1.62-1.07	1.70-1.18	11.9-17.2	.317-.312 / .269-.178	650	69 / 103	Brass / Alum.	417A*	418A*	6.500-3.250	0.010	6.660-3.410	.010	0.080	.010
510		1.45-2.20	20.67-13.62	1.157	25.91	1.60-1.05	1.67-1.18	7.5-10.7		510					5.100-2.550	0.010	5.260-2.710	.010	0.080	.010
430	W	1.70-2.60	17.63-11.53	1.372	21.84	1.61-1.06	1.70-1.18	5.2-7.5	.588-.385 / .501-.330	430	104 / 105	Brass / Alum.	435A*	437A*	4.300-2.150	0.008	4.460-2.310	.008	0.080	.008
340		2.20-3.30	13.63-9.08	1.736	17.27	1.58-1.05	1.78-1.22	3.1-4.5	.877-.572 / .751-.492	340	112 / 113	Brass / Alum.	553*	554*	3.400-1.700	0.005	3.560-1.860	.005	0.080	.005
284	S	2.60-3.95	11.53-7.59	2.078	14.43	1.60-1.05	1.67-1.17	2.2-3.2	1.102-.752 / .940-.641	284	48 / 75	Brass / Alum.	54B / 585A	53 / 584	2.840-1.340	0.005	3.000-1.500	.005	0.080	.005
229		3.30-4.90	9.08-6.12	2.577	11.63	1.56-1.05	1.62-1.17	1.6-2.2		229					2.290-1.145	0.005	2.418-1.273	.005	0.064	.005
187	C	3.95-5.85	7.59-5.12	3.152	9.510	1.60-1.08	1.67-1.19	1.4-2.0	2.08-1.44 / 1.77-1.12	187	49 / 95	Brass / Alum.	148C / 406B	149A / 407	1.872-0.872	0.005	2.000-1.000	.005	0.064	.005
159		4.90-7.05	6.12-4.25	3.711	8.078	1.51-1.05	1.52-1.19	0.79-1.0		159					1.590-0.795	0.004	1.718-0.923	.004	0.064	.004

Band No.	Band	Freq. Range (GHz)								Desig.	Material	Flange						
137		5.85-8.20	5.12-3.66	4.301	6.970	1.47-1.05	1.48-1.17	0.56-0.71	2.45-1.94	106	Alum.	440B	441	1.372-0.622	0.004	1.500-0.750	0.004	0.064
112	X_L	7.05-10.0	4.25-2.99	5.259	5.700	1.49-1.05	1.51-1.17	0.35-0.46	4.12-3.21 / 3.50-2.74	51 / 68	Brass / Alum.	52B / 137B	51 / 138	1.122-0.497	0.004	1.250-0.625	0.004	0.064
102		7.05-10.0								320	Brass	1494	1493	1.020-0.510	0.003	1.148-0.638	0.003	0.064
90	X	8.20-12.40	3.66-2.42	6.557	4.572	1.60-1.06	1.68-1.18	0.20-0.29	6.45-4.48 / 5.49-3.83	52 / 67	Brass / Alum.	40B / 136B	39 / 135	0.900-0.400	0.003	1.000-0.500	0.003	0.050
75		10.00-15.00	2.99-2.00	7.868	3.810	1.57-1.05	1.64-1.17	0.17-0.23		75	Brass	—	—	0.750-0.375	0.003	0.850-0.475	0.003	0.050
62	Ku	12.4-18.0	2.42-1.66	9.486	3.160	1.53-1.05	1.55-1.18	0.12-0.16	9.51-8.31 / 6.14-5.36	91 / 107	Brass 541A / Alum. — / Silver —	419 / — / —		0.622-0.311	0.002	0.702-0.391	0.003	0.040
51		15.00-22.00	2.00-1.36	11.574	2.590	1.54-1.05	1.58-1.18	0.080-0.107		53	Brass	—	—	0.510-0.255	0.0025	0.590-0.335	0.003	0.040
42	K	18.00-26.50	1.66-1.13	14.047	2.134	1.56-1.06	1.60-1.18	0.043-0.058	20.7-14.8 / 17.6-12.6 / 13.3-9.5	53 / 121 / 66	Brass / Alum. / Silver	596A / 598A / —	595 / 597 / —	0.420-0.170	0.0020	0.500-0.250	0.003	0.040
34		22.00-33.00	1.36-0.91	17.328	1.730	1.57-1.05	1.62-1.18	0.034-0.048		34	Brass	1530*		0.340-0.170	0.0020	0.420-0.250	0.003	0.040
28	K_A	26.50-40.00	1.13-0.75	21.081	1.422	1.59-1.05	1.65-1.17	0.022-0.031	21.9-15.0	96	Brass 600A / Alum. — / Silver —	599 / — / —		0.280-0.140	0.0015	0.360-0.220	0.002	0.040
22	Q	33.00-50.00	0.91-0.60	26.342	1.138	1.60-1.05	1.67-1.17	0.014-0.020	31.0-20.9	97	Brass / Silver	383 / —		0.224-0.112	0.0010	0.304-0.192	0.002	0.040
19		40.00-60.00	0.75-0.50	31.357	0.956	1.57-1.05	1.63-1.16	0.011-0.015		98	Brass	1529*		0.188-0.094	0.0010	0.268-0.174	0.002	0.040
15	V	50.00-75.00	0.60-0.40	39.863	0.752	1.60-1.06	1.67-1.17	0.0063-0.0090	52.9-39.1	98	Brass / Silver	385 / —		0.148-0.074	0.0010	0.228-0.154	0.002	0.040
12		60.00-90.00	0.50-0.33	48.350	0.620	1.61-1.06	1.68-1.18	0.0042-0.0060	93.3-52.2	99	Brass / Silver	387 / —		0.122-0.061	0.0005	0.202-0.141	0.002	0.040
10		75.00-110.00	0.40-0.27	59.010	0.508	1.57-1.06	1.61-1.18	0.0030-0.0041			Brass	1528*		0.100-0.050	0.0005	0.180-0.130	0.002	0.040
8		90.00-140.00	0.333-0.214	73.840	0.406	1.64-1.05	1.75-1.17	0.0018-0.0026	152-99	278	Silver Lam.	1527*		0.0800-0.0400	0.0003	0.120-0.080	0.001	0.020
7		110.00-170.00	0.272-0.176	90.840	0.330	1.64-1.06	1.77-1.18	0.0012-0.0017	163-137	276	Silver Lam.	1525*		0.0650-0.0325	0.00025	0.105-0.073	0.001	0.020
5		140.00-220.00	0.214-0.136	115.750	0.259	1.65-1.05	1.78-1.17	0.00071-0.00107	308-193	275	Silver Lam.	1524*		0.0510-0.0255	0.00025	0.091-0.066	0.001	0.020
4		170.00-260.00	0.176-0.115	137.520	0.218	1.61-1.05	1.69-1.17	0.00052-0.00075	384-254	277	Silver Lam.	1526*		0.0430-0.0215	0.00020	0.083-0.062	0.001	0.020
3		220.00-325.00	0.136-0.092	173.280	0.173	1.57-1.06	1.62-1.18	0.00035-0.00047	512-348	3	Silver	—		0.0340-0.0170	0.00020	0.156 dia.	0.001	—

*Contact Flange

† Courtesy of Microwave Development Laboratories, Inc., Needham Heights, Mass.

is analogous to voltage and the magnetic field is analogous to current. There is little practical use for the characteristic-wave-impedance, but it can be determined from the following equations:

$$Z_{\text{TE}} = \frac{\eta}{\sqrt{1 - \left(\dfrac{\lambda}{\lambda_c}\right)^2}} \qquad (2.3\text{-}9)$$

$$Z_{\text{TM}} = \sqrt{1 - \left(\frac{\lambda}{\lambda_c}\right)^2} \, (\eta) \qquad (2.3\text{-}10)$$

where η is the intrinsic impedance of the medium (377 ohms for free space).

Because there are no unique currents and voltages, the characteristic-wave-impedance cannot be determined so easily for waveguide as for a coaxial line. But this is not really a problem, since the process of normalization eliminates characteristic impedance as a requirement for calculations. The basic quantities for waveguide work are reflection coefficient, standing-wave ratio, and propagation constant. From these, the normalized impedance at any point can be determined, and the complete waveguide system can be described in terms of its performance and characteristics.

For coaxial systems the reflection coefficient was defined as the ratio of the reflected signal to the incident signal. It is the same for waveguide.

2.3.4 CYLINDRICAL WAVEGUIDES

Cylindrical waveguides are needed for such devices as rotary joints because rectangular waveguides cannot be used for this purpose without excessive reflections. As was mentioned before, cylindrical waveguides do not have so wide a single mode propagating dominant mode as rectangular waveguide has. A typical cylindrical coordinate system is shown in Fig. 2.3-5. The domi-

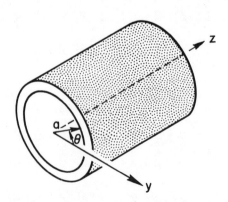

Fig. 2.3-5. Cylindrical waveguide.

nant mode in a cylindrical waveguide is the TE_{11} mode. The indices determining the high-order modes are designed as follows: the first index denotes the total number of full-period variations of either component or field along a circular path; the second index is one more than the total number of sign reversals of either component or field along a radial path. The equations describing the fields use higher mathematics involving Bessel functions. The field configuration of the dominant or TE_{11} mode, which really corresponds to the TE_{10} mode of the rectangular waveguide, is shown in Fig. 2.3-6. Some

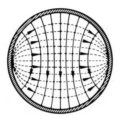

Fig. 2.3-6. Fields of the dominant TE_{11} mode.

of the higher-order modes in the cylindrical waveguide are shown in Fig. 2.3-7. The cutoff frequencies of higher-order modes in a cylindrical waveguide are determined by roots of special functions typically given separately for TE modes and TM modes. The cutoff frequency of the TE_{mn} mode is given by

$$\lambda_{c TE_{m,n}} = \frac{2\pi a}{u'_{m,n}} \tag{2.3-11}$$

where a denotes the inner radius of the guide.

In the denominator, $u'_{m,n}$ are roots of Bessel functions, and these roots are given in Table 2.3-2 for a few lower modes.

Table 2.3-2.

$u'_{0,1} = 3.832$ $u'_{0,2} = 7.016$
$u'_{1,1} = 1.841$ $u'_{1,2} = 5.332$
$u'_{2,1} = 3.054$ $u'_{2,2} = 6.706$
$u'_{3,1} = 4.201$ $u'_{3,2} = 8.013$

The cutoff wavelength for TM modes is given by another formula:

$$\lambda_{c TM_{m,n}} = \frac{2\pi a}{u_{m,n}} \tag{2.3-12}$$

Another set of rules for the special functions is given in Table 2.3-3, to be used with the TM type of modes.

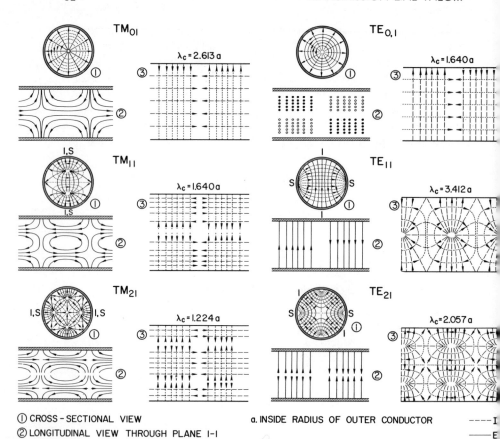

① CROSS-SECTIONAL VIEW   a. INSIDE RADIUS OF OUTER CONDUCTOR   ----- I
② LONGITUDINAL VIEW THROUGH PLANE I-I   ——— E
③ SURFACE VIEW FROM S-S   ····· H

Fig. 2.3-7. Field configurations of some of the modes
in cylindrical waveguides.

Table 2.3-3.

$u_{0,1} = 2.405$	$u_{0,2} = 5.520$	$u_{0,3} = 8.654$
$u_{1,1} = 3.832$	$u_{1,2} = 7.016$	
$u_{2,1} = 5.136$		

2.3.5 ELLIPTICAL WAVEGUIDES

One has to take into consideration the possibilities of machine tolerances and imperfections in cylindrical waveguides. The shape of the circular
cross section can be distorted in handling, making it slightly elliptical. It is
very important to be aware of the difficulties that must be faced if this
happens.

If the dominant (TE_{11}) mode is being propagated on a cylindrical waveguide and ellipticity is encountered, at a random length and random orientation of cylindrical waveguide, the original TE_{11} mode will split into two components in line with the two axes of the ellipsis. Figure 2.3-8 shows an exaggerated picture of this case.

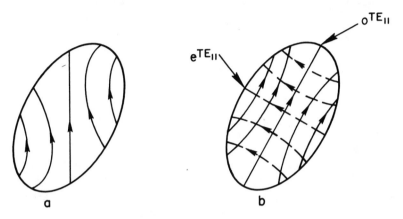

Fig. 2.3-8. Elliptical guide field configuration where dominant mode enters (a) at random orientation but splits (b) into two components of odd and even waves.

The odd and even modes have different propagation characteristics, cutoff frequencies, phase velocities, and attenuation. The fields are described in terms of Mathieu functions. These functions provide even and odd functions, from which the corresponding modes were named even and odd.

2.3.6 PROPAGATION CHARACTERISTICS OF WAVEGUIDES

In all previous discussions about waveguides it was assumed that the waveguides have no losses. This is assumed many times, and most of the time correctly, when small attenuation is not of primary concern. Waveguide is generally used because of its very high power-carrying capabilities, whereas coaxial transmission lines are limited in higher frequencies by their cross-sectional geometry because higher-order modes would be able to propagate. This would limit the area of the wall at higher frequencies, and, consequently, higher losses would be involved due to skin effect. Waveguides in their particular bandwidths have larger dimensions and consequently more surface area in cross section. This allows fewer losses, so they are able to carry higher powers.

The voltage breakdown on waveguide transmission lines is higher than on coaxial lines because they are larger in size in corresponding frequency ranges. Attenuation comes into primary concern when analyzing propagation characteristics of waveguides. Attenuation can be caused by two major contributors, one being the conductor loss and the other the dielectric losses when dielectric fill is used. The attenuation expressions are not so simple as they were in coaxial transmission lines. Attenuation will not only depend upon the frequency increasing; it will depend upon the cutoff frequency, too. In fact, it will depend upon the ratio of the operating frequency over the cutoff frequency. The expressions given for attenuation will apply only for above-cutoff operation. When the frequency of operation gets closer to cutoff, it can be seen that the attenuation will increase very rapidly. This is the reason, which was mentioned before, that waveguides close to their cutoff frequencies are very lossy and their actual useful bandwidth is quite a bit above their cutoff frequencies.

Attenuation due to conductor losses in a rectangular waveguide when the waveguide is operating in the dominant (TE_{10}) mode is given in Eq. (2.3-13), where the waveguide is made of copper and there is no dielectric fill.

$$\alpha_c = \frac{0.01107}{a^{3/2}} \left[\frac{\frac{a}{2b} \left(\frac{f}{f_c}\right)^{3/2} + \left(\frac{f}{f_c}\right)^{-1/2}}{\sqrt{\left(\frac{f}{f_c}\right)^2 - 1}} \right] \text{dB/ft} \qquad (2.3\text{-}13)$$

In the equation, f stands for the operating frequency and f_c is the cutoff frequency; a and b are given in terms of inches. If a metal other than copper is used as a conductor, this formula has to be multiplied by the square root of the ratio of the resistivities, which factor would be the relative loss factor between the particular metal and copper. Many standard waveguides used have specifications of attenuation given in terms of decibels per 100 feet (given in Table 2.3-1). Attenuation as a function of frequency for different modes in a typical copper rectangular waveguide is given in Fig. 2.3-9. Figure 2.3-10 shows attenuation as a function of frequency for some of the lower transmission modes in cylindrical copper waveguide 2 inches in diameter. It is interesting to note from Fig. 2.3-10 that, as the frequency increases, the attenuation will decrease for the TE_{01} mode in cylindrical guide. This is particular only to this mode. Of course, if the smallest ellipticity occurs, this behavior of the T_{01} mode will reverse and will react as all the other modes do. In very high-frequency transmission, the TE_{01} mode is often used for quite narrow band transmission because of this behavior.

Attenuation due to dielectric losses is taken into consideration, as it was in coaxial transmission line cases. The relative dielectric constant is split into two components, one real and one imaginary, where

$$\epsilon_r = \epsilon_r' - j\epsilon_r'' \qquad (2.3\text{-}14)$$

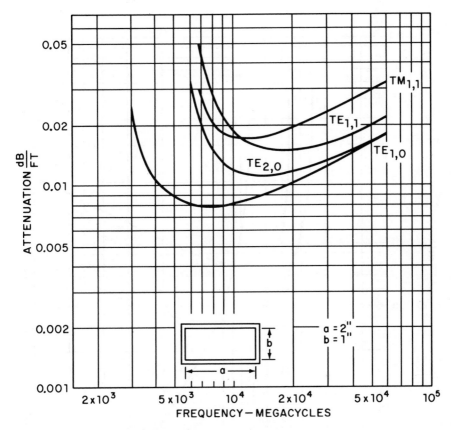

Fig. 2.3-9. Attenuation as a function of frequency for various modes in a typical copper rectangular waveguide.

and the loss tangent

$$\tan \delta = \frac{\epsilon_r''}{\epsilon_r'} \tag{2.3-15}$$

It is important to note that, when waveguide is filled with dielectric material, it will change its cutoff frequency with the square root of the dielectric constant.

$$f_c' = \frac{f_c}{\sqrt{\epsilon_r'}} \tag{2.3-16}$$

In this expression, f_c' stands for the new cutoff frequency and ϵ_r' is the actual relative dielectric constant.

Dielectric material will not only come into consideration in the attenu-

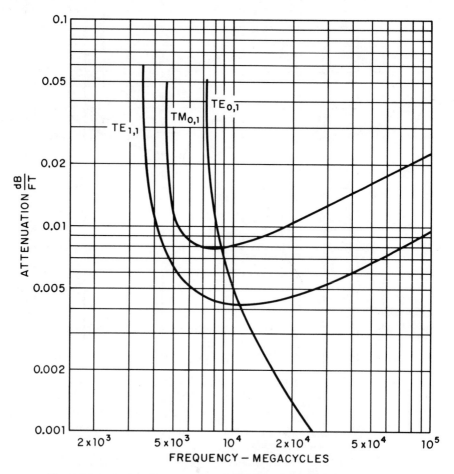

Fig. 2.3-10. Attenuation as a function of frequency for
some of the lower transmission modes in
a circular copper waveguide 2 inches in
diameter.

ation of the dielectric losses, but it will also modify the attenuation due to
conductor losses. The attenuation due to conductor losses is a function of
frequency, cutoff frequency, and the a and b dimensions in the rectangular
waveguide. The fourth root of the relative dielectric constant has to be used as
a multiplication factor, and f_c' has to be used at all places where f_c appears
in Eq. (2.3-13). The modified equation for conductor losses will be

$$\alpha_c = \frac{0.01107\ \epsilon_r^{1/4}}{a^{3/2}} \left[\frac{\frac{a}{2b}\left(\frac{f}{f_c'}\right)^{3/2} + \left(\frac{f}{f_c'}\right)^{-1/2}}{\sqrt{\left(\frac{f}{f_c'}\right)^2 - 1}} \right] \text{dB/ft} \qquad (2.3\text{-}17)$$

The expression for attenuation due to dielectric losses in a waveguide is

$$\alpha_D = 830 \frac{\epsilon''}{\lambda}\left(\frac{\lambda_g}{\lambda}\right) \text{dB/ft} \qquad (2.3\text{-}18)$$

where the wavelength information is given in terms of centimeters. Attenuation due to both conductor and dielectric losses gives the total attenuation, which then becomes

$$\alpha_{\text{total}} = \alpha_C + \alpha_D \qquad (2.3\text{-}19)$$

2.3.7 WAVEGUIDE STRUCTURES, DISCONTINUITIES

Wherever there is a change in the rectangular dimensions of a rectangular waveguide, such as a bend, a load, or a junction, there is a disturbance of the energy storage relationship that exists between the electric and magnetic fields. The effect is equivalent to a susceptive shunt. If the electric field is strengthened relative to the magnetic field, the irregularity in the guide is equivalent to a capacitive shunt. If the magnetic field is strengthened, the irregularity is equivalent to an inductive shunt. The inductances and capacitances are normally quite small and are negligible at low frequencies. However, if the dimensions of the obstacle are appreciable fractions of a wavelength, the energy stored in the distorted field must be considered.

Reactive Obstacles

Among the more common obstacles that may be placed in a waveguide to introduce intentional susceptance are diaphragms or irises, which partially close the guide, and projections into the guide, such as screws or rods. Figure 2.3-11 shows the construction of various types of susceptances that may be placed in a waveguide. If the inductive and capacitive diaphragms are combined to form a window, as in Fig. 2.3-12, the effect is a parallel resonance circuit. It can introduce either capacitive or inductive susceptance, depending on the frequency. At the resonant frequency the equivalent shunt susceptance is zero and the window acts as if it were not present. Off resonance, it may reflect a considerable portion of the incident power. If the window is to be made highly frequency selective, it should be very narrow and about one-half a wavelength long. The easiest susceptance to construct is to allow a screw

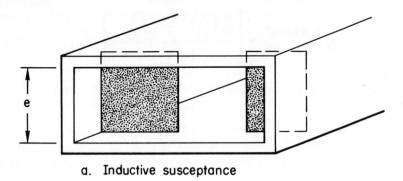

a. Inductive susceptance

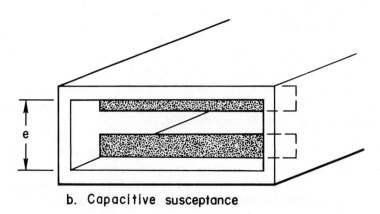

b. Capacitive susceptance

Fig. 2.3-11. Rectangular diaphrams (a) inductive
susceptance (b) capacitive susceptance.

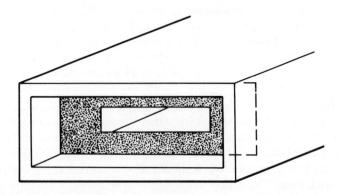

Fig. 2.3-12. Parallel resonant "window."

or other metal probe to protrude into the guide. If the screw is less than a quarter-wavelength long, the effect is a capacitive susceptance. Longer than a quarter-wavelength, the effect is an inductive susceptance. If the screw is a quarter-wavelength long, it produces effects similar to a series resonance shunt, and almost all of the incident power will be reflected. Table 2.3-4 shows the impedances of an aperture in a rectangular waveguide. The impedances refer to a plane of the aperture.

Table 2.3-4.*

Transverse outline of aperture	Circuit character	$\dfrac{Z}{Z_0}$†
a. Symmetrical capacitive aperture		$-j\dfrac{\lambda_g}{4b}\left(\ln\dfrac{2b}{\pi d}+\dfrac{b^2}{2\lambda_g^2}\right)^{-1}\qquad \dfrac{d}{b}\ll 1$ $-j\dfrac{2b\lambda_g}{\pi^2(b-d)^2}\qquad \dfrac{b-d}{b}\ll 1$
b. Asymmetrical capacitive aperture		As above but with λ_g replaced by $\dfrac{\lambda_g}{2}$
c. Symmetrical inductive aperture		$j\dfrac{a}{\lambda_g}\tan^2\dfrac{\pi d}{2a}\left(1+\dfrac{1}{6}\dfrac{\pi^2 d^2}{\lambda^2}\right)\qquad \dfrac{d}{a}\ll 1$ $j\dfrac{a}{\lambda_g}\cot^2 2\pi\dfrac{a-d}{a}\left[1+\dfrac{8\pi^2}{3}\dfrac{(a-d)^2}{\lambda^2}\right]$ $\dfrac{a-d}{a}\ll 1$
d. Asymmetrical inductive aperture		$j\dfrac{2a^3}{\pi^2\lambda_g(a-d)^2}\left[1+\dfrac{\pi^2}{\lambda^2}(a-d)^2\ln\dfrac{\pi}{2}\dfrac{a-d}{a}\right]^{-1}$ $\dfrac{a-d}{a}\ll 1$
e. Small circular aperture		$j\dfrac{2\pi d^3}{3ab\lambda_g}\qquad d\ll b$
f. Resonant aperture		$\dfrac{Z}{Z_0}\rightarrow\infty$ when $\dfrac{a'}{b'}\sqrt{1-\left(\dfrac{\lambda}{2a'}\right)^2}=\dfrac{a}{b}\sqrt{1-\left(\dfrac{\lambda}{2a}\right)^2}$

† Approximate values, under restrictions noted.

* Reprinted from H. A. Atwater, *Introduction to Microwave Theory* (New York: McGraw-Hill Book Company, 1962) p. 112 by permission of the publisher.

Coupling to Waveguides

Because any means used to insert energy into a waveguide serves equally well to extract energy, only the methods of coupling to waveguides need be considered. As might be expected, waveguide can be easily coupled to free space or to a cavity by merely leaving the end of the guide open. Energy will also enter the guide if the open end is in a region where an electromagnetic field of the appropriate waveguide frequency exists. Radiation from and into an open waveguide end can be increased and directed by flaring the guide into a horn which serves as an impedance-matching transition from the guide to free space. Coupling from a coaxial line into a waveguide is one of the most often-used transitions. If a waveguide is considered analogous to a parallel strip line, an obvious means of coupling would be to attach one conductor of the coaxial line to the tip of the guide and the other conductor to the bottom of the guide, as shown in Fig. 2.3-13. The left end of the guide

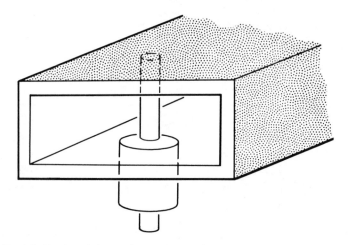

Fig. 2.3-13. Coaxial coupling to waveguide.

is closed by a short-circuiting plate a quarter-wavelength from the point of connection. Transmission of radiation from the wrong direction is prevented. This type of coupling has the disadvantage of not being adjustable. If the center conductor of the coaxial line is terminated in a probe, it does not touch the top of the guide. The device may be adjustable by varying the probe penetration.

An alternate method of coupling to waveguide is to bend the center conductor into a loop that is attached to the outer conductor, where it enters the waveguide. This method provides a rather simple but unadjustable coupling. Ideally, both the coaxial line and the waveguide transmission line should be terminated in its characteristic impedance with no reactive components. In addition, the electric and magnetic fields about the end of the coaxial line must have the same general form as the desired fields in the guide. The greater the similarity, the greater the energy transfer. Careful designs can approach these conditions, but tuning is ordinarily completed by means of tuning screws which project into the guide. At the plane of coupling to a waveguide or at any discontinuity in the guide, the electric and magnetic fields do not have the simple modes, as previously illustrated. The fields in the guide near coupling probes or loops, for instance, are formed laterally by probes or loops and may have little resemblance to the desired waveguide mode. However, any possible mode pattern can be analyzed by means of a Fourier type of approach, which superimposes one field upon another. The analysis results in a fundamental mode and higher modes of appropriate phase and magnitude. The fundamental mode will propagate down the line, but the higher-order modes will die away in a short distance because the cutoff frequencies of these higher-order modes are too high for the waveguide to propagate. Therefore the complicated field exists only in the vicinity of the irregularity, discontinuity.

Waveguide Junctions[4]

Waveguide junctions are useful in a variety of situations. Typical applications include power dividers, mixers, and sampling junctions. The common junction used in waveguide is the waveguide tee. The characteristics of these junctions can be analyzed by considering the behavior of the electromagnetic fields in the junction. If the axis of the arm is parallel to the electric field of the main transmission line, the junction is called an E-plane junction or series tee. If the axis of the arm is parallel to the magnetic field of the main transmission line, the junction is called an H-plane junction or shunt tee. Illustrations of both types are shown in Figs. 2.3-14 and 2.3-15. The progress of a representative line of electric force on the waveform is shown in Fig. 2.3-14b. Note that the line of force bends as it leaves the side arm, and the two resultant fields in arms 1 and 2 are opposite in polarity, or out of phase. They will be equal in magnitude if the junction is completely symmetrical. Similar analysis on the shunt or H-plane tee will show that the wave entering the tee at arm 3 will divide and leave the arms in phase—and equal in magnitude if the junction is symmetrical.

E- and H-plane tees, when constructed of hollow waveguide, are usually poorly matched devices. Normally a tuning device is necessary to provide the necessary impedance matches. A combination of the E- and H-plane tees is a *hybrid junction* known as a magic tee. This term is sometimes reserved

[4] Atwater, *Introduction to Microwave Theory* (New York: McGraw-Hill Book Company, 1962), Chap. 4.

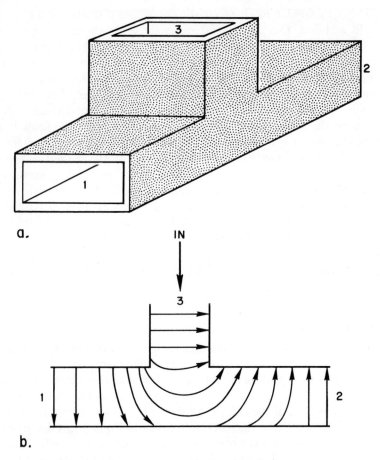

Fig. 2.3-14. E-plane tee.

for such a hybrid junction that has matching structures such as tuning rods in the E- and H-arms. An illustration of this type of junction is shown in Fig. 2.3-16. The geometry of this type of junction is such that the wave entering the E-arm excites equal waves opposite in arms 1 and 2. Similarly, a wave entering the H-arm excites equal waves equal in phase in arms 1 and 2. Because of the symmetry of the device, a wave entering the E-arm does not excite a dominant wave in the H-arm. Reciprocally, there is no direct transmission from the H-arm to the E-arm. If fields equal in magnitude and phase enter arms 1 and 2, the net field in the E-arm will be zero, and the total energy will emerge from the H-arm. The magic tee can be used in microwave receivers for a mixer. If crystal detectors are placed in arms 1 and 2, the local oscillator fed into the E-arm, and the signal frequency fed into the H-arm, mixing and

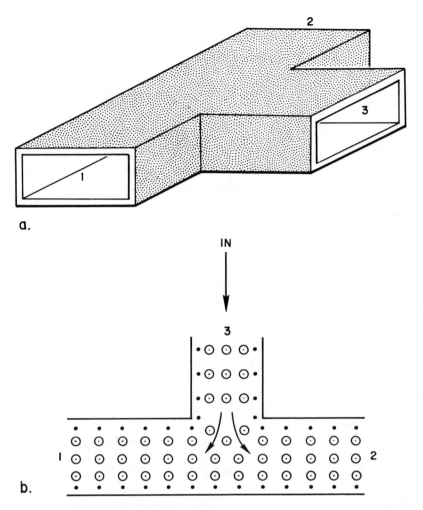

Fig. 2.3-15. H-plane tee.

detection will take place in the tee. The tee junction provides the necessary isolation between two sources of signals which are mixed and detected by the crystals. The magic tee can also be used as a phase shifter by terminating the E- and H-arms with adjustable shorts.

QUESTIONS

1. What is the waveguide wavelength on a waveguide transmission line if the cutoff frequency is 6.56 GHz at 10 GHz?

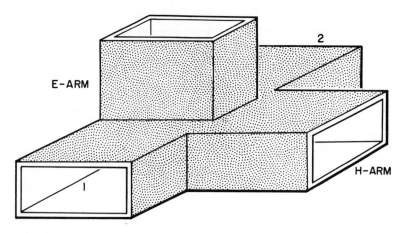

Fig. 2.3-16. Hybrid junction.

2. What is the waveguide wavelength of a dielectric-filled waveguide at 6 GHz if the empty waveguide's cutoff frequency is 8.6 GHz? The dielectric constant is $\epsilon_r = 2.56$.

3. What is the cutoff wavelength and cutoff frequency of a rectangular waveguide in the TE_{32} mode if $a = 0.400$ in. and $b = 0.200$ in.?

4. What is the cutoff frequency of a cylindrical waveguide in the TE_{11} mode if the inner diameter is 1.000 in.?

5. What is the diameter of a cylindrical waveguide if the cutoff frequency in the dominant mode is 9.2 GHz? At what frequency will the TM_{01} mode appear (cutoff frequency)?

6. What is the attenuation in the dominant mode due to conductor losses of a 10-ft-long rectangular waveguide made of copper when the inner dimensions are $a = 2.84$ in. and $b = 1.34$ in.? The operating frequency is 3 GHz.

7. What is the attenuation in the dominant mode due to both conductor and dielectric losses of a dielectric-filled 3-ft-long rectangular waveguide made of copper when the relative dielectric constant of the dielectric is 2.56 and the dielectric loss is 0.001? The waveguide dimensions are $a = 0.500$ in. and $b = 0.250$ in. The operating frequency is 10 GHz.

2.4 RESONANT CAVITIES

Circuit elements at ordinary radio frequencies consist of inductors, capacitors, resistors, and transformers. Resonant devices can be built from these circuit elements. An ordinary parallel resonant circuit is shown in Fig.

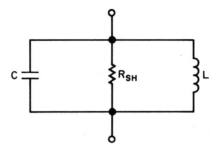

Fig. 2.4-1. Parallel resonant circuit.

2.4-1. Thompson's Formula describes the resonant frequency of such a circuit.

$$f_0 = \frac{1}{2\pi\sqrt{LC}} \qquad (2.4\text{-}1)$$

where the characteristic impedance of such a circuit can be given by

$$Z_0 = \sqrt{\frac{L}{C}} \qquad (2.4\text{-}2)$$

and the input admittance of such a resonant circuit by

$$Y_{\text{in}} = \frac{1}{R_{\text{SH}}} + j\left(\frac{f}{f_0} - \frac{f_0}{f}\right)\sqrt{\frac{C}{L}} \qquad (2.4\text{-}3)$$

From this equation it can be readily seen that the input impedance below resonant frequency will be inductive, since the imaginary part is positive. Also, it will become capacitive above resonant frequency, since the imaginary part of the impedance becomes negative and the impedance at resonance will be the reciprocal value of the shunt resistance. From this, the definition of Q can be given:

$$Q = \frac{R_{\text{SH}}}{2\pi fL} \qquad (2.4\text{-}4)$$

As was mentioned in the discussion of transmission line theory, as frequency increases and the geometrical dimensions of the components of the network in question become comparable with an appreciable portion of the wavelength, lumped-circuit components cannot be used readily. In higher frequencies it has been shown that it is possible to establish such reactances, inductances, or capacitances with small sections of transmission lines. To review this, imagine that energy is applied to a piece of transmission line with one end short-circuited and the other open-circuited. If the piece of line is shorter than one-quarter of the wavelength of the applied signal, it provides an inductance. If it is longer than one-quarter but shorter than one-half wavelength, it becomes a capacitor. At exactly a quarter-wavelength long, the line becomes resonant at that particular frequency; if this piece is doubled and both ends are terminated with either a short circuit or an open circuit, it will still resonate because it is now a half-wavelength long. A quarter-wavelength line terminated at one end with a short circuit and at the other end with an open circuit satisfies boundary conditions for establishing a quarter-wavelength resonant circuit. At the short-circuited end there will be voltage minimum but current maximum, whereas on the open-circuited end there will be a voltage maximum and current minimum.

Figure 2.4-2 shows a quarter-wavelength-long resonant line and its voltage distribution at resonance.

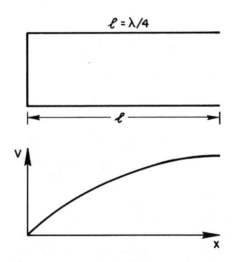

Fig. 2.4-2. Quarter-wavelength resonant line.

Figure 2.4-3 shows a half-wavelength-long resonant line with the voltage pattern at resonance.

If the frequencies have kept increasing, and other resonances can be seen on a quarter-wavelength resonant line, with increasing frequency other resonances can also be observed. The resonant frequencies would be when l, the length of the line, was exactly $\frac{3}{4}$, $\frac{5}{4}$, $\frac{7}{4}$, etc.—odd numbers of quarter-wavelengths.

The half-wavelength resonant line would resonate at full, $1\frac{1}{2}$, 2, $2\frac{1}{2}$, multiples of half-wavelengths. Pieces of transmission lines can be then used as resonant circuit elements in any microwave transmission system. If a variable resonant frequency device (in other words, a tunable resonant circuit) has to be realized at microwave frequencies, a variable length of transmission

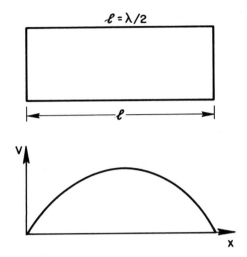

Fig. 2.4-3. Half-wavelength resonant line.

line can be used for this purpose. Resonant devices can be made of any type of transmission line, such as twin lead or Lecher wires, coaxial transmission lines, or even from waveguides. But the resonant wavelength has to be calculated using the applicable wavelength, for instance, in the case of a waveguide, the waveguide wavelength. Since resonators have resonances other than the desired ones, it always has to be taken into consideration that these spurious resonances are to be known and, if it is possible, have to be moved far away from the anticipated resonant frequency to be measured or used. These resonant circuits built of pieces of transmission lines terminated with short circuits or short and open circuits are called resonant cavities. Resonant cavities can be used for building oscillators, establishing resonant networks for oscillators, and as passive devices—for instance, as wavemeters or frequency meters. These types of devices will be discussed in later chapters.

QUESTION

1. What is the input impedance of a resonant circuit having a characteristic impedance of 77 ohms if the resonant frequency is 6 GHz, the operating frequency is 5 GHz, and the Q of the resonant circuit is 3,000?

3
PRINCIPLES OF MICROWAVE MEASUREMENTS

3.1 DETECTION OF MICROWAVE SIGNALS

Measuring any unknown quantity of a certain parameter is always done by comparison of the unknown quantity to a known, which is taken at that time as a standard. For example, if one measures voltage between two conductors with a multimeter while reading the value of voltage, a comparison is made to the standard with which the multimeter was calibrated. In other words, whenever measurement is performed, some kind of substitution or comparing of a value of an unknown with a known is done. These requirements must be met to make measurements on a transmission system. A *signal source* is needed, plus a device that will *detect* or modify the signal and connect it to an instrument which provides *indication* of the presence or behavior of that signal. Signal sources will not be dealt with at this time, since the next chapter is devoted entirely to that subject.

3.1.1 DETECTORS

Detectors are not new to the reader. In the old crystal radio set the crystal performed basically the same role that the detectors in microwaves perform. A detector rectifies the RF signal subjected to it.

Microwave crystal detectors are usually the point-contact type of semiconductor devices. These diodes have low capacitances across the point-contact junction; consequently they are suitable for microwave rectification, as other nonlinear devices are at low frequencies. The semiconductor used is usually silicon or germanium and is doped with certain amounts of impurities.

The diode operates because a contact potential is established between two dissimilar conductors. To understand how this can happen, consider two metals joined as shown in Fig. 3.1-1. If metal B has more free electrons (is more conductive) upon contact, an electron flow takes place predominantly from B to A. After equilibrium is reached, A will apparently be charged more negatively and B will be more positive. Essentially, a potential barrier will be formed at the junction, which effectively provides rectification.

Crystal detectors are widely used in the microwave field because of their sensitivity and simplicity. They are used as video detectors to provide either a dc output when unmodulated microwave energy is applied or a low-frequency ac output up to tens of megahertz or higher when the microwave signal is modulated. They are also used as mixers in superheterodyne systems, especially at microwave frequencies where other mixers, such as vacuum tubes, are insufficient or inefficient.

The essential parts of a crystal detector are a semiconducting chip and a metal whisker, which contacts the chip. A typical microwave crystal

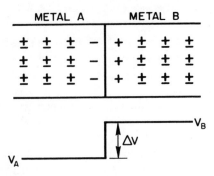

Fig. 3.1-1. Two metals contact with dissimilar conductivity.

detector uses a silicon chip about $\frac{1}{16}$-inch square and a pointed tungsten whisker wire about $\frac{3}{1,000}$ inch in diameter. The other parts of a crystal detector or mount are needed simply to support the chip and the whisker and to couple electrical energy to the detector. Although crystal detectors are successful at microwave frequencies partly because of their extremely small size, these dimensions also limit their power-handling capability; 100 mw is sufficient to damage some crystals.

3.1.2 SQUARE LAW OF CRYSTAL DETECTORS

Crystals as video detectors are very convenient for use as a simple detector on a slotted line or for indicating relative power levels. For these applications the user must usually know the law relating output voltage to applied microwave voltage. One can easily deduce the law for a simple circuit which will illustrate the general behavior in more complicated situations. Consider the idealized circuit shown in Fig. 3.1-2, in which a sinusoidal microwave voltage is applied to the RF terminals. Capacitor C bypasses the RF, leaving dc current to flow to the milliammeter. A typical crystal detector

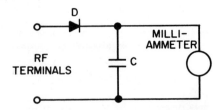

Fig. 3.1-2. Idealized crystal detector circuit.

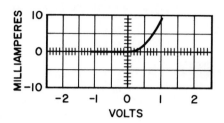

Fig. 3.1-3. Square characteristics of a crystal diode.

has a current voltage characteristic similar to that shown in Fig. 3.1-3. Any such curve can always be approximated by a Taylor series consisting of terms involving powers of v, that is,

$$i = a_0 + a_1 v + a_2 v^2 + a_3 v^3 + \ldots \tag{3.1-1}$$

If the operating point is the origin ($v = 0$, $i = 0$), as it is for Fig. 3.1-2, then a_0 is 0. Let

$$v = A \cos \omega t \tag{3.1-2}$$

where A is the amplitude, ω is equal to $2\pi f$, and f is the microwave frequency. Substituting in Eq. (3.1-1) yields

$$i = a_1(A \cos \omega t) + a_2(A \cos \omega t)^2 + a_3(A \cos \omega t)^3 + \ldots \tag{3.1-3}$$

For extremely small signals, all terms of Eq. (3.1-3) except the first are negligible. And, $i = a_1(A \cos \omega t)$. The current is simply proportional to the applied voltage, and the crystal behaves as a simple resistor with negligible dc current flowing through the milliammeter. However, for somewhat larger signals, the second term must be included to obtain reasonable accuracy.

$$i = a_1(A \cos \omega t) + a_2(A \cos \omega t)^2 \tag{3.1-4}$$

or

$$i = a_1(A \cos \omega t) + \frac{a_2 A^2}{2} (1 + \cos 2 \omega t) \tag{3.1-5}$$

A standard trigonometric identity has been used.

$$\cos^2 \theta = \tfrac{1}{2} + \tfrac{1}{2} \cos 2\theta \tag{3.1-6}$$

The current now includes the dc component $(a_2 A^2)/2$, which flows through the milliammeter, and the second harmonic component $(a_2 A^2)/2 \times \cos 2t$, which flows through C. Thus the milliammeter indication is proportional to the square of the amplitude A of the microwave voltage. At still-higher signal levels more terms of Eq. (3.1-3) must be retained, and the crystal behavior departs from *square law*. Commonly used video crystal detector circuits are more complicated than that in Fig. 3.1-2. They are characterized by a square-law region limited at the low-power end by the noise level of the meter or amplifier connected to the crystal and at the high-power end by departure from square-law behavior.

3.1.3 INDICATORS

Indicators are used primarily for visual presentation of the presence or behavior of the detected signal. Such devices can be oscilloscopes, galvanometers, or special indicator devices like the standing-wave indicator. A standing-wave indicator is a high-gain, tuned amplifier which takes an input from a crystal detector or any audio source. It also has a built-in bias supply to provide bias current for devices like bolometers. The input signal is amplified and applied to a meter calibrated for use with square-law detectors. Figure 3.1-4 shows a block diagram of a standing-wave indicator. Input voltages first go to input switching to provide either the bias supply or an impedance match for the right input characteristics needed. Then the signal is fed to the first section of a range switch and then to the input of an ampli-

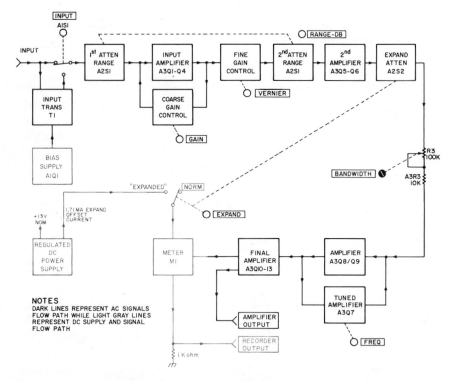

Fig. 3.1-4. Block diagram of a standing-wave indicator.

fier. The second section of the switch is located between the first amplifier and the second amplifier. Switch positions are in 10-dB steps. In the input amplifier, the gain and vernier controls are associated with the amplifier and vary gain over a range of more than 10:1. The gain control is a coarse control to adjust the negative feedback in this amplifier. Vernier is a fine gain control and changes gain in series with the output signal. An ac feedback is provided in the second amplifier for stability and high-input impedance. The output of this amplifier is applied to the expand attenuator and then to the following amplifier. The expand function allows any signal level to be measured on the expanded scale with continuous coverage while maintaining the original reference level. Expansion is accomplished by applying a precise amount of dc offset current to the meter and simultaneously increasing the signal to the third amplifier. This increased gain allows a certain decibel change in signal level to deflect the meter across its full scale. The offset current places the zero signal indication off-scale to the left.

Frequency-Selective Circuits

The frequency response of the third amplifier is shaped by negative feedback. The feedback path includes a Wien bridge and an amplifier. At the null frequency of the Wien bridge, the negative feedback path is open and the gain of the amplifier is maximum. Off center frequency, the negative feedback in the Wien bridge reduces gain. The amount of the off-resonance gain reduction depends on the setting of the bandwidth control. The Wien bridge is set for a sharp null at center frequency, which is adjustable. Actually this control is set for a very slight bridge unbalance to produce just enough positive feedback. Thus, at resonance, negligible signal current flows through the bandwidth control, and the gain is independent of its setting. Center frequency is set by variable resistors. The output amplifier consists of four transistors (usually two output transistors which are operating in a push-pull class-B amplifier). Both collectors are ac grounded. Large negative feedback makes the gain of the output amplifier very nearly unity. As standing-wave indicators, these selective amplifiers are used mostly for standing-wave ratio measurements in conjunction with the slotted line and many other various measurements. This will be discussed in later chapters.

QUESTIONS

1. What provides rectification in a crystal diode? What makes rectification happen at the contact?
2. If a continuous wave (CW) microwave signal is applied to one port of a crystal detector, what kind of signal comes out the other port of the crystal detector?

3. What kind of curve describes the current voltage characteristics of a crystal detector?

4. Standing-wave indicators are calibrated to be used with square-law devices. If one uses a direct signal input to the standing-wave indicator and drops the signal level by 10 dB, how much change, in dB, will be shown by the square-law indicator?

3.2 FLOW GRAPH REPRESENTATION OF MICROWAVE NETWORKS

3.2.1 SCATTERING PARAMETERS

In analyzing microwave transmission line problems, one would like to find some generalized parameters to write for a network in question—parameters which can be measured with reasonable simplicity, even in microwave frequencies. Analysis of the energy flow through a two-port network is one way to do this.

A simple two-port network can be shown as a "black box" (Fig. 3.2-1).

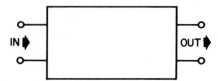

Fig. 3.2-1. Simple two-port network.

We are not interested in knowing what is built into the black box, but only in what it will do to a signal applied to either port. For example, if the black box contains an amplifier and we would like to know the various parameters, we can measure the input impedance while the output is short- and open-circuited and measure the output impedance while the input is short- and open-circuited. This will give us some of the commonly known z, y, and h parameters. However, this technique has some shortcomings at higher frequencies. Some devices may oscillate (probably at some frequency different from the measurement frequency) or have some unwanted, parasitic effects if they are terminated with a short or open circuit.

The ideal case would be to express a set of parameters when the input and output ports are terminated with their own characteristic impedances at all frequencies. The scattering (s) parameters are the set of parameters that are measured under such conditions. An added, inherent advantage of these parameters is that they describe the signal flow within the network. Kuro-kawa,[1] Penfield,[2,3] and Youla[4] studied generalized scattering parameters. Hunton[5] used signal flow to analyze microwave-measurement techniques with s parameters and expressed them with flow graphs, since these parameters relate directly to the signal flow. Kuhn[6] used a topographical approach for resolving these flow graphs.

3.2.2 BASIC FLOW GRAPHS

A flow graph can be drawn to analyze the energy flow of a two-port network. (See Fig. 3.2-2.) A flow graph has two nodes for each port, one for

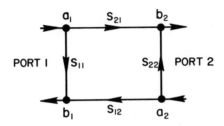

Fig. 3.2-2. Flow graph of a two-port network.

the entering (incident) wave and the other for the leaving (reflected) wave of that port. The incident node is the a node and the reflected node is the b node. In our example of the simple two-port network, when the incident wave enters the device at port 1, part of it will be returned through the s_{11} path and b_1 node. The remaining part of the incident wave goes through the s_{21} path and leaves the network through the b_2 node. If a device that has some reflections is connected to port 2, and if it will reflect part of the wave leaving b_2, this reflection will reenter the network through the a_2 node. Then, part of that may be reflected, passing along the s_{22} path and leaving the network

[1] Kurokawa, K., *IEEE Trans.-MTT*, March 1965, p. 194.

[2] Penfield, P., Jr., "Noise in Negative Resistance Amplifiers," *IRE Trans.-CT*, Vol. CT-7, June 1960, pp. 166–70.

[3] Penfield, P., Jr., "A Classification of Lossless Three Ports," *IRE Trans.-CT*, Vol. CT-9, September 1962, pp. 215–23.

[4] Youla, D. C., "On Scattering Matrices Normalized to Complex Port Numbers," *Proc. IRE*, Vol. 49, July 1961, p. 122.

[5] Hunton, J. K., "Analysis of Microwave Measurement Techniques by Means of Signal Flow Graphs," *Trans. IRE*, Vol. MTT-8, March 1960, pp. 206–12.

[6] Kuhn, Nicholaus, "Simplified Signal Flow Graph Analysis," *The Microwave Journal*, November 1963, pp. 59–66.

through the b_2 node. The other part of the wave passes through the s_{12} path and leaves the circuit through the b_1 node.

Following the arrows in the flow graph, we can write the following equations.

$$b_1 = a_1 s_{11} + a_2 s_{12} \tag{3.2-1}$$

$$b_2 = a_1 s_{21} + a_2 s_{22} \tag{3.2-2}$$

By analyzing these equations, one can see that these parameters can be easily measured under certain conditions.

Assume that there is no signal entering at a_2 node (this can be achieved by terminating port 2 with its characteristic impedance). Then Eqs. (3.2-1) and (3.2-2) become

$$b_1 = a_1 s_{11}, \tag{3.2-3}$$

$$b_2 = a_1 s_{21}. \tag{3.2-4}$$

By reversing the network, that is, terminating port 1 with its characteristic impedance and applying the signal to port 2, Eqs. (3.2-1) and (3.2-2) will become

$$b_1 = a_2 s_{12}, \tag{3.2-5}$$

$$b_2 = a_2 s_{22}, \tag{3.2-6}$$

since $a_1 = 0$.

Expressing the scattering parameters from Eqs. (3.2-3), (3.2-4), (3.2-5), and (3.2-6), we can write the following:

$$s_{11} = \frac{b_1}{a_1} \quad \bigg| \quad a_2 = 0 \tag{3.2-7}$$

$$s_{21} = \frac{b_2}{a_1} \quad \bigg| \quad a_2 = 0 \tag{3.2-8}$$

$$s_{12} = \frac{b_1}{a_2} \quad \bigg| \quad a_1 = 0 \tag{3.2-9}$$

$$s_{22} = \frac{b_2}{a_2} \quad \bigg| \quad a_1 = 0 \tag{3.2-10}$$

Furthermore, these expressions show the means of measuring these parameters: s_{11} can be measured when port 2 is terminated with its characteristic impedance and only the ratio of reflected wave and incident wave has to be measured at port 1. We saw that in Chap. 2, where reflection coefficients were discussed. This means that s_{11} is really the input-reflection coefficient of the device.

s_{22} is measured in exactly the same manner as s_{11}, except that port 1 is terminated with its characteristic impedance and the signal is applied to port 2; s_{22} is the output-reflection coefficient of the network.

s_{21} is measured when port 2 is terminated with its characteristic impedance and the signal is applied into port 1. The ratio of the signals measured at the b_2 and a_1 nodes (voltage between output and input ports) defines the value of s_{21}. Simply, s_{21} is the forward transducer coefficient.

s_{12} is measured by reversing the ports and terminating port 1 in its characteristic impedance and applying the signal to port 2. The ratio of the signals appearing at the b_1 and a_2 nodes will define the value of the s_{12} parameter. s_{12} is the reverse transducer coefficient of the network.

These parameters are vector values, and they have both magnitude and phase information.

It is much easier to make swept-frequency, wideband measurements of s parameters than of h, y, and z parameters, especially above 100 MHz. To use the many design techniques defined in terms of h, y, and z parameters, it is quite simple to convert data to any of these parameters from the scattering parameters. Table (3.2-1) shows the conversion equations for each of these parameters and the scattering parameters.

Table 3.2-1. Conversion Equations Between h, z, y, and s Parameters

$$s_{11} = \frac{(z_{11} - 1)(z_{22} + 1) - z_{12}z_{21}}{(z_{11} + 1)(z_{22} + 1) - z_{12}z_{21}}$$
$$z_{11} = \frac{(1 + s_{11})(1 - s_{22}) + s_{12}s_{21}}{(1 - s_{11})(1 - s_{22}) - s_{12}s_{21}}$$

$$s_{12} = \frac{2z_{12}}{(z_{11} + 1)(z_{22} + 1) - z_{12}z_{21}}$$
$$z_{12} = \frac{2s_{12}}{(1 - s_{11})(1 - s_{22}) - s_{12}s_{21}}$$

$$s_{21} = \frac{2z_{21}}{(z_{11} + 1)(z_{22} + 1) - z_{12}z_{21}}$$
$$z_{21} = \frac{2s_{21}}{(1 - s_{11})(1 - s_{22}) - s_{12}s_{21}}$$

$$s_{22} = \frac{(z_{11} + 1)(z_{22} - 1) - z_{12}z_{21}}{(z_{11} + 1)(z_{22} + 1) - z_{12}z_{21}}$$
$$z_{22} = \frac{(1 + s_{22})(1 - s_{11}) + s_{12}s_{21}}{(1 - s_{11})(1 - s_{22}) - s_{12}s_{21}}$$

$$s_{11} = \frac{(1 - y_{11})(1 + y_{22}) - y_{12}y_{21}}{(1 + y_{11})(1 + y_{22}) - y_{12}y_{21}}$$
$$y_{11} = \frac{(1 + s_{22})(1 - s_{11}) + s_{12}s_{21}}{(1 + s_{11})(1 + s_{22}) - s_{12}s_{21}}$$

$$s_{12} = \frac{-2y_{12}}{(1 + y_{11})(1 + y_{22}) - y_{12}y_{21}}$$
$$y_{12} = \frac{-2s_{12}}{(1 + s_{11})(1 + s_{22}) - s_{12}s_{21}}$$

$$s_{21} = \frac{-2y_{21}}{(1 + y_{11})(1 + y_{22}) - y_{12}y_{21}}$$
$$y_{21} = \frac{-2s_{21}}{(1 + s_{11})(1 + s_{22}) - s_{12}s_{21}}$$

$$s_{22} = \frac{(1 + y_{11})(1 - y_{22}) - y_{21}y_{12}}{(1 + y_{11})(1 + y_{22}) - y_{12}y_{21}}$$
$$y_{22} = \frac{(1 + s_{11})(1 - s_{22}) + s_{12}s_{21}}{(1 + s_{22})(1 + s_{11}) - s_{12}s_{21}}$$

$$s_{11} = \frac{(h_{11} - 1)(h_{22} + 1) - h_{12}h_{21}}{(h_{11} + 1)(h_{22} + 1) - h_{12}h_{21}}$$
$$h_{11} = \frac{(1 + s_{11})(1 + s_{22}) - s_{12}s_{21}}{(1 - s_{11})(1 + s_{22}) + s_{12}s_{21}}$$

$$s_{12} = \frac{2h_{12}}{(h_{11} + 1)(h_{22} + 1) - h_{12}h_{21}}$$
$$h_{12} = \frac{2s_{12}}{(1 - s_{11})(1 + s_{22}) + s_{12}s_{21}}$$

$$s_{21} = \frac{-2h_{21}}{(h_{11} + 1)(h_{22} + 1) - h_{12}h_{21}}$$
$$h_{21} = \frac{-2s_{21}}{(1 - s_{11})(1 + s_{22}) + s_{12}s_{21}}$$

$$s_{22} = \frac{(1 + h_{11})(1 - h_{22}) + h_{12}h_{21}}{(h_{11} + 1)(h_{22} + 1) - h_{12}h_{21}}$$
$$h_{22} = \frac{(1 - s_{22})(1 - s_{11}) - s_{12}s_{21}}{(1 - s_{11})(1 + s_{22}) + s_{12}s_{21}}$$

3.2.3 TOPOGRAPHICAL APPROACH TO RESOLVE FLOW GRAPHS[7]

It was emphasized in the previous section that the scattering parameters are descriptive of signal flow; consequently, signal flow graphs can easily show the scattering parameters as signal flow elements. A two-port network has been described already. The flow graph of a three-port network can be realized in the same manner. Figure 3.2-3 shows such a flow graph.

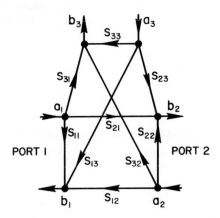

Fig. 3.2-3. Flow graph of a three-port network.

Nodes that represent waves entering and leaving the network are designated a_n and b_n, respectively. There is always a connecting line from an a_n node to a b_n node within the network flow graph, and these connecting lines always go from a to b. They are associated with an s parameter.

Networks can be cascaded one after the other, and their flow graphs can be cascaded similarly, as in Fig. 3.2-4, which shows two two-port networks treated in this way. It is interesting to note that node b_2 and a_1' are synonymous; a_2 and b_1' are also synonymous. In a flow graph, synonymous nodes can be connected with an arrow having a value of "1," meaning that there is no electrical length between them. These two groups of nodes should not be considered identical; the direction of the arrow between b_2 and a_1' is important. Basic transmission line elements can be divided into one-port, two-port, and multiport groups. Every port will have two nodes: one where the wave enters (a) and the other where the wave leaves that port (b).

Flow graph representation of some one-port networks is shown in Fig.

[7] Kuhn, "Simplified Signal Flow Graph Analysis," *Microwave Journal*, November 1963.

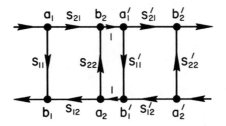

Fig. 3.2-4. Two two-port networks cascaded.

3.2-5. M is the meter reading of an indicator, as shown; K represents the law of the detector and does not change with power level so long as the detector law does not change with power level. Furthermore, M includes the effect of the transmission loss due to the detector's reflection $\sqrt{1 - \rho_D^2}$.

Flow graphs of some two-port networks are shown in Fig. 3.2-6. These flow graphs are only the most-used elements. Remember that Γ stands for

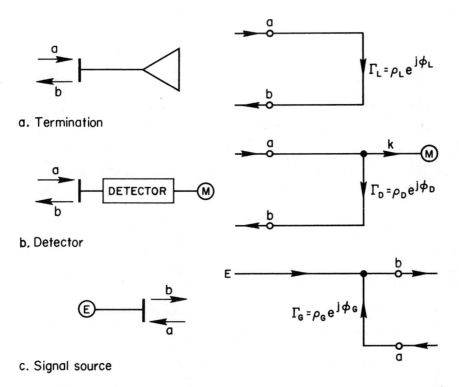

a. Termination

b. Detector

c. Signal source

Fig. 3.2-5. Flow graph representation of some one-port networks.

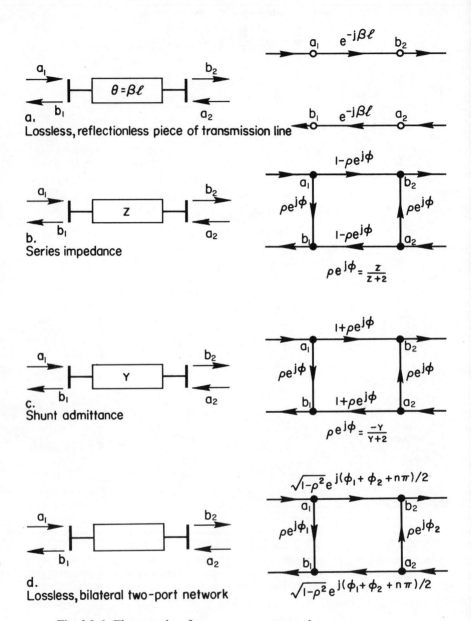

a.
Lossless, reflectionless piece of transmission line

b.
Series impedance

c.
Shunt admittance

d.
Lossless, bilateral two-port network

Fig. 3.2-6. Flow graphs of some two-port networks.

the entire vectorial expression of reflection coefficients, including magnitude and phase information; in other words, Γ describes all the information included in s_{11} and s_{22}. ρ stands for the absolute value of the reflection coefficient.

The three-port device shown in Fig. 3.2-3 is another example for constructing a flow graph. Such a network might be a directional coupler (Fig. 3.2-7) in forward-coupling direction. A directional coupler in the forward

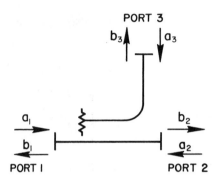

Fig. 3.2-7. Directional coupler in the forward direction.

position splits in two the incoming signal entering at port 1; part of it couples into the auxiliary arm and leaves at port 3, and the remaining part leaves the device at port 2. In other words, the directional coupler in forward direction operates as a power divider. The coupling of the directional coupler will determine how much of the signal entering at port 1 will be coupled into the auxiliary arm and leave at port 3. For example, a 3-dB directional coupler divides the power into two equal parts. Half the power (3 dB is half power) that entered at port 1 will appear at port 3. Consequently, the other half of the power will leave the device at port 2.

Using a 10-dB coupler will mean that, when a signal enters port 1, one-tenth of the power is coupled into the auxiliary arm and leaves the device at port 3, while the rest of the power leaves the device at port 2. (Any power reflecting back from port 3 will be absorbed by a load built into the dead end of the auxiliary arm.) In terms of dB, the level at port 2 will be 0.46 dB less than the signal that entered at port 1 (since power loss of 0.1 is 0.46 dB).

If a directional coupler is reversed, and port 2 is the input, theoretically no power will reach port 3. There is coupling between the through section of the waveguide and the auxiliary arm, but the power that goes into the auxiliary arm from port 2 is absorbed by the load in the dead end. Consequently, even if there are no reflections at the input of port 2 to cause power loss, the power appearing at port 1 will be decreased by this amount of absorbed coupled power. In practice it is not possible to achieve perfect directivity for wideband directional couplers; therefore some signal will get to port 3. The ratio of the signal levels from reverse to forward direction appearing at port 3 is the directivity of the directional coupler and is usually denoted by D.

The flow graph of a directional coupler in forward direction is shown in Fig. 3.2-8. Comparing Fig. 3.2-3 with Fig. 3.2-8, it can be seen that the

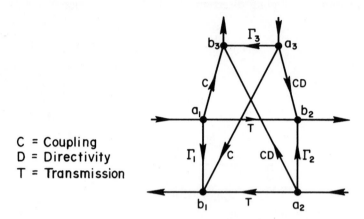

C = Coupling
D = Directivity
T = Transmission

Fig. 3.2-8. Flow graph of a forward directional coupler.

flow graphs are identical, except that the s parameters have been substituted for actual values of reflection and transmission coefficients. Measurement setups are built of components that can be represented with flow graphs. Each component flow graph has input and output ports showing the signal flow. These flow graphs can easily be cascaded, and the complete flow graph of any measurement setup can be drawn. One is usually interested in what happens to the input signal at an output port on the flow graph. Kuhn (see footnote 7) offers a topographical approach to manipulate such flow graphs. This technique is sometimes quite lengthy, but the physical meaning can be seen and approximations made at any step through the manipulation. There are only four rules to keep in mind, and they are easy to remember. These topographical manipulations reduce the complexity of the flow graph through various steps until the solution can be readily seen.

RULE 1. Two branches, whose common node has only one incoming and one outgoing branch (branches in series), may be combined to form a single branch whose coefficient is the product of the coefficients of the original branches. Thus the common node is eliminated (Fig. 3.2-9).

To prove the validity of such reduction, let us write the equations.

$$E_2 = s_{21}E_1$$

$$E_3 = s_{32}E_2$$

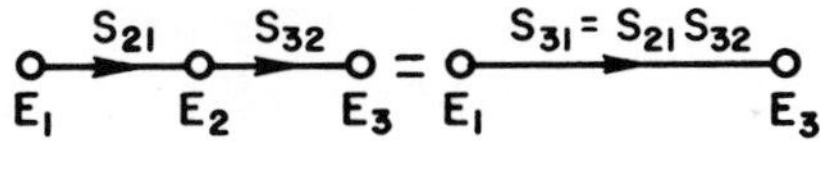

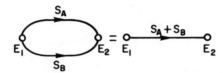

Fig. 3.2-9. Node elimination, rule number 1.

By substitution,

$$E_3 = s_{32}s_{21}E_1$$

proves rule 1 to be correct and allowed.

RULE 2. Two branches pointing from a common node to another common node (branches in parallel) may be combined into a single branch whose coefficient is the sum of the coefficients of the original branches (Fig. 3.2-10).

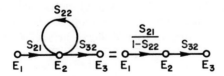

Fig. 3.2-10. Branch elimination, rule number 2.

The proof is obvious from the equation

$$E_2 = s_A E_1 + s_B E_1 = (s_A + s_B)E_1$$

which allows the use of this rule.

RULE 3. When node N possesses a self-loop (a branch which begins and ends at N) of coefficient s_{nn}, the self-loop may be eliminated by dividing the coefficient of every other branch entering node N by $1 - s_{nn}$ (Fig. 3.2-11).

Fig. 3.2-11. Feedback loop elimination, rule number 3.

To prove the reduction of a feedback loop (rule 3), let us write

$$E_2 = s_{21}E_1 + s_{22}E_2$$

and

$$E_3 = s_{32}E_2$$

Solving gives us

$$E_2 - s_{22}E_2 = s_{21}E_1$$

$$E_2 = \frac{s_{21}}{1 - s_{22}} E_1$$

and

$$E_3 = \frac{s_{21}s_{32}}{1 - s_{22}} E_1$$

Applying rule 1 to reduce the series branches in the right side of Fig. 3.2-11 further yields the above expression. It is interesting to note the correspondence of the term $1 - s_{nn}$ with the feedback circuit's term of $1 - \mu\beta$.

The next rule, which is an expanding one instead of a reducing one, enables one to make further reductions in succeeding steps.

RULE 4. A node may be duplicated (i.e., split into two nodes that may be subsequently treated as two separate nodes) as long as the resulting flow graph contains, once and only once, each combination of separate (not a branch which forms a self-loop) input and output branches that connect to the original node. Any self-loop attached to the original node must also be attached to each of the nodes resulting from duplication (Fig. 3.2-12).

To prove Case C in Fig. 3.2-12, let us write

$$E_2 = s_A E_1 + s_B E_1 = (s_A + s_B)E_1$$

$$E_3 = s_C E_2 + s_D E_2 = (s_C + s_D)E_2$$

Hence

$$E_3 = (s_C + s_D)(s_A + s_B)E_1 = (s_A s_C + s_A s_D + s_B s_C + s_B s_D)E_1$$

which proves the validity of this case.

Another example for node duplication is shown in Fig. 3.2-13. Here it will be applied with feedback loops. Notice each step of the manipulation. The original flow graph to be reduced is (a). Since a feedback loop exists between E_2 and E_3 nodes in step (b), a node duplication of E_3 was exercised which results in a duplication of s_{32}. Step (c) shows the loop itself, and step (d) applies rule 3 to eliminate the feedback loop. Both incoming branches were modified in value according to rule 3. Step (e) reapplies the node duplication so that one can write two simple equations for E_4. Step (f) applies rule 1 and the equation

$$E_4 = \frac{s_{2B}s_{32}s_{43}}{1 - s_{23}s_{32}} E_B + \frac{s_{2A}s_{32}s_{43}}{1 - s_{23}s_{32}} E_A$$

Signal flow graph representations of microwave measurement setups allow one to analyze the errors involved in such measurements. If one builds a simple insertion-loss measurement scheme, the basic technique is to take a signal source and a detector connected to an indicator and to note the indicator reading. The unknown device is then inserted between the signal source

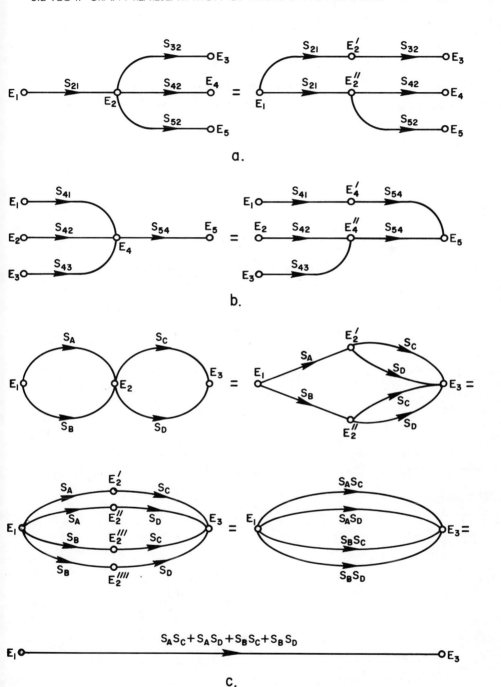

Fig. 3.2-12. Node duplication, rule number 4.

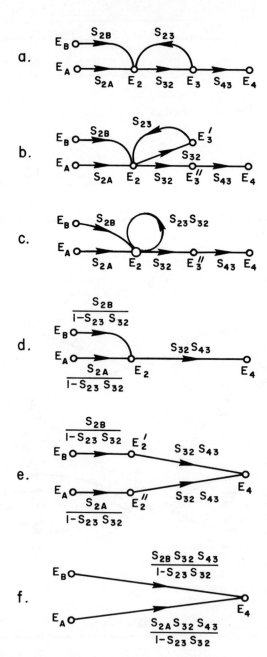

Fig. 3.2-13. Node duplication, applied with feedback loops.

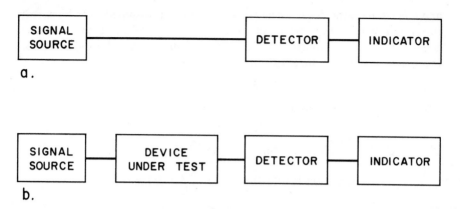

a.

b.

Fig. 3.2-14. Attenuation measurement setup:
(a) calibration (b) test.

and the detector. Insertion loss will be derived from the indicator readings taken before and after insertion of the device. Figure 3.2-14 shows the setup making this measurement.

If the signal source and the detector are perfectly matched, there will be no multiple mismatch errors. Only calibration, repeatability, and drift errors contribute to the inaccuracy of this measurement. But, since the signal source and the detector are seldom perfectly matched, mismatch error will also increase the ambiguity of the measurement. To find the effects of these mismatches, one may draw the flow graphs of the measurement. Figure 3.2-15 shows these flow graphs. T represents the power lost due to absorption

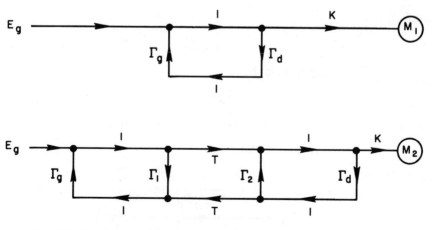

Fig. 3.2-15. Flow graph of the attenuation measurement
setup: calibration and test.

and reflection on the device. The device under test is assumed to be bilateral; therefore T will be the same in both directions of signal flow. If the reflection coefficients of signal source and detector are equal to zero (in other words, the system is perfectly matched), it can be seen that

$$M_1 = KE_g$$

Similarly, for the measurement,

$$M_2 = KE_gT$$

If M_1 and M_2 are voltage readings, the attenuation of the device is

$$\text{Insertion loss (dB)} = 20 \log \frac{M_1}{M_2} = 20 \log \left| \frac{1}{T} \right|$$

However, it is very unlikely that the reflection coefficients of the signal source and the detector will be equal to zero. To see how the expression for insertion loss can be modified, the flow graphs must be reduced. Reduction of the calibration flow graph is shown in Fig. 3.2-16. Reduction of the measurement flow graph is shown in Fig. 3.2-17. The meter reading of calibration is

$$M_1' = \frac{K}{1 - \Gamma_g \Gamma_d} E_g$$

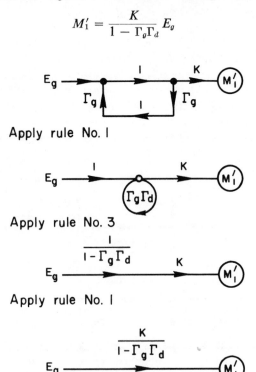

Fig. 3.2-16. Reduction of the calibration flow graph.

The meter reading of measurement (refined) is

$$M_2' = \frac{KT}{1 - \Gamma_g\Gamma_1 - \Gamma_2\Gamma_d + \Gamma_g\Gamma_d\Gamma_1\Gamma_2 - T^2\Gamma_g\Gamma_d} E_g$$

The insertion loss, using M_1' and M_2' expressions if they are voltages, is

$$\text{Insertion loss (dB)} = 20 \log \frac{M_1'}{M_2'}$$

$$= 20 \log \left| \frac{\dfrac{KE_g}{1 - \Gamma_g\Gamma_d}}{1 - \Gamma_g\Gamma_1 - \Gamma_2\Gamma_d + \Gamma_g\Gamma_d\Gamma_1\Gamma_2 - T^2\Gamma_g\Gamma_d} \right|$$

$$= 20 \log \left| \frac{1 - \Gamma_g\Gamma_1 - \Gamma_2\Gamma_d + \Gamma_g\Gamma_d\Gamma_1\Gamma_2 - T^2\Gamma_g\Gamma_d}{T(1 - \Gamma_g\Gamma_d)} \right|$$

Comparing this value with the perfectly matched case, the error term due to multiple mismatches can be found as follows.

$$\text{Error (dB)} = 20 \log \frac{M_1}{M_2} \bigg/ \frac{M_1'}{M_2'} = 20 \log \frac{M_1}{M_2} \frac{M_2'}{M_1'}$$

$$= 20 \log \left| \frac{1}{T} \cdot \frac{T(1 - \Gamma_g\Gamma_d)}{1 - \Gamma_g\Gamma_1 - \Gamma_2\Gamma_d + \Gamma_g\Gamma_d\Gamma_1\Gamma_2 - T^2\Gamma_g\Gamma_d} \right|$$

$$= 20 \log \left| \frac{1 - \Gamma_g\Gamma_d}{1 - \Gamma_g\Gamma_1 - \Gamma_2\Gamma_d + \Gamma_g\Gamma_d\Gamma_1\Gamma_2 - T^2\Gamma_g\Gamma_d} \right|$$

Since the reflection coefficients are complex quantities, the absolute values must be used to take their logarithm. If maximum and minimum error values only are of interest, the numerator can be set to a maximum value while the denominator is set to a minimum (and vice versa) to gain the maximum and minimum error limits, respectively.

3.2.4 ANALYTICAL APPROACH TO RESOLVE FLOW GRAPHS

The "nontouching-loop" rule is an analytical method for solving a flow graph. This technique is fast, but it is very easy to forget and miss a few loops. Although the topographical approach of resolving flow graphs takes longer to do, it is very easy to remember. A few basic definitions have to be understood before the nontouching-loop rule can be learned.

A "path" is a series of branches followed in sequence and in the same direction in such a manner that no node is touched more than once. The "value of the path" is the product of the coefficients of the branches encountered en route.

Figure 3.2-18 shows the flow graph of a two-port network driven with a signal source and terminated with a load. One path goes from the generator to node b_2; its value is s_{21}. There are two paths from the generator to node b_1. The values of these paths are s_{11} and $s_{21}\Gamma_L s_{12}$.

If a path starts and finishes in the same node, it is called a "loop," rather than a path. A "first-order loop" is a path coming to a closure with no node passed more than once. The value of the loop is calculated as the value of the path, or the product of the value of all branches encountered en route.

A "second-order loop" is defined as two first-order loops not touching each other at any node. The value of a second-order loop is the product of

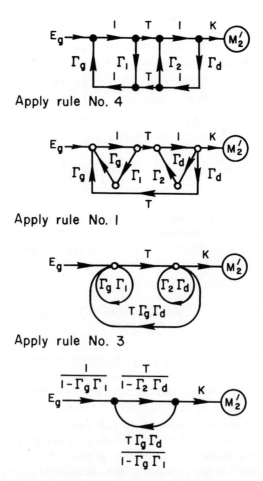

Fig. 3.2-17. Reduction of the measurement flow graph.

Apply rule No. 4

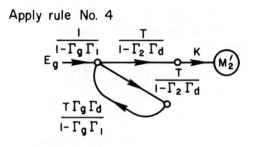

Apply rule No. I

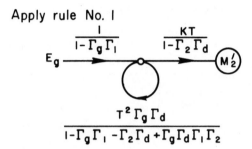

Apply rule No. 3

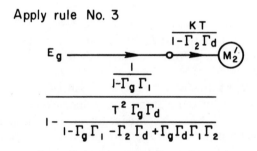

Apply rule No. I

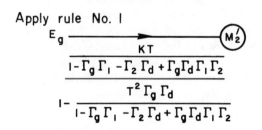

Fig. 3.2-17 (*continued*)

the values of the two first-order loops. Third- and higher-order loops are three or more first-order loops not touching each other at any point. Their values are calculated in the same manner as described above for the second-order loop, that is, by multiplying the coefficients of branches encountered.

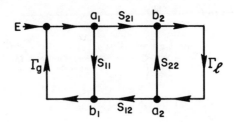

Fig. 3.2-18. Flow graph of a two-port network with a signal source and a load.

For example, in Fig. 3.2-18, there are three first-order loops ($s_{11}\Gamma_g$, $s_{22}\Gamma_L$, and $\Gamma_g s_{21} \Gamma_L s_{12}$) and one second-order loop ($\Gamma_g s_{11} s_{22} \Gamma_L$).

The nontouching-loop rule[8,9] can be applied to solve any flow graph. The equation in symbolic form is

$$T = \frac{\begin{aligned}P_1[1 - \Sigma L(1)^{(1)} + \Sigma L(2)^{(1)} - \Sigma L(3)^{(1)} + \ldots] \\ + P_2[1 - \Sigma L(1)^{(2)} + \Sigma L(2)^{(2)} - \ldots] \\ + P_3[1 - \Sigma L(1)^{(3)} + \ldots] + P_4(1 - \ldots) + \end{aligned}}{1 - \Sigma L(1) + \Sigma L(2) - \Sigma L(3) + \ldots}$$

where $\Sigma L(1)$ stands for the sum of all first-order loops, $\Sigma L(2)$ is the sum of all second-order loops, and so on; P_1, P_2, P_3, etc., stand for the values of all paths that can be followed from the independent variable, in most cases the generator, to the node whose value is desired; $\Sigma L(1)^{(1)}$ denotes the sum of those first-order loops which do not touch the path of P_1 at any node; $\Sigma L(2)^{(1)}$ denotes then the sum of those second-order loops which do not touch the path of P_1 at any point; $\Sigma L(1)^{(2)}$ consequently denotes the sum of those first-order loops which do not touch the path of P_2 at any point. Each path is multiplied by the factor in parentheses which involves all the loops of all orders that the path does not touch. T represents the ratio of the dependent variable in question and the independent variable.

The example shown in Fig. 3.2-18 can be calculated for two dependent variables. One is the reflection coefficient of the two-port network b_1/a_1, and the second is the transmission coefficient b_2/E. In the first case, when b_1/a_1 is to be found, the generator is not involved, so it should be neglected. The solution is

$$\frac{b_1}{a_1} = \frac{s_{11}(1 - s_{22}\Gamma_L) + s_{21}\Gamma_L s_{12}}{1 - s_{22}\Gamma_L}$$

s_{11} is the first path, P_1, which has to be multiplied with $1 - \Sigma L(1)^{(1)}$. $s_{22}\Gamma_L$ is

[8] Lorens, C. S., "A Proof of the Nonintersecting Loop Rule for the Solution of Linear Equations by Flow Graphs," Res. Lab. of Electronics, M.I.T., Cambridge, Mass., *Quarterly Progress Report*, January 1956, pp. 97–102.

[9] Happ, W. W., "Lecture Notes on Signal Flow Graphs," from *Analysis of Transistor Circuits*, Extension Course, University of California, Catalog 834AB.

the only first-order loop not touching the P_1 path; higher-order loops not touching the P_1 path do not exist. Path number two, P_2, will be $s_{21}\Gamma_L s_{12}$; since there are no first-order or any higher-order loops not touching this path, it will be multiplied by 1. The denominator shows the only first-order loop, $s_{22}\Gamma_L$, subtracted from unity.

The entire flow graph, including the generator, is needed to write the solution for the transmission coefficient.

$$\frac{b_2}{E} = \frac{s_{21}}{1 - \Gamma_g s_{11} - s_{22}\Gamma_L - \Gamma_g s_{21}\Gamma_L s_{12} + \Gamma_g s_{11} s_{22}\Gamma_L}$$

Because there is only one possible path from E to the b_2 node, and there are no loops not touching this path, only s_{21} will stay in the numerator. It can be seen that there are three first-order loops and a second-order loop in the denominator.

It would be interesting to see the attenuation measurement flow graph discussed in the topographical approach as another example. Figure 3.2-15 shows the flow graphs in question. Equations have to be written for M'_1/E_g and M'_2/E_g; the values of M_1/E_g and M_2/E_g have already been found analytically.

$$\frac{M'_1}{E_g} = \frac{k}{1 - \Gamma_g \Gamma_d}$$

since k is the only path and $\Gamma_g \Gamma_d$ is the only loop.

$$\frac{M'_2}{E_g} = \frac{kT}{1 - \Gamma_g \Gamma_1 - \Gamma_2 \Gamma_d - \Gamma_g T^2 \Gamma_d + \Gamma_g \Gamma_1 \Gamma_2 \Gamma_d}$$

Again the only path is kT, and all loops touch this path. Three first-order loops and a second-order loop can be found in the denominator.

It is worth mentioning that third- and higher-order loops can usually be neglected after careful analysis of the values of various coefficients in question. This is because values smaller than unity multiplied with each other become even smaller. This point will be emphasized later in the text.

Nontouching-Loop Rule

$$T = \frac{\begin{array}{c} P_1(1 - \Sigma L(1)^{(1)} + \Sigma L(2)^{(1)} - \Sigma L(3)^{(1)} + \ldots) \\ + P_2(1 - \Sigma L(1)^{(2)} + \Sigma L(2)^{(2)} - \ldots) + P_3(1 - \Sigma L(1)^{(3)} + \ldots) + \ldots \end{array}}{1 - \Sigma L(1) + \Sigma L(2) - \Sigma L(3) + \ldots}$$

$\Sigma L(1)$	Sum of all first-order loops
$\Sigma L(2)$	Sum of all second-order loops
P_1, P_2, P_3	Values of paths corresponding to indices
$\Sigma L(1)^{(1)}$	Sum of those first-order loops which do not touch P_1
$\Sigma L(2)^{(1)}$	Sum of those second-order loops which do not touch P_1
$\Sigma L(n)^{(m)}$	Sum of those n-order loops which do not touch P_m path
T	Ratio of dependent variable in question and independent variable

QUESTIONS

1. Convert the given *s* parameters of a two-port network to the equivalent *z* parameters: $s_{11} = 0.12$, $s_{12} = 0.02$, $s_{21} = 1.2$, $s_{22} = 0.15$.

2. Solve the following flow graph for M_2/M_1 using the topographical approach.

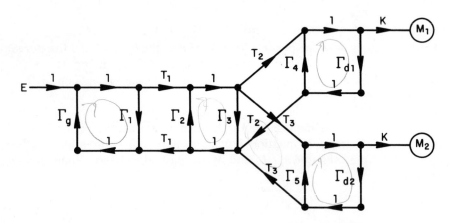

3. Solve the following flow graph for M/E using the analytical approach.

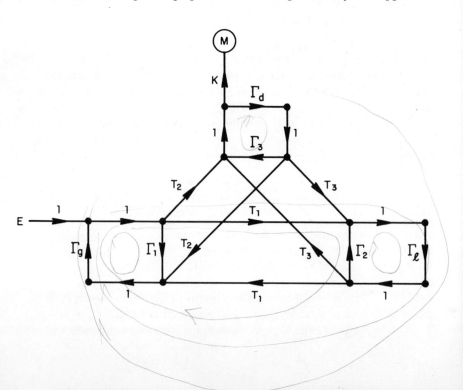

4
LABORATORY SOURCES
FOR
MICROWAVE SIGNAL GENERATION

In general, signal sources can be divided into two basic categories: single-frequency and swept-frequency types. Many further divisions can be made within these categories. Three basic requirements distinguish signal generators from other oscillators: frequency dial calibration, calibrated output, and modulation capability.

As the state of the art advanced, and as more sophisticated communication systems and the advent of radar increased bandwidth requirements, scientists began to look for ways to utilize frequencies higher than the customary short waves. As the need for higher frequencies arose, higher-frequency oscillators were needed. In the late 1930's, scientists reached the UHF band and found that the performance of their standard tube-type lumped-circuit element oscillators was very marginal at this range.

4.1 TRIODE OSCILLATORS

A conventional triode tube-type oscillator is shown in Fig. 4.1-1. A resonant-tank circuit between the cathode and the grid is inductively coupled

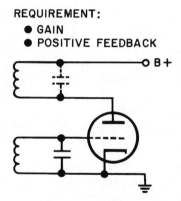

Fig. 4.1-1. Regenerative triode oscillator.

with the inductor of the plate circuit and connected in such a manner that positive feedback will occur. The signal in the plate circuit is 180° out of phase with the input. The windings of the plate circuit coil are connected in such a way that positive feedback from the plate circuit back into the resonant-tank circuit is assured. As the frequency range increases, the size of L and C components must decrease. These components become so small that the capacitances of the tube and the leads connecting to them are very significant. In fact, their reactances become dominant around the 500-MHz region, so that it is impractical to develop any type of conventional coil or capacitor. Lengths of transmission lines such as the Lecher wires (twin leads) are substituted for lumped-element tank circuits. Dual-triode oscillators such as those shown in Fig. 4.1-2 were used where the coupling capacitor was not really needed be-

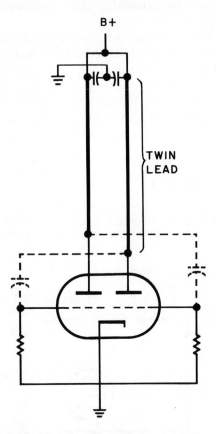

Fig. 4.1-2. Dual triode push-pull oscillator employing
twin lead resonant circuit.

cause the tube had enough coupling within itself to establish oscillation. The length of the Lecher wire coupled to the plate and the tubes' capacitance determine the resonant frequency. This technique allowed scientists to develop higher-frequency oscillators, but at about 1 GHz further problems occurred within the tube.

In a triode, an electron cloud leaves the cathode that has negative potential and is attracted to the plate that has positive potential on it. The amount of time it takes for the electrons to travel from the cathode to the plate is called the transit time. If the transit time is appreciably long compared to the cycle time of the frequency, positive feedback (which is necessary to establish oscillation) will no longer be assured. The oscillator will be very unstable, and oscillation will not occur due to the lack of positive feedback. This transit time can be shortened in two ways. The first is to increase the plate voltage, which increases the velocity of the electrons traveling from cathode to plate. The second is to shorten the distance, moving the plate closer to the cathode. Of course, there are technical limitations to doing this. These techniques were used to obtain further gains in frequency coverage. Furthermore, the shape of electron tubes was changed to accommodate the transmission line.

The lighthouse tube, or pencil triode, was developed. Figure 4.1-3 shows

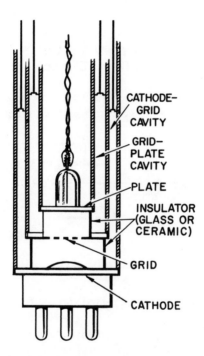

CATHODE–
GRID
CAVITY

GRID–
PLATE
CAVITY

PLATE

INSULATOR
(GLASS OR
CERAMIC)

GRID

CATHODE

Fig. 4.1-3. Triaxial lighthouse tube oscillator.

a lighthouse tube placed into a tunable cavity that is really a triaxial structure having two simultaneously tunable cavities. The inner cavity is the grid-plate cavity, and the other one is the cathode-grid cavity. The ends of each of the conductors are directly in contact with the electrodes of the tube, which are brought out from the tube in a coaxial manner. It is quite obvious that the movable short circuits which provide the tuning to each of the cavities must not make ohmic contact with the walls, since this would short out the cathode to the grid and the grid to the plate. These short circuits have to be of the capacitive or resonant-choke types, which will be explained later in the text. This kind of oscillator is still being made in the lower end of the microwave spectrum. However, as frequency requirements kept increasing, these oscillators became inadequate in higher frequencies because the electron transit time from the cathode to the plate was still too long.

QUESTION

1. What is the principal limiting factor of a triode in making higher-frequency oscillators?

4.2 DENSITY MODULATION AND VELOCITY MODULATION OF ELECTRON BEAMS

Let us see how modulation occurs in a triode. Figure 4.2-1 shows how the cathode, grid, and plate are spaced from each other. An electron cloud

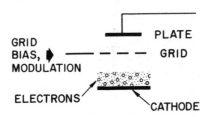

Fig. 4.2-1. Triode, density modulation of electrons.

emerges from the cathode when it is heated and is attracted to the plate, which is on a positive potential. The velocity of the electrons traveling toward the plate will depend entirely upon the positive plate voltage. The bias on the grid or the modulation applied to the grid will vary only the density of the electrons traveling toward the plate; it will not affect the transit time of the electrons from cathode to plate. (Only the potential between the cathode and the plate will determine the velocity of the electrons traveling from cathode to plate.) This is called the DENSITY MODULATION of electrons.

VELOCITY MODULATION of electron beams is achieved as follows. An electron beam is generated by an electron gun similar to the type used in a CRT but usually smaller. This beam is directed through a cavity resonator with an opening in the center, and the electric field is axially aligned with the beam. Microwave energy coupled into the cavity causes it to resonate, and the electric field interacts with the electron beam. Half the time, when the electric field is opposed to the electron beam, the electrons will be retarded. When the field is aligned with the travel of the electron beam the other half of the time, the speed of the electrons will increase. Consequently some of the electrons will be retarded and some advanced, resulting in a bunching effect. This effect is shown in Fig. 4.2-2.

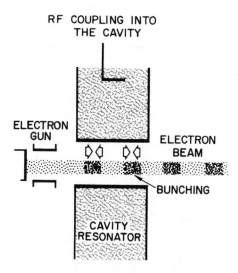

Fig. 4.2-2. Velocity modulation of electron beams.

QUESTION

1. How is microwave energy sustained when it is amplified in electron tubes using velocity modulation of the electron beam?

4.3 KLYSTRONS

4.3.1 CHARACTERISTICS OF KLYSTRONS

Russell and Siegurd Varian* placed another cavity behind the first one, as shown in Fig. 4.3-1, and perfected the klystron amplifier. In the first cavity, where the RF energy is coupled in, the electron beam is velocity modulated, and the second cavity is tuned to the same frequency. In the second cavity the RF energy is coupled through the electron beam by placing the second cavity into the proper position at an optimum distance. The RF interacting with the electron beam causes a kinetic energy loss from the beam that results

---

*Inventors of the klystron (1938).

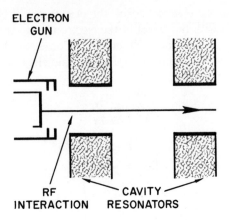

Fig. 4.3-1. Klystron amplifier.

in gain. This was the first use of velocity modulation of an electron beam. Frequency multiplication also can be obtained by tuning the second cavity into a harmonic of the first cavity's frequency. Several cavities can be placed one after the other to achieve higher gain and narrower bandwidth.

This technique met the first requirement for an oscillator: the gain. Positive feedback could be established by connecting the second cavity back into the first cavity. To assure a wide tuning capability of this dual cavity requires more consideration about phase relations involved. Using the klystron as an oscillator would require a variable-phase shifter between the first and second cavities to assure positive feedback at all frequencies.

Figure 4.3-2 shows the REFLEX KLYSTRON in which oscillation can be achieved within a single cavity. There are three basic regions of a reflex klystron: the *cathode-anode region,* or the electron gun; the *RF structure* where interaction takes place, causing electron bunching; and the *drift space.* The repeller, an electrode with a more negative potential, is placed at this point where the electrons are returned. After they turn around, they may arrive back at the interaction space when bunches are being formed. If their travel time in the drift space, coming back into the RF structure, is such that they are in phase (in other words, when the electrons are being bunched), they will provide the necessary positive feedback to establish oscillation. Oscillation can be obtained as the repeller voltage is varied. The transit time from interaction space into the drift space and back to the interaction space is varied. If bunches returning from the drift space are not in phase with bunches being formed, oscillation will not occur. The bunches will be out of phase. Variation of transit time by the adjustment of repeller voltage will make the returning bunch meet in phase with another bunch that is being formed. Then oscillation will occur again. This means that oscillation can occur not only in one repeller voltage at each frequency; several repeller voltages will provide

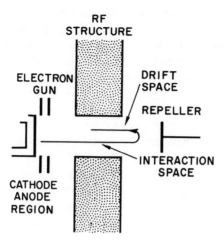

Fig. 4.3-2. Reflex klystron.

oscillation. These are called the different repeller modes. The higher the repeller voltage, the shorter the transit time from interaction space out into the drift space and back into the interaction space. In the highest-voltage repeller mode, a bunch is formed, travels out through the drift space, and then, coming back, hits the next bunch being formed. The lowest-voltage repeller mode is very close to zero volts. It is not a good practice to go too low with the repeller voltage; the tube is not designed to carry any current to the repeller, and all the electrons are repelled, turning back into the interaction space where they will terminate on the RF structure.

Figure 4.3-3 shows the power supply requirements of a reflex klystron.

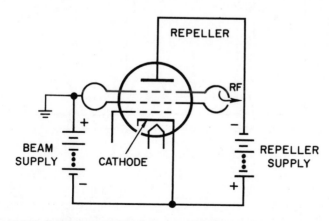

Fig. 4.3-3. Power supply requirements of reflex klystrons.

It can be seen that the RF structure, the cavity itself, is the most positive, and it is grounded. The cathode is negative. The beam voltage will be measured between the cathode and the RF structure. Another negative voltage, the repeller voltage, is connected between the cathode and the repeller. This voltage is usually variable to allow the operator to tune in the right repeller mode to establish oscillation. While turning on a reflex klystron, it is always a good habit to turn on the repeller power supply first and then the beam supply to be sure that no positive voltage goes to the repeller. Commercially available klystron power supplies have a safety device built into the repeller supply that will never allow the repeller to go positive; in fact, it will never let the repeller go down to zero voltage. If the resonant cavity is tuned to any one frequency, a number of repeller voltages will provide the same frequency oscillation. These are, as was mentioned before, the different repeller modes. Figure 4.3-4

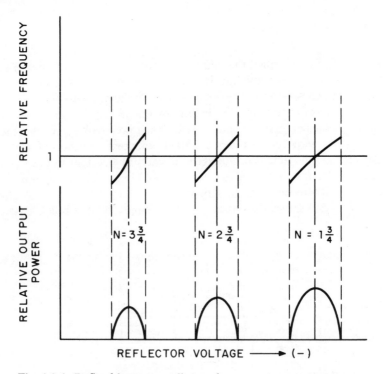

Fig. 4.3-4. Reflex klystron repeller modes.

shows a plot of those repeller modes—power output versus repeller voltage. Another plot right below the power output shows the frequency variation with repeller voltage. This is quite obvious if one thinks how the electron bunches are returned from drift space through different repeller voltages. Their transit times are slightly different, which means slightly different frequencies. This could be used as a fine frequency control of the reflex klystron. As the negative repeller voltage is increased, the transit time gets shorter, and the electron bunches arrive back at the interaction space earlier. This will increase the frequency. If the negative repeller voltage decreases, the frequency will also decrease because the transit time gets longer and the electron bunches will arrive back into the interaction space later, giving a longer cycle time. If one looks at the power plot, he will see that the highest power output within a mode from the klystron will occur at the center of the mode. That is why this technique of frequency variation is only used for very slight variations.

4.3.2 MODULATION OF KLYSTRONS

Because of the nature of the reflex klystron, amplitude modulation is impractical. Any voltage variation on the reflex klystron will result in frequency modulation. If the repeller voltage is varied, frequency change will occur. If the electron beam voltage is varied, of course, the velocity of the electron beam will be changing, which will also result in frequency variation. Only frequency modulation is possible on a reflex klystron; amplitude modulation would result in unwanted frequency modulation.

Pulse modulation can be achieved quite simply by turning the oscillation on and off. The electron beam can be turned on and off, but this is not done customarily. The way pulse modulation, or square-wave modulation, is achieved is to modulate the repeller voltage. It is shown in Fig. 4.3-5 that the repeller voltage is varied rapidly between the repeller mode regions and the center of the repeller modes. This will establish a simple pulse modulation effect. Care should be taken as far as the rise time and the decay time of the pulse are concerned, since they will result in an inherent frequency modulation while the voltage is being changed at a finite time from one value to the other and is going through that part of the repeller mode where frequency variation will occur. This effect is undesired. It is a good practice to clip part of the signal modulating the repeller of the reflex klystron. This assures the fastest rise and decay times and no overshoots. Overshoots may result in getting into another repeller mode's area, which will result in a dual-frequency oscillation.

QUESTIONS

1. What are the regions of a reflex klystron?
2. What happens to the oscillation of a reflex klystron when the repeller voltage is varied?
3. Why is it impractical to amplitude-modulate a reflex klystron directly?

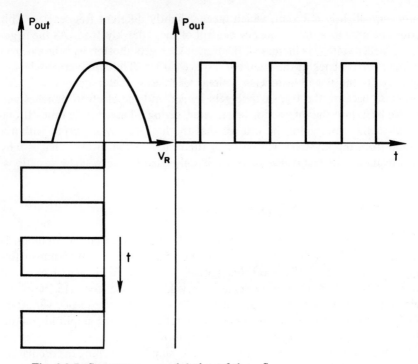

Fig. 4.3-5. Square-wave modulation of the reflex klystron.

4.4 TRAVELING-WAVE TUBE AMPLIFIER

Another concept for obtaining microwave amplification, based on electron bunching, was developed in the early 1950's. In contrast to the klystron amplifier, this device, the traveling-wave tube amplifier (TWT), does not use resonant cavities; it employs a direct interaction between a nonresonant transmission line and an electron beam. Consequently the TWT provides wide frequency band gain without any mechanical adjustments, such as the cavity resonant-frequency tuning that is necessary to change the resonant frequency of the cavities in the klystron amplifier.

The importance of these amplifiers lies in their high gain, generous power output, linear amplification characteristics, extremely versatile modulation properties, and very wide bandwidths. These exceptional features can be used to solve many difficult problems in high-frequency work.

A traveling-wave tube amplifier can faithfully amplify many broadband signals such as those employed in television relay and broadband microwave carrier systems. In addition to this broadband feature, the traveling-wave tube amplifier has a linear amplification characteristic and relatively flat frequency response over the band.

Since the traveling-wave tube amplifier does not require tuning, it is exceptionally easy to operate and is capable of solving many high-frequency problems in a very simple, straightforward manner. Some specific applications for the traveling-wave tube amplifier's broadband-amplification properties are:

1. investigation of information-handling capacity in broadband microwave communication systems;
2. preamplification of low-level input signals to wideband microwave receivers;
3. amplification of low-frequency harmonics to produce frequency markers used in microwave frequency measurements.

Many more applications have been found in recent years.

In many narrow-band applications, the traveling-wave tube amplifier's great bandwidth permits shifting the narrow band over a considerable frequency range to avoid noise, interference, etc., without changing basic amplifier circuitry. However, one strong objection to the use of a broadband amplifier for narrow-band amplification is the noise amplified over the greater bandwidth. In such cases, noise can be greatly reduced by using a narrow-band filter in the amplifier output.

Traveling-wave tube amplifiers can also be used as moderate-power, wideband signal sources by amplifying the low-power output typical of many klystron signal generators. Thus a signal generator/traveling-wave tube amplifier combination can be used in many applications requiring large amounts of microwave power, such as wide-range antenna measurements to plot patterns and determine efficiency and directivity. It can be used as low-cost, portable means of providing moderate-power microwave signal sources for field-siting microwave installations.

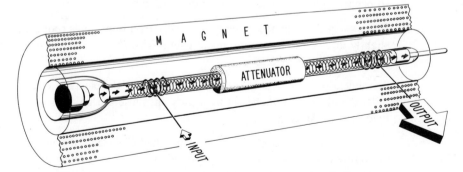

Fig. 4.4-1. Cutaway view showing important elements
of a traveling-wave tube amplifier.

Key to Fig. 4.4-2

1. Electron beam directed through center of helix.

2. CW signal coupled into helix. Arrows in detail show direction and magnitude of force exerted on the electron beam by the CW signal.

3. Electron bunching caused by the electric field of the CW signal (See detail).

4. Amplification of signal on helix begins as the field formed by the electron bunches interacts with the electric field of the CW signal. The newly formed electron bunch adds a small amount of voltage to the CW signal on the helix. The slightly amplified CW signal then produces a denser electron bunch, which in turn adds a still greater voltage to the CW signal, and so on.

5. Amplification increases as the greater velocity of the electron beam pulls the electron bunches more nearly in phase with the electric field of the CW signal. The additive effect of the two fields exactly in phase produces the greatest resultant amplification.

6. Attenuators placed near the center of the helix reduce all the waves traveling along the helix to nearly zero. This prevents undesired waves, such as waves reflected from mismatched loads, from returning to the tube input and causing oscillation.

7. Electron bunches travel through attenuator unaffected.

8. Electron bunches emerging from attenuator induce a new CW signal on helix. New CW signal is the same frequency as the original CW signal applied.

9. Field of newly induced CW signal interacts with bunched electrons to begin the amplification process over again.

10. For a short distance the velocity of the electron bunches is reduced slightly due to the large amount of energy absorbed by the formation of the new CW signal.

11. Amplification increases as the greater velocity of the electron beam pulls the electron bunches more nearly in phase with the electric field of the CW signal.

12. At point of desired amplification the amplified CW signal is coupled out of the helix. *Note that the "amplified" CW signal is a new signal whose energy is wholly supplied by the bunched electron beam.*

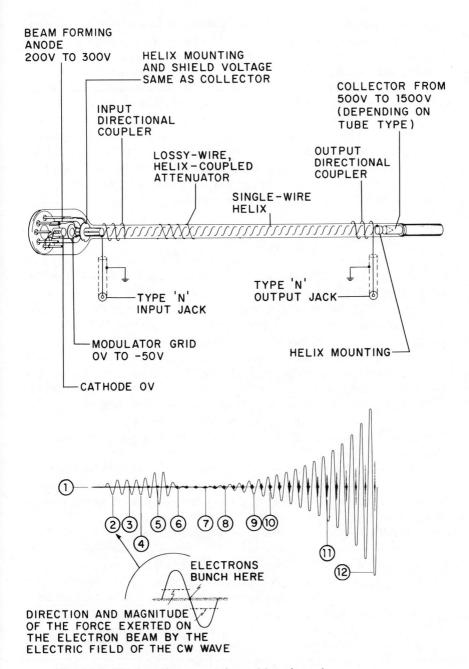

Fig. 4.4-2. The traveling-wave tube and how it works.

The traveling-wave tube amplifier also acts as a buffer between a microwave signal source and an external load. As a buffer, it isolates load reflections from the signal source and eliminates the ill effects that occur when the source is modulated directly.

The basic traveling-wave tube amplifier consists of an electron gun which projects a focused electron beam through a helically wound coil to a collector electrode, as shown in Fig. 4.4-1. The focused electrons are held in a pinlike beam through the center of the helix by a powerful magnetic field around the full length of the tube. The magnetic field can be obtained through an electromagnet. This approach was used in the earlier TWT's. Later models use permanent magnets preset at the factory for optimum focusing. Periodic permanent magnets (PPM) provide even better focusing.

4.4.1 SLOW-WAVE STRUCTURES

A CW signal coupled into the gun end of the helix travels around the turns of the helix and thus has its linear velocity reduced by an amount equal to the ratio of the length of wire in the helix to the length of the helix itself. This is called the slow-wave structure. In higher-frequency TWT's, the helix structure has been replaced with some other types of slow-wave structures (e.g., interdigital slow-wave structures).

The electron beam velocity, determined by the potential difference between the cathode and the helix, is adjusted so that the electron beam travels a little faster than the CW signal. The electric field of the CW signal on the helix interacts with the electric field created by the electron beam and increases the amplitude of the signal on the helix, thus producing the desired amplification.

Figure 4.4-2 shows the principal elements of a typical traveling-wave tube in the upper portion and the important steps in the amplification process in the lower portion. The steps should be followed by referring to the numbered captions.

A TWT is a linear amplifier at its small-signal region. As the input signal increases and the output power reaches a certain level, saturation distorts this linearity. When a TWT amplifier is operated close to its maximum output power region, the gain will no longer be constant. The amplifier is not linear; it will become dependent upon output power. Figure 4.4-3 shows a typical TWT power in–power out curve at a particular frequency. This curve may vary quite drastically from one frequency to another. In Fig. 4.4-3, the straight line proves constant gain in the small-signal region. With increasing power input, a maximum output power is reached. Then, further increase of input power will result in a drop of output power from maximum value. Gain is proportional to the tangent of the curve at any point.

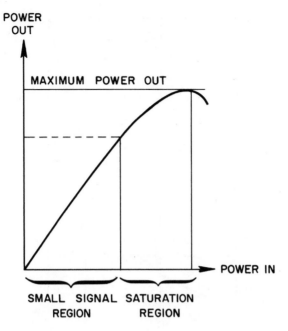

Fig. 4.4-3. Power curve of a typical TWT.

4.4.2 USING TWT'S TO MODULATE KLYSTRONS

The traveling-wave tube amplifier is particularly suitable for use as the power amplifier of an *amplitude-modulated* master-oscillator-power-amplifier (MOPA) system. This feature opens new fields of application, since it is not possible to amplitude-modulate a reflex klystron directly. Furthermore, using the traveling-wave tube amplifier as a power amplifier means that the RF output from a microwave oscillator can be sine-wave modulated, pulse modulated, or pulse-train modulated without the starting delays and jitter generally present when the oscillator itself is modulated. Thus, in addition to amplification, the traveling-wave tube provides a simple system of amplitude modulation.

Amplitude modulation is accomplished by varying the electron beam current while a CW signal is being amplified. Beam current can be changed by varying the potential applied to one of the electrodes in the electron gun—for example, varying the modulation grid potential. However, amplitude-modulated signals produced by varying the grid potential have attendant phase modulation. The graph in Fig. 4.4-4 shows that, over the linear portion of the grid-voltage-versus-output-voltage curve, about a 10-dB amplitude

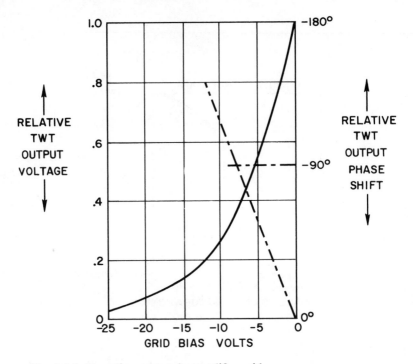

Fig. 4.4-4. Traveling-wave tube amplifier grid
modulation characteristics.

modulation produces approximately 90° phase modulation. Although this degree of phase modulation may limit the usefulness of grid modulation, this method can be used in applications where phase relationships are not involved in the particular application.

    The traveling-wave tube amplifier can also be used as the amplifier of a *phase-modulated or frequency-modulated* master oscillator-power amplifier system. The phase-modulation characteristic is nearly linear and permits many unique and specialized applications.

QUESTIONS

1. Where is the energy taken from in the TWT amplifier in the process of amplification?
2. What is the purpose of the attenuator built into the TWT slow-wave structure?
3. What unwanted effect occurs when grid-amplitude modulation is used in a traveling-wave tube amplifier?

4.5 THE BACKWARD-WAVE OSCILLATOR (BWO)

The backward-wave oscillator (BWO) provides a versatile source of microwave energy that can be voltage-tuned over bandwidths as high as 5:1. The output frequency of the BWO is determined by a frequency-selective feedback and amplification process rather than by resonant circuits, as used in conventional microwave oscillators.

The BWO tube consists of an electron gun, a helix structure, and a collector at the far end of the helix (see Fig. 4.5-1). Physically, the backward-wave

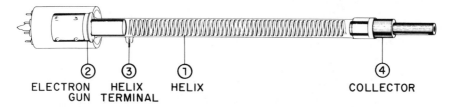

ELECTRON　HELIX　HELIX　　　　　　　　　COLLECTOR
GUN　TERMINAL

Fig. 4.5-1. Backward-wave oscillator tube (without a focusing magnet).

oscillator resembles the traveling-wave amplifier tube; however, for comparable frequencies it is larger in diameter and somewhat shorter in length. Another difference, not apparent from a visual inspection of the tube, is that the helical BWO uses a hollow electron beam with a strong concentration of the electrons near the helix. This hollow electron beam is focused along the length of the helix by a strong magnetic field supplied by an axial solenoid surrounding the tube or by a permanent magnet, as described for the TWT.

The RF output of the BWO is a result of the interaction between the electron beam and the electric fields accompanying a microwave signal present on the helix. The term "backward-wave oscillator" is quite appropriate for this tube, since the RF energy moves and builds up in a direction opposite to that of the electron beam and is coupled out at the gun end of the tube via the helix terminal.

What is the nature of this wave that can go in one direction and still be in step with electrons going in the opposite direction? The operation of the BWO tube can be explained in terms of a series of feedback loops similar to those common to low-frequency electronics circuits. Each of these regenerative loops can function as an amplifier or an oscillator and is designed so that the phase shift around the loop is one cycle. One of these feedback loops is shown in Fig. 4.5-2, where, using conventional terminology, the forward or μ circuit consists of a section of transmission line and the backward or β circuit is a unilateral amplifier connecting the output of the transmission line to

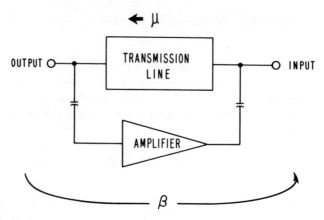

Fig. 4.5-2. A single regenerative loop.

the input. Positive feedback will occur in this circuit when the amplifier gain becomes sufficiently high to overcome the loss in the transmission line, and the $\mu\beta$ loop will oscillate at a frequency for which the total phase delay is one cycle. If the amplifier is designed for limited high-frequency response, oscillation will occur only when the phase delay is one cycle, and the frequency of oscillation can be shifted by changing the phase delay in the amplifier. The essential feature of the voltage-tuned backward-wave oscillator is that the frequency of oscillation can be changed electrically by changing the phase delay in the amplifier.

Figure 4.5-3 shows a chain of identical regenerative feedback loops. Along the top of the chain is a series of transmission line sections that will support a wave moving either to the right or to the left. Along the bottom of the chain is a series of unilateral amplifiers in which signals can pass only in

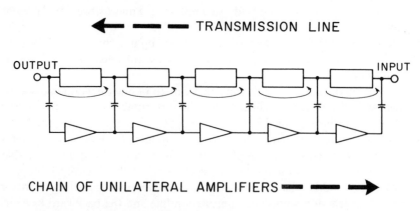

Fig. 4.5-3. A chain of regenerative loops.

the left-to-right direction. Each loop then consists of a transmission line, two coupling capacitors, and an amplifier transmitting from left to right. In operation, positive feedback, which leads to regenerative amplification or oscillation, occurs. It then utilizes a wave going from right to left on the transmission line when the phase delay in a single loop is just one cycle. The total phase delay around a group of n loops will then be n cycles.

For low values of amplification, the chain of loops will act as a regenerative amplifier operating at the frequency which provides positive feedback. However, oscillations will start if the transmission line is terminated in its characteristic impedance at the input, and the amount of amplification is increased. The frequency of oscillation will be controlled by the phase delay in the amplifier chain.

With this background we can now examine the actual functioning of the backward-wave oscillator tube. Figure 4.5-4 shows a cross section of the helix

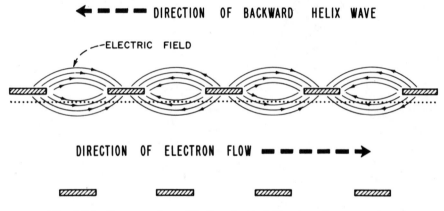

Fig. 4.5-4. Cross section of helix, electron beam, and helix wave.

and a portion of the electron beam. The helix structure consists of a cylindrically wound flat-wire tape; the electron beam is hollow and passes very close to the helix. The strong axial magnetic field focuses the electrons in the beam and allows movement only in the direction of the axis of the tube. The lines of force of the electric fields associated with an RF wave traveling along the helix are also shown in Fig. 4.5-4. These fields rotate around the helix at the velocity of light equal to the ratio of the turn-to-turn spacing of the helix divided by its circumference. The axial electric fields will be strong between helix turns and very weak under the turns, since electric fields cannot exist parallel to a conductor. The strong effect of these fields between helix turns on the velocity of the electrons in the beam produces an interaction process which is represented by the capacitive coupling between the transmission line and the amplifier chain shown in Fig. 4.5-3. In this way feedback loops are formed between the midpoints of adjacent helix gaps.

Although the concept of discrete feedback loops is a useful device for explanation, the backward-wave interaction is actually a continuous process. The maximum coupling between the helix wave and the beam will occur midway between gaps and gradually taper off to a minimum directly under the helix turns. One of these regenerative loop chains exists at each angular position around the helix. Each of these regenerative loop chains is independently coupled to the helix transmission line, so the net effect is a continuous amplification and feedback process occurring down the entire length of the tube.

The basic mechanism of amplification is a velocity-modulation process which causes the electrons to bunch in the beam. Figure 4.5-5 shows the

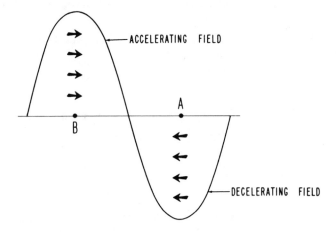

Fig. 4.5-5. Axial electric field of helix wave that provides velocity modulation and bunching of electron beam.

sinusoidal variations in amplitude of the electric field at the midpoint between helix turns. The phase relationship between the backward wave on the helix and the velocity of the electron beam is such that each specific portion of the electron beam will be affected by an electric field of the same phase as it passes successive gaps down the helix. Referring to Fig. 4.5-5, an electron at point A experiencing the decelerating effect of the field at the first gap in the helix will experience a continuous decelerating effect caused by fields of the same phase and direction of force as it proceeds down the tube. In a like manner, an electron at point B will be continuously accelerated in its journey down the tube. In this way some parts of the electron beam are slowed down while others are advanced, and the net effect is a bunch formed at the midpoint of

Fig. 4.5-5 between the accelerating and decelerating fields. This situation is shown in Fig. 4.5-6. The spiral form of the bunched electron beam is caused

Fig. 4.5-6. Helix showing bunching of electron beam.

by velocity modulation which occurs at different RF phases at various angular positions around the spirally wound helix.

At this point it should be mentioned that the average electron velocity of the beam is slightly faster than the effective phase velocity of the amplifier chain. This means that the electron bunches will advance a quarter of a cycle as they approach the collector end of the tube; thus they encounter the full decelerating effects of the electric field and give up a maximum amount of kinetic energy to the wave on the helix.

Figure 4.5-7 shows that the density of the electron bunches increases

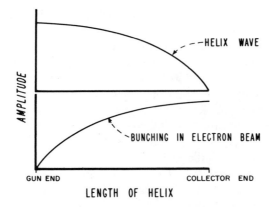

Fig. 4.5-7. Relative amplitude of the envelope of the helix wave and the degree of electron bunching along the helix.

according to a sine-wave relationship; the envelope of the bunching, rather than instantaneous amplitudes, is shown, since many RF cycles exist along the length of the BWO tube. The wave on the helix moves from right to left toward the gun end of the tube and gains amplitude between each turn according to the degree of electron bunching in the beam. In this way the envelope of the wave on the helix shown in Fig. 4.5-7 is the integral of the bunching envelope, so the maximum energy transfer from the beam to the wave on the helix occurs at the collector end of the tube.

Now that a correspondence between the chain of lumped regenerative loops and the helix of the backward-wave oscillator tube has been established, it can be seen that, if the velocity of the electron beam is varied, the phase delay around each of the regenerative loops will be changed. Also, if the electron beam current is high enough, the chain of regenerative loops will oscillate at a frequency where the phase delay of each loop is equal to one cycle.

The following explains the effect of axial electric fields, charge distribution, and energy exchange on oscillation in a BWO. For interaction with electrons, we are concerned with the electric field component of the helix wave that is in the direction of the tube axis. The strong magnetic field prevents electron motions transverse to the axis but allows accelerations and motions along the tube axis. When we have an RF wave traveling along the helix, the electric field on the axis of the helix is a smoothly varying sine wave at any moment in time. But, as we move away from the axis and nearer to the helix, which is wound of wire tape, we find strong electric fields between the wires but no axial electric field directly under the wires, since an electric field cannot be parallel to an electrical conductor. The RF wave will move along the helix with ripples due to the shadow of the helix that remains stationary. Since these field ripples are stationary in space, they can be considered as a standing wave, which we know can be further decomposed into two running waves going in opposite directions. The microwave signal close to the helix always consists of three components: the usual one indicating which way the main signal is going and carrying its power, plus these two oppositely directed waves associated with the field ripples next to the helix. The oppositely directed waves have higher harmonics, but they are not used here, so they will be ignored. The backward-wave interaction is between the electron beam and the ripple wave that is going opposite in direction to the main signal. In waveguide we are used to a number of modes of propagation. These are individual waves and can exist separately. However, the group of three waves discussed here cannot exist individually but must all stick together; if energy is fed into one of them, all three share it.

The wavelength of the field ripples is approximately equal to the turn-to-turn spacing of the helix. Thus if ripple wavelength, frequency, and synchronism velocity are related, the equation is that the product of the wavelength and the frequency must be equal to velocity. Thus, with the wavelength almost constant, the frequency of interaction can change only if the velocity changes. This is the frequency-determining mechanism.

The basic phenomenon in the backward-wave tube is the mechanism of amplification. If we launch a microwave signal on the helix at the collector end, it moves along the helix, velocity-modulating the electron beam, which then produces bunches that reinforce the microwave signal on the helix. This, of course, can occur only if the electrons are in step with one of the helix waves—in this case, the backward-moving wave in the group of three discussed above. The electron bunches start at the gun end of the tube. They form around a point of zero electric field where the wave ahead has been decelerated and the subsequent wave is accelerated to obtain maximum amplification. This phase position that produces the electron bunch should go a little faster than the wave on the helix, so that, by the time bunches are formed, they are synchronized with the decelerating wave of the helix.

This amplification mechanism also includes a regenerative or feedback effect. Starting with an electric field at a point on the helix near the electron gun, this field velocity modulates the electron beam, which moves toward the collector and forms a bunch. This bunch supplies energy to the helix wave that is carrying its power in the gun direction. Thus the amplified wave gets back to the original point, completing the feedback loop. This is a continuous group of feedback loops. When there is sufficient current in the electron beam, this feedback mechanism produces oscillations at the frequency where the electron velocity is in step with the backward wave of the helix. The current at the threshold of operation is called the starting current of the tube. These oscillations build up from noise in the electron beam and also from thermal noise existing on the helix at the synchronized frequency until limiting occurs and henceforth a constant level of oscillation is maintained. The electron bunching, as shown in Fig. 4.5-8, increases sinusoidally from the gun to the collector, whereas the resulting helix wave increases in the same manner from the collector to the gun end. Where the electric field is strong on the helix near the gun, the electron bunches increase rapidly. Conversely, where the electron bunches are strong at the collector end, the wave on the helix builds up faster. As in the traveling-wave tube, a portion of the energy of the high-speed electron beam is converted into a useful microwave signal when the bunches are decelerated by the electric field of the helix wave, which transfers the beam energy to the helix wave. The resulting action of this group of three waves associated with the ripple field near the helix is that the electron bunches pass between windings when the helix is decelerating for electrons and pass under the windings when the electron bunches could absorb energy from the helix wave. Figure 4.5-8 shows that the transfer of power is predominantly from the beam to the helix. However, at 14b, it is seen that the helix wave is accelerating the electron bunch momentarily, but this lasts only for a short time before the bunch passes under a winding where no acceleration can take place.

Since the electrons at different angular positions around the beam encounter the ripple wave bunching at different times, an electron bunch that has a spiral form eventually results. The bunching at an angular position denoted by AA' is 90° out of phase with the bunches formed along BB'.

The backward-wave tube can be used as a voltage-tunable band pass

Key to Fig. 4.5-8

1. *Helix*—a slow-wave transmission line that provides interaction with the electron beam.

2. *Electron gun*—generates a hollow electron beam that travels through the helix very close to the helix turns.

3. *Helix terminal*—RF energy output terminal and helix potential terminal.

4. *Collector*—collects the used electron beam.

5. Cross-section of hollow electron beam.

6. Cross-section of helix showing proximity of beam to helix.

7. *Expanded view of tape-wound helix*—transmits the RF signal which interacts with the electron beam.

8. Smooth electron beam entering the helix where velocity modulation begins.

9. Axial electric field of RF wave traveling around the helix turns.

10a. Axial electric field strength between helix turns that decelerates electrons.

10b. No axial electric field directly under the helix turns.

10c. Axial electric field strength between helix turns that accelerates electrons.

11. Velocity modulation of the electron beam causes bunching. The ripple of the RF wave and the helix structure produce a helical-shaped bunch.

12. Electron current density seen at the top of the beam, along A–A'.

13. Electron current density seen at the same time along B–B'.

14a. Power exchange between electron bunch and helix wave. Helix wave decelerates high-speed electron bunch, which transfers energy from the beam to the helix wave. The power transfer is proportional to the product of electric field strength and electron beam current density.

14b. Helix wave losing energy by accelerating the electron bunch. This action occurs for a very short time only, and very little energy is transferred in this direction.

14c. Under the helix wires, the electric field of the helix wave is perpendicular instead of axial; therefore, there is no power exchange.

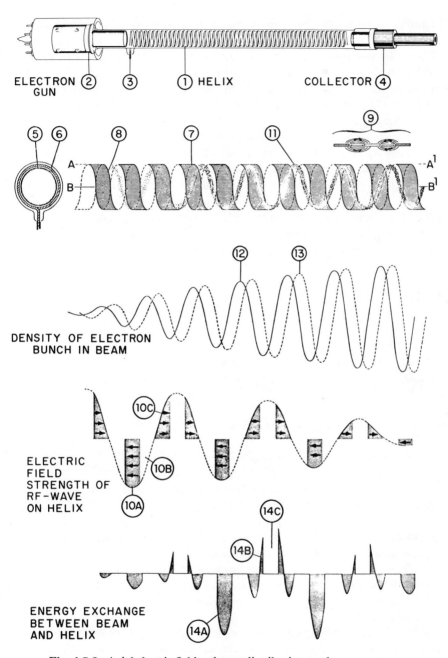

Fig. 4.5-8. Axial electric fields, charge distribution, and
energy exchange in a helical backward-wave
oscillator tube.

amplifier or as a voltage-tuned oscillator. In a sweep oscillator, the tube is used as a voltage-tuned oscillator. This allows a very flexible microwave oscillator. In the lower-frequency range, we are used to mechanically tuned oscillators and reactance-tuned oscillators, but nowhere is such an easily tuned oscillator with a wide bandwidth available. In practically any other tuned oscillator many electrical or mechanical parts are involved. Here, however, we are dealing only with a single vacuum tube. The electrical sweep feature allows a very wide range of sweep rates that lead to an unsurpassed versatility. One other feature is that the amplitude of the oscillation can be decreased continuously to zero. This feature is unusual in oscillators and can be readily used in the microwave region to provide amplitude modulation (which does, unfortunately, have some attendant frequency modulation). This does not prevent it from being useful for providing a constant output oscillator when we wish to sample the output signal and provide a feedback control to maintain constant output. This subject will be discussed later in this chapter.

QUESTIONS

1. What is the basic mechanism which makes a backward-wave oscillator oscillate?
2. How is the oscillation frequency of a BWO varied?
3. Can the BWO be amplitude-modulated? If so, how?

4.6 PIN DIODE MODULATORS

Electronic amplitude modulation of reflex klystrons is impractical. However, if the CW output of the reflex klystron is connected through a variable attenuator, amplitude modulation can be obtained by varying attenuation. The PIN diode has opened the way to a new method of modulating microwave signals. Placed across a transmission line, the PIN diode becomes an absorption-type attenuator which permits sine-wave, square-wave, and pulse modulation without frequency pulling.

The limitations of present amplitude-modulation methods are becoming more evident with the advance of microwave technology. For example, it has been nearly impossible to amplitude-modulate a klystron oscillator directly with anything but a square wave or pulse. Sine-wave modulation traditionally results in significant frequency shifts with changes in amplitude. In addition, conventional klystron oscillators have relatively slow rise and decay times and poor frequency stability, and jitter in the RF pulse is too high for precise pulse measurements.

The PIN modulator is a high-speed, current-controlled, absorption-

type attenuator. A simplified illustration of a PIN modulator is shown in
Fig. 4.6-1. Each PIN unit includes one low-pass filter, two high-pass filters, a

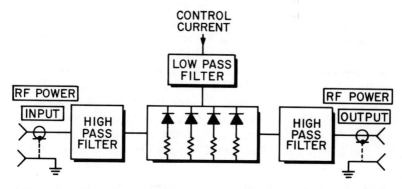

Fig. 4.6-1. Simplified block diagram of PIN modulator.

number of PIN diodes, and a 50-ohm strip transmission line (ridged waveguide in the higher-frequency units). The PIN diode is a silicon junction
diode whose P and N trace regions are separated by a layer of intrinsic (I)
semiconductor (silicon). Thus we have the name PIN diode.

4.6.1 OPERATION AND USES OF PIN ATTENUATORS, MODULATORS

At frequencies below about 100 MHz, the PIN diode rectifies as a junction diode. However, at frequencies above 100 MHz, rectification ceases due
to stored charge resistance by conducting current in both directions. This
equivalent resistance is inversely proportional to the amount of charge in the
I layer. An increase in forward bias current (current at a negative voltage)
increases the stored charge and decreases the equivalent resistance of the PIN
diodes. When reverse bias is applied, reverse current flows until the stored
charge is depleted, at which time equivalent resistance becomes a maximum
in the order of thousands of ohms.

To understand how a PIN modulator works, consider the following:
the PIN diodes are mounted as shunt elements between the RF transmission
path and ground. The transmission path has a characteristic impedance of
50 ohms. When the PIN diodes are forward biased, the equivalent diode
resistance is about 30 ohms, and most of the RF energy is absorbed by the
diodes instead of propagating down the 50-ohm transmission path. However, when the diodes are reverse biased, the equivalent diode resistance is in
the order of thousands of ohms, and the microwave currents will flow down

the transmission path because diode resistance compared to the 50-ohm path impedance is negligible.

PIN modulators are three-port devices that accept RF power at either end port and, depending on the bias signal applied, provide a modulated RF output at the other end port. As an operating device, the PIN modulator should be thought of as a variable RF attenuator whose attenuation level is controlled by dc current and voltage. By applying a +5-volt dc potential to the bias input connector, RF signals will be passed through the PIN modulator with only minimum residual attenuation. By applying a current at about 0.7 volt dc at the bias input, attenuation is provided at all points across the frequency band. The dc current at a negative voltage is forward biased. By varying the forward bias current between the rated value and zero, any level below maximum rated attenuation can be established. Figure 4.6-2 shows

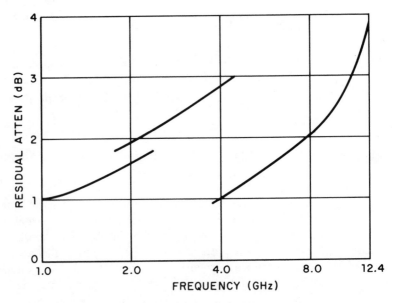

Fig. 4.6-2. Typical residual attenuation of different PIN modulators.

typical residual attenuation curves of some PIN modulators having a maximum of 80 dB on-off ratio.

Attenuation increases with forward bias current almost linearly with dB until the diodes saturate. Attenuation sensitivity varies from unit to unit and also with time for individual units. For this reason the PIN modulators are not suitable for use as programmed attenuators. The range of expected sensitivity variation between units, and for a single unit with time, is shown in the

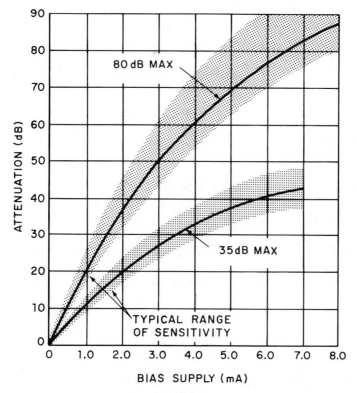

Fig. 4.6-3. Typical sensitivity of attenuation (plotted with frequency constant).

shaded areas of the typical 35-dB and 80-dB attenuation plots in Fig. 4.6-3. Attenuation with a constant forward bias varies with frequency. This variation is small at low levels of attenuation and more noticeable at higher values of attenuation. In addition, it should be noted that attenuation is a minimum at each end of the frequency band. Thus the frequency band can be exceeded if a degradation in attenuation range as well as other specifications is acceptable. Exceeding the band limits by a large degree is impractical.

The PIN modulator has a good impedance match at all levels of attenuation over its frequency range. This match is illustrated in Fig. 4.6-4, which shows the SWR of a typical PIN modulator measured under both zero-bias and maximum forward-bias conditions. Hence, with this constant match, frequency-pulling effects due to the PIN modulator are negligible. If the PIN modulator is used outside its specified frequency range, SWR characteristics will be degraded along with the other operating specifications. Also, spurious responses can result when the specified frequency range is exceeded.

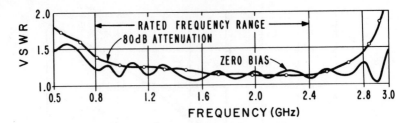

Fig. 4.6-4. Typical SWR of a PIN modulator.

The attenuation variation with applied bias current and bias voltage at different temperatures is illustrated, over a limited range, in Fig. 4.6-5. Attenuation in dB varies almost linearly with current, whereas variation with voltage follows an almost perfect exponential curve. If a constant voltage is maintained, the attenuation rises with increasing temperature. With a constant current, the attenuation drops with increasing temperature. Since the constant-voltage temperature coefficient of attenuation is the opposite of the constant-current coefficient, the proper selection of voltage and series

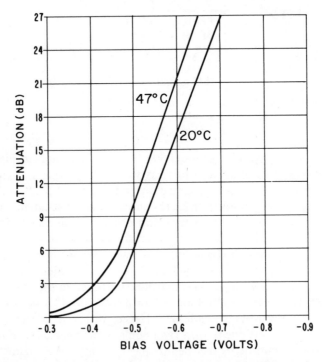

Fig. 4.6-5. Typical attenuation variation at two different ambient temperatures.

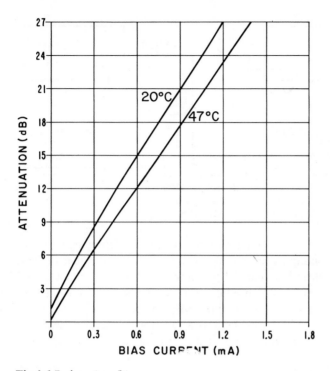

Fig.4.6-5. (*continued*)

resistance between the source and the PIN modulator will give temperature compensation over a limited operating range.

The PIN modulator can be used to amplitude-modulate an RF signal with almost any time-varying signal. Modulation is accomplished with a dc bias current to obtain a specific attenuation level and then superimposing a time-varying current upon the bias current. The specific attenuation level upon which the modulating signal is superimposed must be equal to or greater than the peak amplitude of the modulating signal, or peak clipping will occur. A typical setup for modulating applications is illustrated in Fig. 4.6-6.

4.6.2 MODULATION DISTORTION

When a PIN modulator is used in an AM system, some envelope distortion occurs. This distortion is a function of the peak attenuation and total attenuation range covered by the modulating signal. To minimize RF envelope distortion, the reference level upon which the modulating signal is superimposed should always be a minimum (i.e., reference level should be set

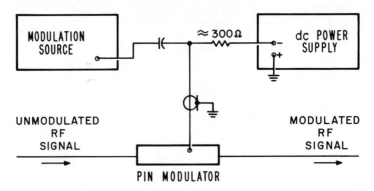

Fig. 4.6-6. Typical general modulation setup.

at a point only slightly greater than the peak-modulating signal amplitude). Under these conditions a 50% sinusoidal modulating signal (9.5 dB total swing) will result in less than 6.1% distortion. A plot of measured AM distortion for a typical PIN modulator is shown in Fig. 4.6-7. (*Note:* AM distortion was measured under the optimum conditions described above. Since distortion, as described here, is a function of the nonlinear sensitivity relationship, shaping circuits may be incorporated to limit overall distortion.)

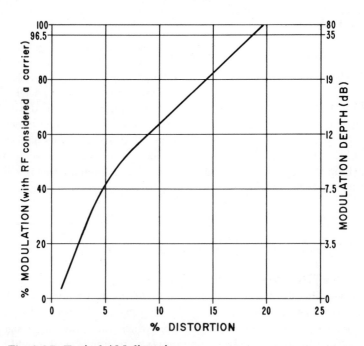

Fig. 4.6-7. Typical AM distortion.

4.6.3 AMPLITUDE MODULATION
LIMITATION

In any AM system the modulating equipment limits RF-modulating capabilities almost entirely. For example, the power supply and modulation signal source must both be capable of supplying sufficient current to the low impedance bias input. See Fig. 4.6-3 for typical currents necessary for desired attenuation levels. In addition, the PIN modulator itself limits the total per-cent of modulation obtainable. With 35 dB attenuation, AM percent of modu-lation is limited to a maximum of about 96.5%; with 80 dB attenuation, the percent of modulation is essentially 100%.

The PIN modulator can be used as a pulsing or switching device for RF power levels. Modulation is accomplished by applying rated forward bias for a maximum attenuation. Once maximum attenuation level is established, the RF power may be pulsed "on" by applying a constant $+5$ to $+6$ volts to the bias input (voltage must be referenced to PIN modulator ground). At the end of desired pulse width, the $+5$- to $+6$-volt dc potential must be switched to a -0.8-volt level with rated bias current so that RF power level is pulsed "off."

To obtain pulsing with rise and fall times in the order of 15 to 40 nano-seconds, the PIN modulator must be biased with a specially shaped impulse waveform, such as that illustrated in Fig. 4.6-8. However, if rise and fall times in the order of 100 to 300 nanoseconds are satisfactory, a setup such as that shown in Fig. 4.6-6 may be used. (*Note:* In any modulation system, modula-tion capability depends on the modulating waveform at the bias input con-nector. Hence lead lengths should be as short as possible to avoid capacitive cable effects.)

QUESTIONS

1. What does the I stand for in PIN diode?
2. Why doesn't the PIN diode work as a voltage-controlled resistor in low frequencies?
3. In which direction must the PIN diode be biased to decrease its equivalent circuit resistance?
4. What kind of distortion occurs in PIN diode modulators when they are amplitude-modulated?

4.7 LEVELING OF SIGNAL SOURCES

Varying the frequency of a signal source will provide a quite randomly varied output signal level. When using a signal source to perform varied

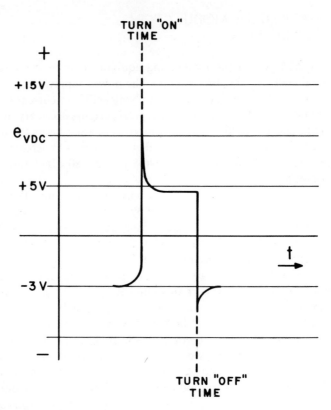

Fig. 4.6-8. Dynamic bias switching waveform.

measurements, it is often quite cumbersome to measure the output level of the source and to reset an attenuator connected after the source in order to keep the output level constant with varying frequency. A feedback circuit, the so-called automatic leveling circuit (ALC), has been developed to provide a constant output level. A directional coupler, explained in detail in another chapter, provides a continuous sample of the power level flowing in the incident direction out of the source. Theoretically, a directional coupler only couples energy from the line flowing in one direction; the signal flowing in the reverse direction does not couple. In practice, this is not true, but quite large differences in coupling can be maintained between forward and reverse coupling. Figure 4.7-1 shows the flow graph of a directional coupler, where C stands for forward coupling and D stands for directivity, or the difference from forward to reverse coupling.

An ALC system can be used if one takes a signal source in which the output power level can be electronically adjusted by applying voltage or current. If no adjustment of this kind can be arranged, as in the case of the reflex

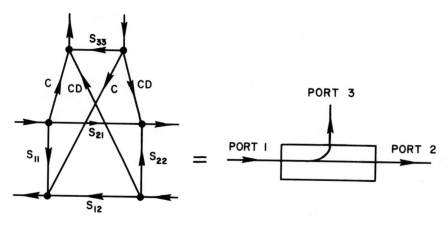

Fig. 4.7-1. Flow graph of a directional coupler.

klystron, a PIN diode modulator can be used as a current-controlled level set attenuator. The circuit in Fig. 4.7-2 shows how automatic level control can be achieved.

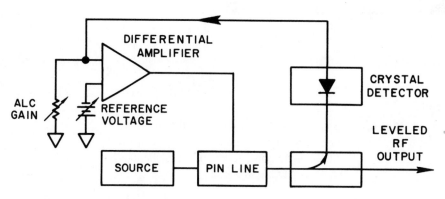

Fig. 4.7-2. Basic leveling circuit.

The ALC input is a dc feedback voltage derived from a crystal detector monitoring the RF level at some point in the measurement system. In a basic leveling system, as shown in Fig. 4.7-2, the feedback loop is arranged so that forward power is held constant with frequency and load impedance variation. The leveling amplifier has a differential input with an adjustable reference voltage that sets the operating power level. Increasing this voltage calls for more detector voltage, and thus more RF output, to satisfy the leveler amplifier. The overall loop gain of the ALC system is adjusted by a variable resistance at the other input to the differential amplifier. Increasing the gain control increases negative feedback, reducing power peaks and improving leveling. Excessive gain causes the feedback loop to be unstable, and oscillations can occur.

Because the ALC system utilizes a feedback amplifier, the circuit must be properly adjusted to the leveling components for maximum gain without oscillations in the feedback loop. If loop gain is too low, the leveling circuit will be ineffective, as shown by the power variations in Fig. 4.7-3(a). If loop

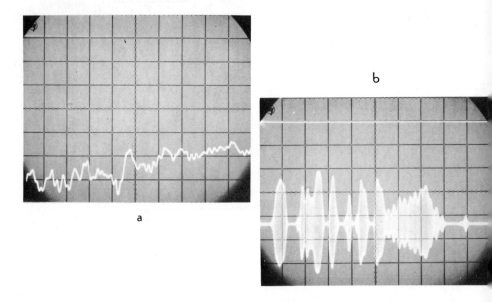

Fig. 4.7-3. Oscilloscope patterns (a) insufficient ALC gain-power output unleveled (b) excessive ALC gain for power level setting; oscillations occur in ALC feedback loop.

gain is too high, the system will become unstable, and oscillations, such as shown in Fig. 4.7-3(b), will occur. Improper operation of the ALC loop can be readily observed in measuring systems using oscilloscope readout.

Once a system has been adjusted, readjustment of the gain control is normally not required unless different couplers or detectors are substituted in the leveling loop. The calibrated power level control may be simply adjusted each time the system is turned on.

Leveling not only provides flat power level output with frequency; it also makes the signal source look as if it were very well matched—an effect

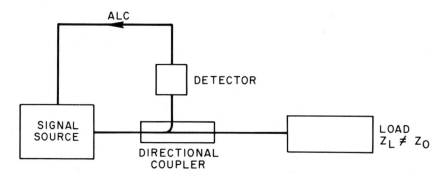

Fig. 4.7-4. Leveled signal source terminated with a reflecting load.

many times more important than flatness. This effect can be easily under-stood by assuming a leveled source terminated with a load reflecting part of the power applied to it (Fig. 4.7-4). If part of this incident signal is reflected from the load, and it travels backward in the direction of the source, it passes the directional coupler. Since the directional coupler couples energy from only one direction into the auxiliary line (to the detector), and in this case that direction is the forward one, it will not sense the reflected signal. Con-sequently the ALC loop will not feel any change from the reflected wave coming from the load. This signal entering the source will rereflect from the discontinuity of the source and produce a change in the forward-going signal. This will either add to or subtract from the incident signal, depending upon its phase. If this signal level change is forward going, it will be sensed by the ALC loop and consequently corrected.

Essentially, the ALC loop makes the source look perfectly matched. Since directional couplers are not perfect, they have directivity errors, and the correction will only be as good as the directional coupler's directivity. For instance, if a directional coupler has 40 dB directivity, the coupler will make the source look as if it were matched to 1.02:1 VSWR, since this ratio relates to a 40 dB return loss.

There will be other sources of mismatches occuring besides the direc-tivity. These sources are also vectorial quantities adding vectorially to direc-tivity, such as all discontinuities past the plane of coupling in the directional coupler. The discontinuities on the remainder part of the mainline in the coupler including the connector of the coupler, which is usually considered to be the plane of the leveled signal source, are also considered to be lumped into these sources of mismatches.

QUESTIONS

1. What does the source match depend on in a leveled source?
2. What adjustment controls the leveled output of a leveled signal source system?

4.8 HARMONIC GENERATORS, FREQUENCY MULTIPLIERS

Frequency multiplication is often used to produce signals at a frequency higher than that produced by state-of-the-art signal generators. When working with lower frequencies, particularly in frequency standards labs, frequency multiplication is often the most desirable way to raise the frequency of an input signal. A tuned amplifier having an input circuit resonant at frequency F_1 can, through a push-pull circuit, provide an output at frequency $2F_1$ if the output circuit is tuned to that frequency. Higher multiples of the input frequency can be easily achieved through similar techniques. The frequency-multiplying chains needed to generate standard frequency sequence signals are discussed in Section 5.2.2. This operation consists simply of generation of harmonics of the fundamental signal.

Harmonic generation can be achieved by means of nonlinear devices, such as crystal diodes.[1] During the past few years the step recovery diode has found increasing application in the generation of microwave power by means of single-stage multiplication of higher harmonic order. These multipliers offer reduced complexity and increased versatility in multiplying by orders of 2 to 100 over multistage solid-state chains using Varactor diodes. Conversion efficiencies in excess of $1/n$, where n is the harmonic number, are easily achievable by this technique.

Step Recovery Diode Principles

The step recovery diode is a silicon PN junction diode with a dynamic performance that depends very much on minority carrier lifetime and junction retarding field. The dynamics of diode behavior have been discussed in detail in papers by Moll, Krakauer, and Shen,[2] among others.

The important mechanism for harmonic generation is the behavior of the diode during the intermittent stages of forward conduction, reverse

---

[1] Hall, R., HPA Application Note #2, Harmonic Generation with Step Recovery Diode.

[2] Moll, Krakauer, and Shen, "P-N Junction Charge Storage Diodes," *Proc. IRE*, Vol. 50, January 1962, pp. 43–53.

conduction, and the abrupt termination of reverse conduction. During forward conduction, charge is stored in the diode by the minority carriers. The amount of charge stored is determined by the minority carrier lifetime of the diode. Upon application of a reverse bias voltage, charge flows out of the diode as reverse current. During the initial phase of reverse recovery, the conductivity of the diode will remain essentially the same as its forward conduction value, and the impedance of the diode therefore remains low during this interval.

The reverse current ceases to flow when the stored minority carriers have been depleted by the flow of reverse current and by minority carrier recombination. This causes an abrupt transition or "step" in the reverse recovery waveform. This fall time, called "transition time," is determined by the junction retarding field and can be controlled during manufacture from less than 100 picoseconds to several nanoseconds. Upon completion of this rapid transition time, the RF impedance of the diode becomes that of the reverse-bias diode capacitance.

This abrupt transition during reverse conduction is shown in Fig. 4.8-1(a) for a diode with sinusoidal excitation. This unique waveform is the result of optimizing the step recovery diode for finite controlled storage, coupled with a very abrupt transition from reverse storage conduction to cutoff.

It is in the analysis of this waveform that we determine the harmonic generating capability of this diode. The harmonic spectrum that is associated with the waveform can be evaluated by conventional Fourier analysis. S. Krakauer[3] performed this analysis in demonstrating the fundamental differences in the step recovery diode as a harmonic generating element versus other widely used semiconductor diodes.

Semiconductor diodes have been widely used as harmonic generating elements because of their nonlinear behavior. Profound generic differences can exist in the character of this nonlinearity, depending upon both the design of the diode and its operating conditions. These differences in nonlinear behavior lead to fundamental differences in harmonic generating capability.

Conventional harmonic generating diodes, such as Varactors, are operated within their reverse saturation region so as to avoid both forward conduction and avalanche breakdown. Under these conditions the diode will act as a voltage variable capacitor having some dissipation. This smooth capacitance voltage characteristic does not naturally generate high-order harmonic power. Conversion efficiency in this mode has been found to drop off at the rate of $1/n^2$ (where n is the harmonic number) because the available

[3] Krakauer, S., "Harmonic Generation, Rectification and Lifetime Evaluation with the Step Recovery Diode," *Proc. IRE*, Vol. 50, July 1962, 1665–76.

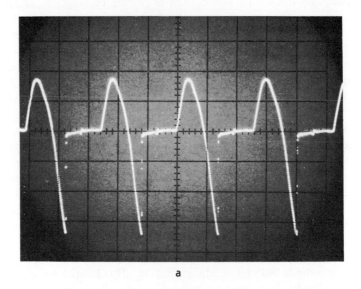

a

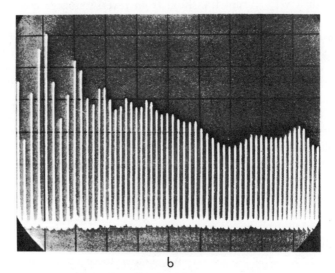

b

Fig. 4.8-1. (a) The picosecond transition of the step recovery diode after a period of reverse condition. A 10-MHz sine wave has been applied across the diode and the current through the diode is monitored. (b) This transition can be used as a very rich source of high order harmonics.

harmonic power output drops off so rapidly. For this reason, efficient harmonic generation with Varactors usually requires a cascade of doublers and triplers.

The step recovery diode has a reverse-bias capacitance which varies slightly with bias voltage; hence the contribution of this mechanism to harmonic power may be neglected. However, Fourier analysis shows that the conversion efficiency of the step recovery diode varies as $1/n$ versus the $1/n^2$ of a Varactor, where n is larger than 5. Hence the effect of stored charge due to operation in the forward direction increases high-order efficiency over operation in the reverse saturation region where minority carrier lifetime approaches zero. Therefore the step recovery diode as a single-stage multiplier is a very rich source of higher-order harmonics, as shown in Fig. 4.8-1(b).

Design Criteria for Step Recovery Diode Multipliers

The functional block diagram of a single-stage multiplier is shown in Fig. 4.8-2(a), and a typical physical realization of that block diagram is shown in Fig. 4.8-2(b).

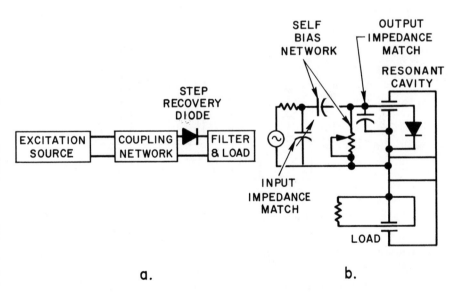

a.   b.

Fig. 4.8-2. (a) Functional block diagram, (b) typical physical realization.

The design of the step recovery diode harmonic generator circuit should:

1. provide a broadband reactive termination for unwanted harmonics;
2. resonate the diode at the output frequency;

3. match input power down to the diode impedance (1–10 ohms);
4. tune out the circuit susceptance of the input frequency; and
5. provide high-Q energy storage at the output frequency.

These criteria are provided for in either of the two alternative designs shown in Figs. 4.8-3 and 4.8-4. In Fig. 4.8-3, a broadband input circuit is utilized to enable changes in frequencies without returning the input. The resonant

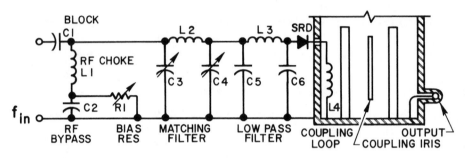

Fig. 4.8-3. Broadband input circuit.

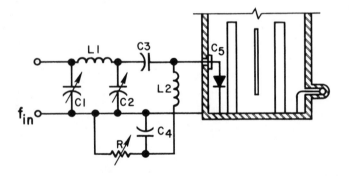

Fig. 4.8-4. Resonant input circuit.

input circuit of Fig. 4.8-4 is simpler for fixed-frequency applications and still decouples harmonics efficiently from the source. In either circuit, self-bias of the diode is satisfactory, but in some applications it may be desirable to use fixed bias and to control the bias resistance.

In Fig. 4.8-3, capacitors C1 and C2 are large so that their reactances are small at the input frequency. The same is true of C3 and C4 in Fig. 4.8-4. In both circuits, a PI section (C3, L2, C4 in Fig. 4.8-3 and C1, L1, and C2 in Fig. 4.8-4) is used to match the source resistance down to the diode resistance. These transforming sections have 90° phase shift and are analogous to a quarter-wavelength transformer. For matching between two resistance values, R_s of the source and R_d of the diode, the three reactances of the PI section are all equal to R_sR_d at the input frequency. By making the capacitances adjustable, the diode reactance may also be compensated at the input frequency. The adjustable resistors R will usually be in the range of 1 to 10 kΩ for drive levels of 100 to 1000 mw.

In Fig. 4.8-3, a low-impedance, low-pass filter is used to prevent loading of harmonics by the input circuit. This filter, consisting of C5, L3, and C6, should have a cutoff frequency about 1.25 times the input frequency and may have more elements if desired. It can be omitted with some slight sacrifice in efficiency, since the low-pass matching network also cuts off harmonics. This was done in a 100-MHz to 5-GHz multiplier, which attained up to 15% efficiency despite some leakage of harmonics through the input matching network and about 2.5 dB circuit losses.

In Fig. 4.8-4, harmonics are decoupled from the input circuit by a low-reactance resonant tank circuit, L2, C5. The Q of these elements should be at least 100, so that their reactances may be 1 to 2 ohms at the input frequency. The semipictorial representation of C5 shows that it may be constructed in the form of a dielectric (ceramic) washer mounted on the wall of the output tuned cavity. This geometry is free from undesirable resonances across the extreme frequency ranges of the higher-order multipliers. In like manner, the diode coupling in Fig. 4.8-3 may be done the same way.

The diode in its associated package and the output circuit should be resonant at the output frequency, a condition normally difficult to calculate since it depends upon geometric details of diode mounting, the diode parameters, the output circuit, and the driving reactance. The diode should not be resonant at other harmonic frequencies. In practice, one usually provides some reactive tuning device at the diode to adjust for best operation. One such device is shown in Fig. 4.8-5.

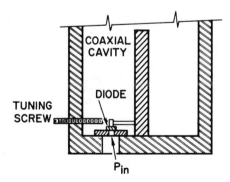

Fig. 4.8-5. Diode resonance adjustment. A tuning screw
can be used as shown to adjust the diode to
resonate at the output frequency.

Energy storage at the output frequency is needed to develop voltage on the diode for maximum efficiency and to eliminate undesired adjacent harmonics. In the examples of Figs. 4.8-3 and 4.8-4, a double-tuned, quarter-wave stub, iris-coupled cavity is indicated. Any of many other configurations of waveguide or coaxial cavity could be used. The cavities should have high Q to reduce insertion loss at the desired output frequency while simultaneously eliminating adjacent harmonics. An alternative description of the requirement is to say that enough energy should be stored so that it is not appreciably depleted by the load in the interval between impulses from the drive frequency. This means suppressing the amplitude modulation of the output at the drive frequency. The degree of suppression desired will depend upon the application and will determine the amount of output filtering required. For example, to reduce neighboring sidebands in a $\times 20$ multiplier by 20 dB, a loaded Q of about 300 is needed.

Adjustment of multipliers using the step recovery diode is a simple procedure. The output resonator should be prealigned to resonate at the proper frequency, and the input frequency and power level should be set. The diode self-bias resistance, the output circuit resonance with the packaged diode, and the input matching circuit are then adjusted for maximum power output and efficiency. Under improper adjustment, parametric oscillations may be observable. If they are present, the circuit will jump suddenly in and out of oscillation, and the output power will fluctuate as the adjustments are made. It should be possible to find a broad region of adjustment for efficient operation where no such variations occur. Inability of the circuit to achieve such operation indicates the existence of resonances at different frequencies. Good geometrical circuit layout and construction practices should be employed to eliminate such resonances.

A Typical Single-stage X20 Multiplier

Let us discuss a single-stage $\times 20$ step recovery diode multiplier designed using the principles cited. An efficiency in excess of 10%, or $2/n$, where n is the harmonic number, can be achieved by this technique using various step recovery diodes. The circuit of the multiplier and the associated filter are shown schematically in Fig. 4.8-6.

The input signal is fed through a coupling capacitor, $C1 = 0.1 \ \mu F$, which, in conjunction with a choke, $L1 = 5 \ \mu H$, is used to separate the input power from the bias voltage developed across the adjustable resistor, $R1 = 50$ kΩ. The input impedance is matched by a network consisting of variable capacitor, $C3 = 14$–150 pF, $L2 = 70$ nH, and the bypass capacitor, $C4 = 70$ pF, which also decouples the higher harmonics from the input. The input impedance of 50 ohms is matched down to a range of 5 to 15 ohms by this circuit to couple efficiently into the diode.

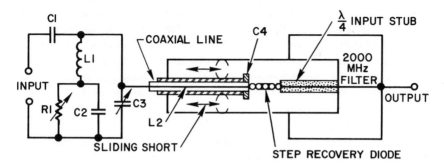

Fig. 4.8-6. Step recovery diode multiplier circuit.

The step recovery diode is placed in series with an input resonator of the shorted type. The only adjustment of the diode reactance and microwave circuit impedance required is by use of the sliding short shown in the figure. The 2000-MHz output filter is a six-resonator, interdigital structure with a bandwidth of 20 MHz and a 2-dB insertion loss. The estimated total circuit losses of the circuit described are approximately 2.5 to 3 dB.

The operation of the circuit is simple, and replacing a diode requires a minimum of retuning to achieve a stable maximum output power. The tuning requires only that R1 and C3 be adjusted for best input match and the sliding short be positioned for maximum output power.

Various HPA step recovery diodes were substituted in this ×20 multiplier, and Table 4.8-1 shows a summary of the average experimental results obtained with the different types of diodes.

Table 4.8-1. Average Operating Results

×20 Multiplier

Input Frequency 100 MHz–Output Frequency 2000 MHz

HPA Diode Type	P_{out}(mw)	Diode Conversion Efficiency (in percent)
0112	20	15
0151, 0251	5	25
0152, 0252	5	20
0153, 0253	15	30
0154, 0254	15	15
0132	80	10

Determination of Diode Parameters

The results of Table 4.8-1 indicate a variation in conversion efficiency and power output as a result of substituting various HPA step recovery diodes. The choice of a diode to obtain a desired output power for a pre-scribed frequency and conversion efficiency depends upon the interrelation of the significant diode parameters. The parameters of interest are τ, minor-ity carrier lifetime; t_t, transition time; C_j, junction capacitance; R_s, series resistance; and BV, breakdown voltage. Certain of these parameters conflict in the process of selecting a suitable diode for a given application.

Minority carrier lifetime τ is determined by the desired input frequency and determines the amount of stored charge available for conversion to higher-order power. For a large-amplitude transition of the reverse-current waveform, the lifetime should be longer than the period of the input signal. A lifetime at least three times longer minimizes the stored charge lost by recombination within the diode and enhances efficient multiplication.

The conversion efficiency depends on the speed and amplitude of the transition time. Step recovery diodes with transition times less than one-half of the period of the desired output frequency achieve efficiencies in excess of $1/n$. The efficiency increases as the transition time becomes short com-pared to the period of the output frequency, until it is one-tenth or less. Although theoretical predictions other than the 100% efficiency permitted by the Manley-Rowe relations are lacking, diode efficiencies up to 30% have been achieved experimentally, in the previously cited $\times 20$ multiplier. Circuit losses limit overall efficiency to less than this value. The probable practical limit of efficiency for high-order multiplication obtained by extrapolating data to the limit of zero transition time is approximately 40%.

Figure 4.8-7 shows experimental results using the $\times 20$ multiplier from a 100-MHz input to a 2000-MHz output. The efficiency at the diode is plotted against transition time, t_t, measured in units of the period, T, of the

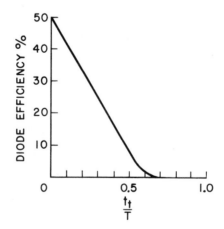

Fig. 4.8-7. Efficiency of $\times 20$ multiplier versus transition-time-to-period ratio.

output frequency. The region $t_t/T < 0.2$ is extrapolated. Approximately 40 different diodes were used to obtain the data for this graph. Data on many more diodes in C-band multipliers confirm the general relation shown here.

Low series resistance, R_s, and junction capacitance, C_j, are required to obtain good circuit efficiency. The series resistance in HPA step recovery diodes is relatively low—typically 1 to 10 ohms—as shown by the high forward current achievable 40 to 400 milliamperes at 1.0 volt bias. The junction capacitance is a function of the junction area and differs for each class of step recovery diode. For a given output frequency, it is desirable to keep R_s and C_j as low as possible. Figure 4.8-8(a) shows the diode equivalent

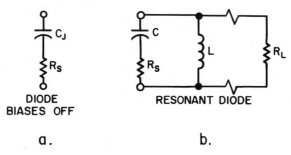

DIODE
BIASES OFF

RESONANT DIODE

a.   b.

Fig. 4.8-8. Equivalents circuits (a) diode biases off
(b) resonant diode.

circuit after transition to the zero reverse current condition as consisting of the R_s and C_j.

Figure 4.8-8(b) shows the equivalent circuit of the diode when resonated at the output frequency. The unloaded circuit Q is equal to

$$\frac{1}{\omega R_s C_j}$$

and the resistance at resonance is approximately

$$\frac{1}{\omega^2 R_s C_j 2}$$

The resonant resistance should be large compared to the load resistance, R_L, to minimize circuit losses at the output frequency. As an example, at 10 GHz with $C_j = 0.5$ pF and $R_s = 1$ ohm, the resistance is approximately 1 kΩ. A loss of only 5% in power is realized with a 50-ohm R_L under these conditions. Power-handling capability depends to a great degree on breakdown voltage rather than diode heating until the burnout level is reached. Figure 4.8-9 shows the relationship of power output versus breakdown voltage

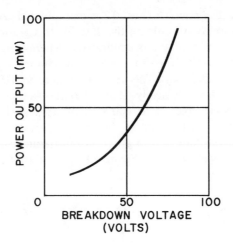

Fig. 4.8-9. Effect of diode breakdown voltage on output
power capability.

from data obtained on various diodes in the ×20 multiplier at an output
frequency of 2.0 GHz.

The relationship shown in Fig. 4.8-9 is consistent with results obtained
from a C-band multiplier. Thus increased power output is obtainable in two
ways: by increasing the diode breakdown voltage and by reducing the circuit
voltages. In circuits of approximately 50 ohms characteristic impedance,
input power levels up to one watt may be safely employed for various HPA
diodes presently available.

Optimizing the important diode parameters presents somewhat conflicting requirements on the diode construction. Figures 4.8-7 and 4.8-9 indicate that very high breakdown voltage and fast transition times lead to
greater output power capability. These may be conflicting requirements in a
diode. Likewise, low junction capacity and high minority carrier lifetime
for high-Q resonance and lower input frequency requirements may not be
achievable, since they both depend in conflicting ways on junction size. However, compromises can be affected in diode parameters to optimize the power
output and efficiency for a given application.

As an example, in many applications it is desirable for the power output
of a multiplier to be independent of the power input over some reasonable
range. Step recovery diodes may be operated in this manner by proper choice
of breakdown voltage. For example, in multiplying to frequencies in the 6- to
7-GHz range with multiplication orders of 10 to 20 times, fairly linear operation was obtained up to about 300 mw driver. Figure 4.8-10 shows typical
results for a diode with breakdown at 28 volts. The saturated output power
will depend upon breakdown voltage, as previously shown.

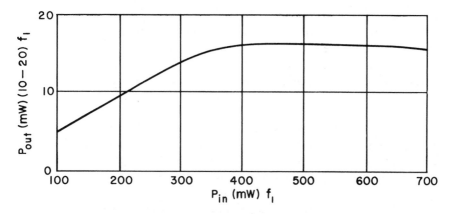

Fig. 4.8-10. Variation of output power versus input power for an HPA 0253 diode with V_{VR} of 28 volts.

Since diodes with breakdown voltages in the 10- to 20-volt range will generally have shorter transition times than higher-voltage units, higher efficiencies are generally achievable with lower input power at a given frequency. Thus the choice of breakdown voltage and transition time for a particular application depends on the power level and efficiency required and on the desirability of saturated operation. Appropriate choice of these parameters can be guided by the graphs in this section, but, since the results are also circuit dependent, adjustment may be needed after initial experimental results have been obtained.

Comparison with Other Techniques

The step recovery diode harmonic generator offers a distinct and prime advantage over all other techniques of multiplication: simplicity. The reduced complexity increases reliability, yields greater circuit flexibility, and allows ease of adjustment.

High-order multipliers using these diodes are the most practical means of generating low-noise, high-stability signals for low-power microwave applications. In broadband output applications requiring more than 1% or 2% instantaneous bandwidth, lower multiplication orders—less than ×20 to ×30—are necessary to eliminate neighboring undesired harmonics from the passband. In broadband Varactor multiplier chains, obtaining reasonably flat output power across the desired frequency band is difficult. This problem is further aggravated by the necessity of using idlers in these chains to increase efficiency and the inherent difficulty in their adjustment. However, the efficiency of multiplication of the step recovery diode permits these idlers to be eliminated. Therefore, as expected, the step recovery diode multiplier gives a flatter output across the band in experimental comparison with a Varactor multiplier chain of the same multiplication order.

Idlers are required in a Varactor multiplier of even moderate order because its smooth capacitance-voltage characteristic does not naturally generate high-order harmonic power. Step recovery diodes, being rich in higher-order harmonics, multiply to the 100th order with better than $1/n$ efficiency. Clearly, in a multiplier of such high order, the many idlers indicated for a Varactor multiplier cannot be included. The resulting circuit simplicity from omitting these idlers gives ease of adjustment. Multiplier circuits of the order of ×10 to ×20 have been adjusted after replacing various diodes in an average tuneup time of 5 minutes per diode.

By omitting idlers, one also eliminates many of the possible modes for parametric oscillation. Clean output power free from spurious frequencies and noise is easily obtainable by using a double- or triple-tuned output circuit or a cavity with sufficiently high Q to suppress neighboring harmonics. A side-by-side comparison of a single-stage HPA step recovery diode multiplier and a conventional Varactor multiplier chain has demonstrated the superior noise performance of the single-stage approach. Parametric up-conversion of noise, except for noise on the driver signal, can be almost eliminated by proper tuning of the Varactor chain. Noise must be removed from the driver by narrow-band filtering at the multiplier input in both cases. After all precautions have been taken and parametric oscillations have been eliminated from the Varactor chain, the noise advantage of the single HPA step recovery diode multiplier relative to a Varactor chain will still be important for critical systems such as microwave relay links and CW detection. The noise output per cycle bandwidth of the single-stage multiplier has been reported to be 120 to 130 dB below the desired output, even within the bandwidth of the circuits. This level is exceedingly difficult to measure.

Using the best reliably reported results, the state of the art may be summarized for step recovery diode multipliers as shown in Table 4.8-2.

Table 4.8-2. Present State of the Art in High-Order Multiplication

F_{out}(GHz)	Multiplication Order (n)	Maximum P_{out}(mw)	Conversion Loss (dB/n)
1–2	10–30	100	0.5
2–4	10–30	75	0.75
4–8	10–40	20	1.0
8–12	9–84	4.0	0.65

The tabulated efficiencies include circuit losses which in some cases are known to be excessive. The variation in performance from band to band is thought to reflect the levels of design effort expended to date.

This tabulation should soon become obsolete as a result of work in progress at Hewlett-Packard Associates and elsewhere. Step recovery diodes should soon become available which will be capable of increasing the output power available to the order of 1.0 watt in the range of 1.0 to 2.0 GHz and 50 to 100 mw in the range of 8.0 to 12.0 GHz at comparable efficiencies.

With the use of harmonic generators, so-called comb generators can be realized. A comb generator generates a train of sharp pulses at the repetition frequency of 1, 10, or 100 MHz supplied internally or at the frequency of an external oscillator. The frequency spectrum of the output is a comb with spectral lines spaced by the repetition frequency, which could be 1, 10, or 100 MHz, or the frequency of an external oscillator. Figure 4.8-11 is a

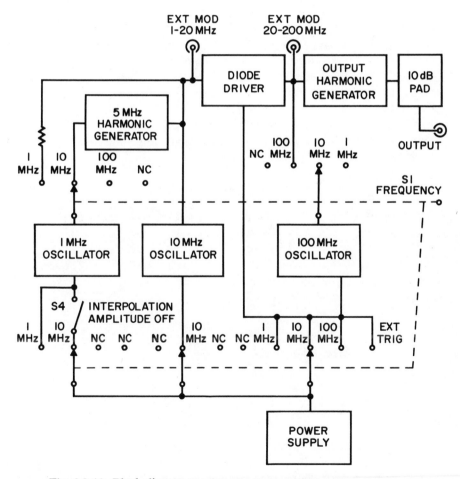

Fig. 4.8-11. Block diagram.

block diagram which shows the interconnections between the main sections of such an instrument. Note that only one oscillator is on at any one time, except when a 1-MHz interpolation oscillator is used to interpolate between the main spectral lines of the 10-MHz oscillator. In the case of the 1-MHz and 10-MHz oscillators, the signal is passed through a diode to the driver before it is applied to the output harmonic generator. Since low-frequency signals do not generate harmonics in sufficient amplitude when applied directly to the output harmonic generator, the diode driver sharpens the transition so that higher-amplitude harmonics are generated. The 100-MHz oscillator amplifier generates high-level harmonics without shaping and thus triggers the step recovery diode directly. A frequency comb generator provides a simple and convenient means of obtaining highly accurate absolute frequency calibration of signals displayed. This instrument is also useful for checking calibration and linearity of backward-wave oscillators and even spectrum analyzers, which will be discussed in a later chapter. Utilizing crystal-controlled oscillators and step recovery diode harmonic generators, such instruments produce frequency markers with 1-, 10-, and 100-MHz spacing. The marker combs are usable from the fundamental marker frequency to beyond 5 GHz and have an accuracy in the order of .01%.

Figure 4.8-12 illustrates a 1-GHz portion for the 100-MHz comb with amplitude versus frequency plotted on an oscilloscope. For interpolation convenience a primary-secondary comb, a combination of the 10- and 1-MHz combs, also can be generated, as shown in Fig. 4.8-13. The frequency of

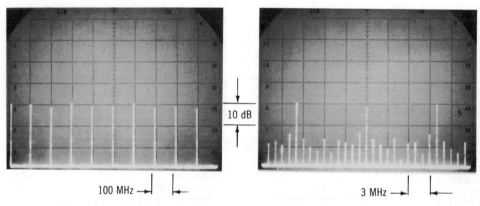

10 dB

100 MHz ⟶ ⊣ ⌊⟵   3 MHz ⟶ ⊣ ⌊⟵

Fig. 4.8-12. Amplitude versus frequency plot of a 1-GHz-wide comb using a 100-MHz signal repetition rate.

Fig. 4.8-13. Spectrogram showing primary-secondary comb produced by combining 10- and 1-MHz combs.

signals falling anywhere between the markers can be determined easily and accurately through the use of an external oscillator. The external signal phase modulates the comb to produce sidebands displaced from the comb marker by the frequency of the modulation signal. In another mode of operation, external signals can be used to generate combs with fundamental frequencies anywhere between 1 and 200 MHz. In other words, marker spacings are equal to the external signal frequency. The presence of accurate frequency combs over such a wide frequency range beyond 5 GHz also makes the frequency comb generator a convenient marker generator for use in rapid calibration of frequency meters.

QUESTIONS

1. What causes the abrupt transition or step in a step recovery diode?
2. What generally is the conversion efficiency of Varactor-type harmonic generators, and what is it for step recovery diodes?

4.9 SOLID-STATE MICROWAVE SOURCES

As the state of art forges ahead, transistor technology is nearing the microwave frequency spectrum, and more and more solid-state devices are discovered oscillating and amplifying in this frequency region.

4.9.1 TUNNEL DIODES

The Esaki (tunnel) diode is really an abrupt-junction diode, made with a very narrow barrier layer, in the order of 200 Å wide. These diodes exhibit a negative resistance region under small forward-bias conditions. Current is carried by tunneling majority carriers through that thin barrier layer of the junction. The voltage-current characteristic of such tunnel diodes is given in Fig. 4.9-1. The current first rises to a maximum and then decreases rapidly to a minimum. Further forward bias will make the minority carriers exceed tunneling current, providing a standard diode I-V region.

Basically there are three regions of the I-V curve shown in Fig. 4.9-1: the first region, where very small bias provides large positive conductance; the second region, where the bias is intermediate, providing large negative conductance; and the third region, which gives standard forward-biased junction diode behavior.

Negative resistance is one of the requirements of oscillation. Frequency-selective circuits connected to tunnel diodes can form useful oscillators. At

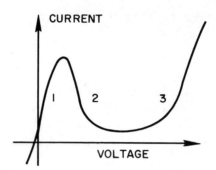

Fig. 4.9-1. Voltage-current characteristics of a tunnel diode.

the present state of art, tunnel diodes are used quite often in amplifier and oscillator applications.

4.9.2 YTTRIUM-IRON-GARNET (YIG)-TUNED OSCILLATORS

We will now describe an electrically tunable filter using the recently developed yttrium-iron-garnet filters.[4,5]

Highly polished spheres of single-crystal YIG, a ferrite material, when placed in an RF structure under the influence of a dc magnetic field, exhibit a high Q resonance at a frequency proportional to the dc magnetic field. To understand the phenomenon of ferrimagnetic resonance, consider diagrams (a) through (e) of Fig. 4.9-2. In the ferrite with no dc magnetic field applied, there is a high density of randomly oriented magnetic dipoles, each consisting of a minute current loop formed by a spinning electron. Viewed macroscopically, there is no net effect because of the random orientations. When a dc magnetic field, H_0, of sufficient magnitude is applied, the dipoles align parallel to the applied field, producing a strong net magnetization, M_0, in the direction of H_0. If an RF magnetic field is applied at right angles to H_0, the net magnetization vector will precess, at the frequency of the RF field, about an axis coincident with H_0. The precessing magnetization vector may be represented as the sum of M_0 and two circularly polarized RF magnetization components m_x and m_y. The angle of precession $\emptyset$, and therefore the magnitudes of m_x and m_y, will be small, except at the natural precession

---

[4] "How a YIG Filter Works," *HP Journal*, Vol. 19, No. 5 (January 1968).

[5] For a detailed treatment see P. S. Carter, Jr., "Magnetically Tunable Microwave Filters Using Single-crystal Yttrium-Iron-Garnet Resonators," *IRE Transactions on Microwave Theory and Techniques*, Vol. MTT-9, No. 3, May 1961.

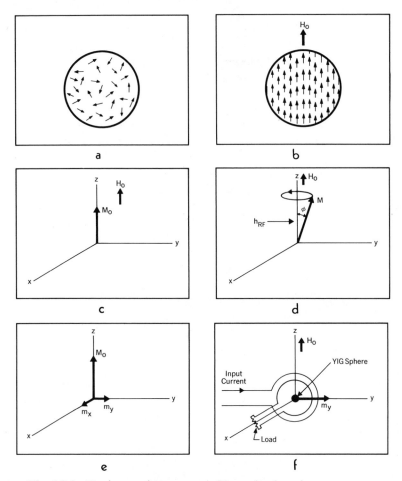

Fig. 4.9-2. Ferrimagnetic resonance (a) randomly oriented magnetic dipoles in the unmagnetized ferrite, (b) magnetic dipoles aligned under the influence of a magnetic field, (c) an equivalent representation of *b* showing the combined effect of the aligned dipoles, (d) precession of the net magnetization vector due to RF magnetic excitation, (e) equivalent representation of precessing magnetization vector, and (f) tuned bandpass filter consisting of YIG sphere at center of two mutually orthogonal loops.

frequency. This frequency, known as the ferrimagnetic resonant frequency, is a linear function of the dc field H_0.

Diagram (f) shows the basic elements of a YIG bandpass filter. The filter consists of a YIG sphere at the center of two loops, whose axes are perpendicular to each other and to the dc field H_0. One loop carries the RF input current, and the other loop is connected to the load. When H_0 is zero, there is large input-to-output isolation, since the two loops are perpendicular. With H_0 applied, there is a net magnetization vector in the direction of H_0. The magnetic field h_x produced by the RF driving current in the input loop causes the net magnetization vector to precess about the Z-axis. The resulting RF magnetization component, m_y, induces a voltage into the output loop. At frequencies away from the ferrimagnetic resonant frequency, m_y and the voltage it induces are small, so input-to-output isolation is high. When the input current is at the ferrimagnetic resonant frequency, $\emptyset$ and m_y are maximum. There is a large transfer of power from input to output, and insertion loss is low. Thus the filter center frequency is the ferrimagnetic resonant frequency and can be tuned by varying H_0. Commonly, the YIG sphere and RF structure are located between the poles of an electromagnet, and tuning is accomplished by furnishing a control current to the magnet coils.

A typical YIG filter to increase the offband isolation can be made by using two YIG spheres, as shown in Fig. 4.9-3. The heart of the YIG filter is an electronically tunable filter. Two spheres of YIG crystal material in a magnetic field are placed in the path of the RF signal. RF can only pass through the YIG filter when the spheres are at resonance. The frequency of resonance is a linear function of the magnetic field strength produced by the

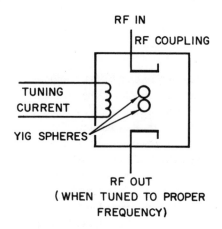

Fig. 4.9-3. YIG tuning circuit.

YIG tuning coil. (See Fig. 4.9-3.) The coil's field is proportional to the tuning current; thus the resonant frequency of the YIG filter is directly proportional to the tuning current supplied to the YIG filter. Insertion loss of the YIG filter at resonance is less than 5 dB. Off resonance, the YIG filter acts like a short circuit reflecting most of the RF energy back to the source.

If a filter like the one described above is connected to the feedback circuit of an oscillator as a tunable bandpass filter, a voltage (current)-tuned signal source can be achieved. Tunnel diodes, transistor oscillators, and even comb generators can be connected with these filters to provide single-frequency, harmonic-free microwave signal sources.

4.9.3 NEW APPROACHES FOR MICROWAVE SIGNAL GENERATION

In recent years solid-state semiconductor devices have replaced electron-tube-type signal sources and amplifying devices to provide manifold improvements in reliability as well as other advantages. Although major difficulties prevented this changeover from occurring simultaneously in the microwave frequency range, two significant discoveries have been reported in the last few years.[6] In 1958 Read[7] reported on his discovery of the negative resistance diode and its possible use in microwave oscillators. In 1965 Johnston, DeLoach, and Cohen[8] discussed their practical realization of avalanche transit-time devices in silicon. In 1964 Gunn[9,10,11] reported his discovery, the so-called "Gunn Effect" in gallium-arsenide.

These techniques depend on "hot" electrons whose energy is large in comparison to kT—possibly in the order of tenths of an electron volt. Both phenomena use hot electrons in a two-port negative conductance form. Negative conductance is due to a large phase shift (90°–180°) between voltage and current in the avalanche transit-time devices made from PN and even PIN junctions when reverse biased near the avalanche breakdown range. At this junction two different phase shifts will occur, one due to the delay of the electron transit time and the other due to the multiplying behavior of the avalanching. The negative conductance of the Gunn Effect device occurs within the gallium-arsenide crystal, where the local current density decreases whenever local electric field increases beyond a certain threshold.

[6] Kroemer, Herbert, "Negative Conductance in Semiconductors," *IEEE Spectrum*, Vol. 5, No. 1, January 1968, pp. 47–57.

[7] Read, W. T., "A Proposed High-frequency Negative-resistance Diode," *Bell System Technical Journal*, Vol. 37, March 1958, pp. 401–46.

[8] Johnston, R. L., DeLoach, D. C., and Cohen, B. G., "A Silicon Diode Microwave Oscillator," *Bell System Technical Journal*, Vol. 44, February 1965, pp. 369–72.

[9] Gunn, J. B., "Microwave Oscillations of Current in III-V Semiconductors," *Solid State Communication*, Vol. 1, September 1963, pp. 88–91.

[10] ———, "Instabilities of Current in III-V Semiconductors," *IBM Journal of Research and Development*, Vol. 8, April 1964, pp. 141–59.

[11] ———, "Instabilities of Current and Potential Distribution in GaAs and InP," Symp. Plasma Effect Solids, Dunod, Pan's, 1964.

Transferred Electron Oscillators (Gunn Effect)[12]

The observations by Ridley and Watkins[13] in 1961, that a carrier-velocity-versus-electric-field characteristic of the type shown in Fig. 4.9-4 is

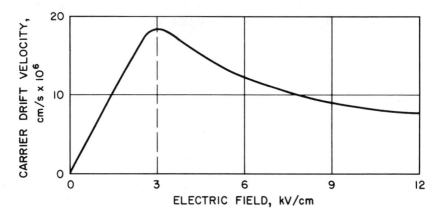

Fig. 4.9-4. Drift velocity of conduction-band electrons in GaAs versus electric field. The ac resistivity is negative when the electric field is biased above 3000 v/cm.

achievable in the bulk of certain semiconductors, is destined to have a profound impact on future sources of microwave and millimeter wavelength power. Because of the speed of the carrier response and the generation of energy by direct conversion from dc, electrically tunable sources promise to be exceptionally simple and reliable.

The existence of a velocity-electric field characteristic of the type shown in Fig. 4.9-4 throughout a volume of semiconductor material having ohmic contacts leads to an electrically and thermodynamically unstable condition with more than about 3200 v/cm applied across a gallium-arsenide device. As a result, the mobile electrons are rapidly redistributed to minimize the total energy and thus reach quasiequilibrium with the formation regions or domains of high and low electric fields which are bounded by layers of charge. Since the charge layers are mobile in the applied electric field, current oscillation occurs having a period determined by the transit time of electrons between the device contacts. This is the phenomenon first observed by Gunn[9] in the course of doing high-electric-field noise experiments on gallium-arsenide.

[12] Private conversation with Del Hanson, Hewlett-Packard Co., July 1968.

[13] Ridley, B. K., and Watkins, T. B., "The Possibility of Negative Resistance in Semiconductors," *Proc. Phys. Soc.* (London), Vol. 78, August 1961, pp. 293–304.

The physical model for GaAs which gives rise to the decrease in average drift velocity at a critical electric field is shown in Fig. 4.9-5, as presented

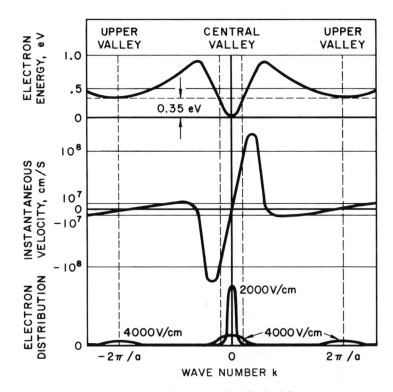

Fig. 4.9-5. Two valley model. As the electric field increases, electrons transfer from high-velocity states in the central valley to low-velocity states in the upper valleys, causing the average velocity to decrease.

by Copeland.[14] At room temperature and with no applied field, most electrons reside near the bottom of the central conduction band valley (at $k = 0$) with zero average velocity. At a critical applied electric field (i.e., 3200 v/cm), electrons gain enough energy to transfer to the upper valleys, where their mobility is reduced, and hence the device current is reduced with increasing applied voltage. The lower portion of Fig. 4.9-5 shows how the carrier density is redistributed as the field is increased.

[14] Copeland, J. A., "Bulk Negative-resistance Semiconductor Devices," *IEEE Spectrum*, Vol. 4, May 1967, pp. 71–77.

The formation of high-field domains and the resultant transit-time current oscillations which occur in bulk GaAs as a result of the transferred electron mechanism are illustrated in Fig. 4.9-6. It is noted that at "snapshot" (b) the device current reaches a maximum, as shown in (g), at the critical field E_c. Due to the resultant instability, as described above, a high-field dipole domain forms, with a resultant decrease in current. With the domain fully formed at "snapshot" (d), the device current remains constant

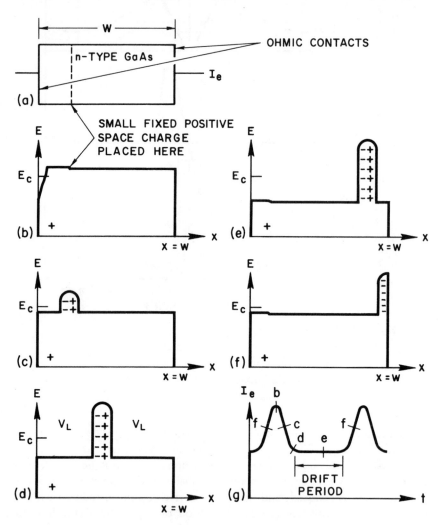

Fig. 4.9-6. Nucleation, growth and propagation of a dipole layer in a perturbed medium of negative conductivity (dipole mode).

until the domain reaches the anode at $X = W$. At this time the device field again reverts to that shown in "snapshot" (b), and the cycle repeats itself.

An oscillation frequency which is completely controlled by the device parameters, as in the above transit-time case, is not particularly attractive for microwave source applications. However, if such a device is placed in a microwave circuit in such a way that the superimposed microwave field across the device terminals can be used to control the formation and quenching of the high-field domain, shown in Fig. 4.9-6, it is possible to achieve circuit-controlled oscillation over more than an octave tuning range.

For instrumentation and various communication and radar applications, it is imperative to have electrical tunability over a broad frequency range. In the frequency range above 4 GHz, the two alternatives are Varactor tuning and YIG tuning. The latter is the preferred approach when octave tunability, high spectral purity, and substantial power output are desired.

Figure 4.9-7 is a schematic drawing of a YIG-tuned, transferred-electron

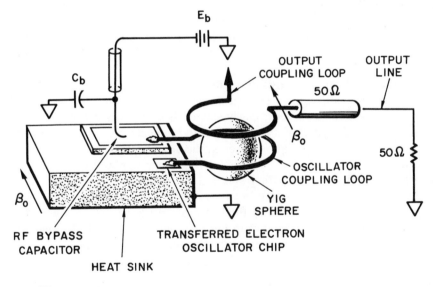

Fig. 4.9-7. YIG-tuned transferred electron oscillator.

oscillator which yields well over an octave continuous tunability in the region above 4 GHz with no spurious responses in the band. In designing a high-performance microwave oscillator of this type, it is necessary to consider simultaneously the synthesis of an optimum microwave circuit, magnetic circuit, and thermal circuit. The necessity of keeping all electrical dimensions substantially less than a wavelength while maintaining a low thermal resistance sink for the GaAs chip and a minimum gap for the magnetic circuit invariably requires the high resolution and reproducibility of thin-film circuit technology. This is particularly vital to minimize the electrical length of the YIG sphere coupling loop while maintaining uniformly close coupling to the uniform rotational mode of the sphere.

The synthesis of an optimum microwave circuit impedance for a bulk GaAs oscillator over its broad range of tunability requires a knowledge of the microwave circuit impedance directly at the device contacts,[15] since the oscillation condition of the device cannot be stabilized at the operating point. This information is used directly to design the YIG tuning circuit to yield maximum efficiency and tuning range.

4.9.4 LASERS (LIGHT AMPLIFICATION BY STIMULATED EMISSION OF RADIATION)

Basically, a laser is an oscillator providing energy at light frequencies. The basic difference between a common oscillator and the laser is that the laser does not take its energy from an electron beam; it takes energy directly from the atom. Most commonly used lasers can be categorized as crystal, gas, liquid, or semiconductor.

The most important quality of the laser is that its signal is coherent. Coherency means that the signal waves generated are all in phase with one another. These signals are extremely stable. Since the signal is coherent, it has modulation capability and consequently is detectable; it is capable of carrying intelligent information. Even in the heavy atmosphere of the earth, the wave within a laser beam does not diverge appreciably. This well-aimed directivity of signal propagation makes it a very efficient transmitter. It is very high in frequency as well as stable, and the coherent signal is pure. The visible light spectrum is in the order of 10^{15} hertz, which allows the beam to carry very wide bandwidth information if modulated. Bandwidths of modulation in the gigahertz region are common with lasers.

Stimulated Atoms[16]

The inherently unique properties of laser light have their origin in basic atomic theories governing electrons, photons, and their energy levels. The same facts apply to the *maser*, or microwave version of the laser, but for clarity only the visible light spectrum will be used in this explanation.

[15] Hanson, D. C., and Rowe, J. E., "Microwave Circuit Characteristics of Bulk GaAs Oscillators," *IEEE Trans. on Electron Devices*, Vol. ED-14, No. 9, September 1967, pp. 469–75.

[16] Reprinted from "A Review of the Laser: A Prospect for Communications," *The Lenkurt Demodulator*, Vol. 14, No. 11, November 1965, by permission of the publisher.

Atoms in nature are usually in a relatively undisturbed or *ground* state. The energy of orbiting electrons is balanced with the energy in the nucleus of the atom. These electrons occupy specific orbits determined by their own energy, but when "excited" by an outside source of energy, may jump to a higher second level raising the total energy of the atom (Fig. 4.9-8). In lasers, adding this outside energy is known as *pumping*.

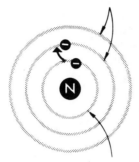

**EXCITED OR
HIGH ENERGY ORBITS**

**"NORMAL" OR
GROUND STATE ORBIT**

Fig. 4.9-8. Atomic electrons normally occupy an orbit
around the nucleus, representing a certain
fixed energy level. A stimulated atom ac-
quires energy as one of its electrons jumps
to a higher level.

The excited state for the atom is unnatural, and it will tend to relax to its ground state. As this happens, the stored energy is dissipated by emitting a photon of radiant energy. The energy of the photon is exactly proportional to its frequency—the higher the energy, the higher the frequency.

The common neon tube is an example of this action. Molecules of gas are excited to upper energy states by high voltage. As the atoms drop to their ground state, they emit light of a characteristic color or frequency— various gases produce various colors.

If left uncontrolled, the atom's spontaneous relaxation occurs in a random manner and results in incoherent light. But during the period when the atom is still excited, it is possible to stimulate the drop to ground level by striking the atom with an outside photon of the same energy it would have otherwise emitted spontaneously. Relaxation is no longer random, and the emitted photons leave the system as coherent light.

Another remarkable feature of laser action results when an emitted photon strikes another excited atom within the laser. As that atom returns to its ground state, another photon is added to the stream exactly in phase with the first, producing amplification.

Various methods have been discovered to improve the efficiency of the lasing action. The *three-level* method pumps atoms not to the second energy level, but to a still higher third level (Fig. 4.9-9). The atoms are very un-

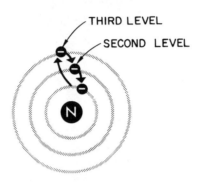

Fig. 4.9-9. Atoms can be pumped to higher third level, allowed to fall back and accumulate at second level, then controlled in their return to ground state. More atoms available at second level improves laser action.

stable at this level and quickly fall back to the intermediate, or second level. Here, by the nature of the laser material, the atoms tend to accumulate and are available in greater numbers for outside stimulation. Cooling the material, say to liquid nitrogen temperature (70°K), increases the effect. These techniques are typical of crystal lasers such as ruby. Gas lasers of the helium-neon type rely on the difference in energy levels between the two gases to provide for more effective pumping.

The Wave Grows

As emitted photons of light travel along the laser tube bumping into more excited atoms, the light wave continues to grow (Fig. 4.9-10). These waves are reflected back and forth inside the tube by mirrors, one of which is slightly transparent. Forming a resonant cavity at light frequencies, the laser now builds up standing waves which continue to multiply on each pass through the cavity. When the gain is strong enough to overcome the loss

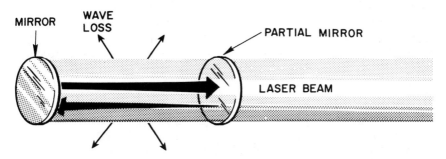

Fig. 4.9-10. Growing photon stream bounces back and forth, emerging as brilliant, coherent light. Extraneous waves are lost through laser walls.

in the mirrors, an intense beam is emitted from the partially transparent mirror. As the light bounces back and forth, any waves moving at angles to the axis between the mirrors will soon leave the system through the walls of the tube. Therefore the output beam of the laser will be extremely parallel.

Because of the coherence of laser light, it is possible to construct extremely efficient optical lenses to further focus the beam. A laser can be focused into a spot no wider than a wavelength of light, or about 0.0001 cm. The result is intense heat at the focal point, useful as a precision cutting tool; for microwelding; in delicate surgery, such as eye operations; and in chemistry, where individual molecules may be subjected to the unique radiant energy. The laser should also provide an invaluable laboratory tool for research in optical physics.

Communications Problems

The communications industry, faced with an already overflowing radio spectrum and the ever-increasing demand for more and more communications service, quickly became interested in the laser's ultrahigh information-carrying potential. But engineers found that before the laser could compete with microwave radio for point-to-point communications on earth, many problems remained to be solved.

The most serious of these problems is finding a suitable transmission path. The earth's atmosphere presents many natural barriers to light—haze, rain, snow—which do not seriously affect microwaves. Several studies have been undertaken to establish path reliability over relatively short ranges. To date, television pictures have been transmitted over a few miles without serious losses, but any system matching microwave reliability and quality over 25-mile links seems improbable with current techniques.

A number of *covered* transmission paths have been proposed, including the use of fiber optics, shiny hollow tubes, and tubes with regularly spaced lenses to redirect the beam around mild corners. Among these is a unique suggestion for a continuous gas-lens tube. A simplified arrangement (Fig. 4.9-11) would pump a steady stream of cool gas into a warm tube. The

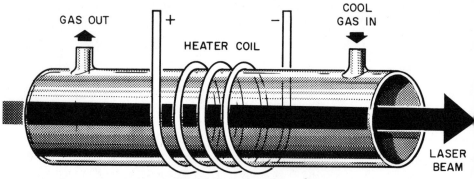

GAS OUT   +   −   COOL GAS IN

HEATER COIL

LASER BEAM

Fig. 4.9-11. Cool gas forced through a hot tube can form a continuous positive lens, suggested as a means of "piping" laser beams between cities.

heated gas near the surface of the tube would be less dense than the cooler gas at the center. This difference, represented by a varying refractive index, would be enough to produce a positive lens for the laser beam.

The slight divergence of the laser's beam—advantageous for keeping energy concentrated theoretically over vast distances—does become a problem in transmitter-to-receiver visual-path alignment. Experimenters have plotted the expansions and contractions of a building caused by heat from the sun by monitoring changes in signal strength between a laser mounted in the building and a receiver nearby. In the communications field this could be a definite problem. But to others this sensitivity to angular change makes the laser an extremely good device for measuring minor physical displacements. For instance, a system using lasers has been proposed to record instantaneously the land shift around California's infamous San Andreas fault. The same capability is being used to detect micromovement in laboratory experiments.

The laser's prime contribution to communications may come in space, where interference from a *dirty* atmosphere ceases to be a problem. Optical links from known positions, such as earth-orbiting space stations and the moon, could carry tremendous quantities of information on a single laser channel. Tracking of vehicles moving freely in space could be more of a problem for the thin-lined laser beam. But techniques developed here could also apply to laser radar, producing a highly sensitive system with greater resolution than ever before possible. Military applications now being tested include a laser fire control radar system to permit low-flying aircraft to see targets normally obscured by ground clutter on microwave radar.

The basic component needs of a laser communications system are the same—in name at least—as any similar radio device. The information to be transmitted must be amplified, modulated, demodulated, and recreated in its original form. The techniques needed here are not all new. For example, earlier developed masers operating at microwave frequencies have for some time been employed as amplifiers in radio astronomy because of their ability to provide high gain and very low noise.

Modulation

Apart from the extensive pure research being done with lasers, considerable effort is being put into finding suitable modulators and demodulators. Modulation of the relatively new semiconductor or injection-type laser is easily accomplished by simply controlling the pumping current. However, semiconductor lasers, such as gallium-arsenide (GaAs), have a low output power and are not so coherent as other sources. Also, the output is a less desirable flat sheet of light as opposed to the solid round beam of other lasers. For the present, communicators are putting more faith in gas and crystal lasers for communications purposes, with further study being given liquid lasers.

One of the first successful devices for amplitude-modulating a laser beam uses the polarization properties of the clear crystal potassium dihydrogen phosphate (KDP). As illustrated in Fig. 4.9-12, the KDP device amplitude-modulates a laser beam projected through it. The first polarizer blocks all light wave polarizations except, for example, the vertical.

The resulting beam may be thought of graphically as a number of ribbons of light, all parallel to each other and at right angles to the direction of propagation. For simplicity of illustration, this example treats the polarized light as only one *ribbon*. It is characteristic of the KDP crystal to shift polarization in a circular direction proportionate to a stimulating voltage. For a higher voltage, there will be more circular change in polarization. If a signal is applied to the KDP crystal and the now vertically polarized laser beam shone through it, the beam will be what might be called *polar modulated*. In the diagram, the ribbon will be twisted in accordance with the modulating signal. The second polarizer, known as the analyzer, will sense this twist as a decrease in amplitude. The output intensity will then vary in relation to the signal, hence, amplitude modulation.

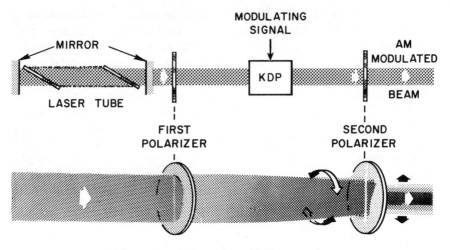

Fig. 4.9-12. As beam passes through an AM laser modulator, it may be visualized as a ribbon
of light twisting proportional to a signal
applied to the KDP crystal.

Bandwidth in the order of 200 mc can be achieved with amplitude
modulation, but this by no means takes full advantage of the capabilities of
coherent laser light. On the other hand, frequency and phase modulation
methods are producing bandwidths of over 1 GHz. One such device, known
as a wideband traveling-wave phase modulator, consists of two parallel brass
rods about one meter in length, with an electro-optical material (such as
KDP) sandwiched between. A microwave signal voltage applied to the device
varies the velocity of light in the crystal, resulting in phase modulation. The
length of the modulator allows longer interaction time between signal and
light beam and hence greater depths of modulation.

Laser Modes

It should be noted that since the laser cavity is thousands of times
longer than any wavelength at light frequencies, a number of frequencies will
resonate in the tube at the same time. This results in the laser having in its
output a number of separate and distinct frequencies or modes (Fig. 4.9-13).
The separation of these modes is determined by the mirror placement in the
laser and may be calculated by the formula

$$df = \frac{c}{2L}$$

where df = the difference frequency between modes,
 c = speed of light, and
 L = length between the mirrors.

Since the obvious desire is to transmit only one frequency, power distributed in modes other than the one to be used is wasted. Likewise, each mode acts as a carrier frequency for any modulation. As sidebands are added to each mode (Fig. 4.9-13(a)), it can be seen that the bandwidth of modula-

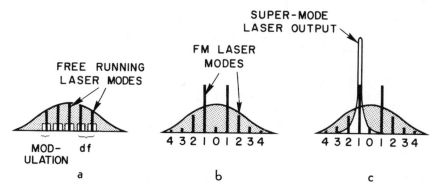

Fig. 4.9-13. Laser modes representing the (a) free-running laser, (b) laser with FM modulator inside the cavity, and (c) super-mode laser, where power in the FM sidebands is combined into one output frequency.

tion on any one mode is limited by the difference in frequency between the modes.

One solution is the supermode laser (Fig. 4.9-14). By inserting a phase modulator inside the laser tube, driven at a frequency nearly equal to the difference frequency between modes (df), the output is converted to a typical FM configuration with sidebands occupying the positions formerly held by the various modes (Fig. 4.9-13(b)). By using this method, the bandwidth limitation in the mode structure has been eliminated. Another phase modulator outside the tube will additionally affect the beam by compressing all the modes together into a single frequency. The final output contains most of the power formerly held in the many modes, plus the highly desirable single frequency (Fig. 4.9-13(c)). This "supermode" beam can be successfully modulated by any chosen method with much superior performance.

A similar means of arriving at the same end is found by placing a device known as a *Fabry-Perot etalon* (not shown) inside the tube, along with the FM modulator. The output frequency is determined by the spacing of two

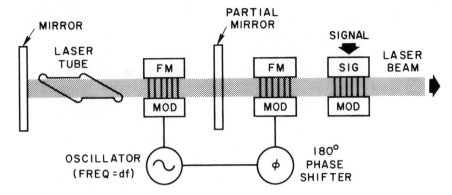

Fig. 4.9-14. Super-mode laser has single frequency output of comparatively high energy. Oscillator frequency is nearly equal to the difference frequency between free-running laser modes.

highly reflective mirrors forming the etalon, creating a resonance at the desired frequency. The beam leaving the tube will be of single frequency if the etalon is tuned to one of the FM sidebands and will be at the laser's full power. The device combines the power of the other sidebands into the output. This can be done because the FM laser modes are sidebands of a single carrier, rather than a set of independent oscillations.

Demodulation

Demodulation of optical radiation is typically accomplished with either a photomultiplier tube or a microwave phototube, each relying on the secondary emission of electrons from a cathode when struck by light photons. The photomultiplier technique redirects emitted electrons onto other secondary-emitting surfaces, producing considerable amplification. The current is eventually collected on an output electrode. The photomultiplier tube has a range from dc to many megahertz, thereby detecting signals directly to baseband frequencies. The microwave phototube (Fig. 4.9-15) is designed with a traveling wave tube helix output and is effective at the higher microwave frequencies. A modification of the microwave phototube, known as the crossed-field electron multiplier, amplifies the signal before the electrons reach the helix. In both cases, since light frequencies are outside the bandwidth capabilities of the phototubes, the electron stream represents only the original modulation placed on the laser beam.

Optical heterodyning is also possible using the photomultiplier tube, as seen in Fig. 4.9-16. A laser local oscillator beam beats with the incoming

Fig. 4.9-15. The microwave phototube converts laser
light to electrons, bunched proportionate
to the original modulation. The current induced in the helix is then processed by usual
microwave techniques.

laser signal in the phototube, resulting in an IF frequency equal to the difference between the two light frequencies. This IF signal is typically in the
microwave region and may be amplified and demodulated by conventional
methods. A discriminator supplying a control signal to the laser local
oscillator maintains frequency stability.

Other Applications

Interest has been shown in using lasers, possibly of the semiconductor
type, for intercomponent communication in high-speed computers. The technique would eliminate the delay associated with placement of components
and connections and would offer much faster switching speeds than now possible with ordinary electronic computer circuits.

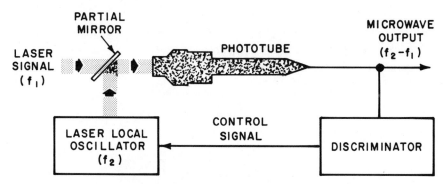

Fig. 4.9-16. Optical heterodyning combines laser signal
with that of laser local oscillator at phototube. Microwave output at IF frequencies
is then amplified and demodulated.

Closer to the communications field, coherent laser light has opened a "magic window" to successful wavefront reconstruction photography, known as *holography*. Holograms record both amplitude and phase variations of laser light reflected from a subject, allowing near-perfect three-dimensional reproduction. Interference patterns between the reflected light and a reference beam are recorded on film without the use of lenses. Applications of the future may include three-dimension color television and medical or industrial X-ray holograms for studying the interior of an object.

Conclusion

Although the laser is becoming a valuable tool in many pure-science areas and may have useful military and industrial applications, its high information-carrying potential for communications seems barely tapped. Techniques for the use of the laser in communications are gaining in sophistication, but a great number of obstacles must be overcome before a practical system becomes feasible. Only a continuing refinement of the art will take the laser out of the laboratory and into the field.

BIBLIOGRAPHY

1. Brotherton, M., *Masers and Lasers*, McGraw-Hill Book Company, 1964.
2. Gordon, J. P., "Optical Communication," *International Science and Technology*, No. 44 (August 1965), pp. 60–64.
3. Leith, Emmett N., and Juris Upatnieks, "Photography by Laser," *Scientific American*, Vol. 212, No. 6 (June 1965), pp. 24–35.
4. Reed, John S., "The Wonderful World of the Laser," *Telephone Engineer and Management*, Vol. 68, Nos. 6, 8, 9 (March 15, 1964, pp. 43–47; April 15, pp. 64–67; May 1, pp. 40–43).
5. Schawlow, Arthur L., "Optical Masers," *Scientific American*, Vol. 204, No. 6 (June 1961), pp. 52–61.

QUESTIONS

1. Which region of the I-V curve of the tunnel diode provides large negative conductance?
2. What parameter has to be varied on a YIG filter to vary its passband frequency?
3. From what source does the laser beam get its energy?

4.10 SIGNAL GENERATORS

It is already well known from low-frequency techniques that several criteria have to be met to make a signal source become a signal generator. In the early World War II days and shortly after, according to C. G. Montgomery,[17] a signal generator was defined as a signal source with which a variety of receiver measurements could be made. He further classified them into two main categories, the modulated and unmodulated CW signal generators.

A signal generator generally consists of a CW oscillator, an attenuator of the variable-calibrated kind, a power monitor or a power meter, and some means of accurate frequency indication: a frequency meter or well-calibrated frequency dial. Further requirements involve leakproof packaging. This problem in the early days constituted one of major difficulty, since the signal level in the generator usually was 100–150 dB higher than what was needed for receiver-type measurements. Shielding of the sources was a major concern, since leakage paths were detected not only through power supplies but in many other routes. Attenuators used in signal generators often have proven unsatisfactory, although the problem was not really due to the design of such attenuators but due to leakage, where a higher level of signal came to the generator output connector (through some leakage path) than was available through the attenuator.

With the advancement of the state of the art and the technology available today, present-day signal generators have stable and clean signals, with very small incidental frequency or amplitude modulation on them. Their frequency dials are calibrated to better than 1% accuracy. The output power at any frequency can be determined to within quite respectable accuracies, depending on the type of design approach taken and the required frequency band. Modulation capabilities are quite good, including calibrated AM, FM, and pulse modulation.

A stable oscillator must be used for a signal source of a signal generator. For quite some time the reflex klystron proved to be an ideal source to fulfill this role in the microwave frequency region. Basically, the instrument includes an RF oscillator, PIN diode modulator, automatic-leveling circuit, modulation circuits, and power supply, as shown in Fig. 4.10-1. The RF oscillator is a reflex klystron which always operates in CW. The PIN diode modulator is a current-controlled device that attenuates RF power up through 20 dB or more. The control circuits provide the modulation currents required by the PIN modulator. The power supply provides the regulated dc voltages required to operate the circuits in the instrument. The RF oscillator which

[17] Montgomery, C. G., "Technique of Microwave Measurements," *M.I.T. Radiation Lab Series*, Vol. 11, Chap. 4.

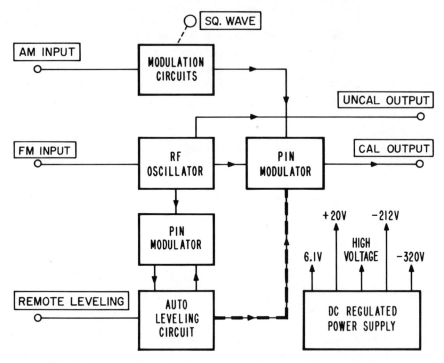

Fig. 4.10-1. Circuit block diagram.

generates the RF power consists of a velocity-modulated tube operating in an external resonant cavity. The tube is a reflex klystron operating in the $1\frac{3}{4}$ and $2\frac{3}{4}$ modes. The RF power output from the oscillator, which may be CW or CW with FM, is obtained from the resonant cavity by means of a pickup probe located in small sections of waveguide which open into the resonant cavity. In this particular design, one of these probes delivers RF power directly to the uncalibrated RF output connector, and the other two deliver RF power to the PIN modulator. With the use of the PIN diode modulator built into the unit, external pulse and internal pulse modulation and internal square-wave modulation can be achieved. Furthermore, external AM can also be achieved all the way up to 100% modulation. With the PIN modulator devices, unfortunately, high-percentage modulation will not be distortionless (it could have as high a distortion as 5% to 20%), but with about 30% modulation the distortion is almost unnoticeable.

In this particular signal generator, internal metering and automatic level control (ALC) is performed with the simple circuit shown in Fig. 4.10-2. The meter amplifier is a dual-function circuit performing both leveling and/or

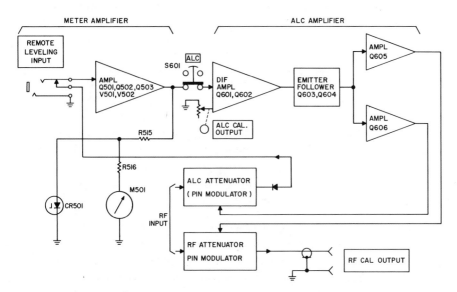

Fig. 4.10-2. ALC and meter circuit.

a power-output-monitoring function. RF power is taken from the klystron cavity through the ALC attenuator assembly (part of the PIN diode modulator) and delivered to the ALC circuit. The meter amplifier monitors the power level, and the leveled operation with the ALC amplifier maintains a constant RF output. Actual operation is as follows: RF power from the klystron is coupled from a fixed probe in the klystron cavity to the ALC attenuator (part of the PIN diode modulator). The RF power is delivered through a high-pass filter to the ALC diode attenuator and then through another high-pass filter to a crystal detector. The detected signal from the crystal detector is then delivered to a low-pass filter and to the ALC circuit. The crystal detector is arranged so that the detected signal is negative in polarity. The increase in RF level as the klystron is tuned across the band will cause a more negative output. A decrease in RF power from the klystron causes a less negative output. Detected RF power level from the crystal detector is then delivered to the base of the first amplifier (Q501A).

Consider the circuit operation when the RF level from the klystron increases. An increasing klystron output level causes a more negative signal on the base of Q501A. The conduction of Q501A decreases, causing the collector of Q501A to go in a positive direction. The positive signal goes through a cathode follower and is applied to the base of Q502, decreasing its conduction. The collector of Q502 goes more negative. A portion of the negative-going signal from the collector of Q502 is applied to the base of Q501B as negative feedback. The feedback factor is determined by the ratio of two resistors in the circuit. The open-loop gain of the meter amplifier is sufficiently high so that the closed-loop gain is essentially a function of the feedback factor and is therefore less dependent upon the normal aging effects on the tubes and transistors in the circuit. The negative-going signal from Q502 is also applied to the meter M501 for output indication. The meter is protected from over-load by the breakdown diode CR501. If the internal ALC switch is on, the negative-going output is applied to the base of the differential amplifier Q601, causing a decrease in conduction. The collector will go more positive, causing an increase in conduction of the emitter follower Q603 and Q604. This causes the emitter of Q604 also to become more positive. The positive-going signal is applied to the bases of Q605 and Q606, increasing their conduction and causing both collectors to become more negative. The collectors of Q605 and Q606 appear as constant current sources, so that the decreasing collector potential causes current to be drawn from the PIN diodes. This increased bias current (increased forward bias) reduces the RF power output to its original level. The negative-going output from Q605 is delivered to the RF PIN diode attenuator, allowing less RF to pass through it also. The net result is that the increase in klystron output causes an increase of forward bias on the PIN diodes which decreases the RF output. For accurate leveling, the ALC and RF PIN diode attenuators must track together as far as attenu-ation and frequency are concerned. The adjustment of the resistors involved provides for matching the attenuator characteristics. The RF output can be controlled by adjusting the front panel ALC calibrator output control, which varies the bias on the base of the differential amplifier, which in turn changes the bias on the PIN diode attenuator.

The simplified diagram (Fig. 4.10-2) of the ALC circuit also shows how external leveling can be done. The operation of external leveling is the same as that described for internal leveling, with two exceptions. The operation of the ALC circuit is such that the ALC attenuator (part of the PIN diode modulator) will no longer be part of the circuit; therefore, since the ALC attenuator is removed from the overall circuit, the meter will indicate an RF power level but not an accurate measure of the calibrated RF output power.

Figure 4.10-3 shows a cutaway view of a klystron cavity and a klystron assembly. The resonant circuit of the RF oscillator klystron includes a reso-nator grid capacitance and inductive impedance of the external cavity. The cavity is a shorted coaxial transmission line, one cylinder within another. The cavity is fitted with a movable plunger (wiper contacts that short-circuit the line at the opposite end of the cavity from the tube), which changes cavity dimensions. This changes the resonant frequency of the oscillator circuit and thus the frequency of oscillation.

Figure 4.10-4 shows the equivalent circuit of a reflex klystron oscillator. The RF oscillator tube is a reflex klystron operating in a tunable cavity reso-

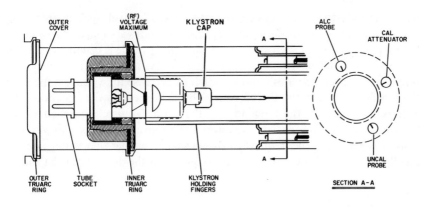

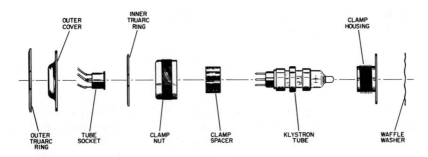

Fig. 4.10-3. Cutaway view of klystron cavity
and klystron assembly.

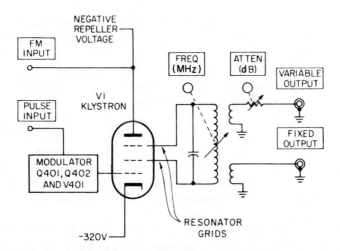

Fig. 4.10-4. Equivalent oscillator circuit.

nator. The klystron cavity system operates on the $\frac{3}{4}$-wavelength cavity mode, and oscillations on both the $1\frac{3}{4}$ and $2\frac{3}{4}$ repeller modes are employed to cover the frequency range of such an instrument. The $1\frac{3}{4}$ mode is usually used from low-frequency end up to about midfrequency. At about midfrequency, the tuning mechanism actuates a mode switch to decrease the voltage applied to the repeller by about 160 to 200 volts. This action places the system on the $2\frac{3}{4}$ repeller mode for the remainder of the band. Voltage is applied to the klystron repeller from variable resistors. The movable arm of this variable resistor is ganged to the frequency drive in such a manner that voltage on the repeller is automatically tracked with frequency in the desired repeller mode.

QUESTION

1. What are the basic requirements that a signal source must satisfy to be considered a signal generator?

4.11 SWEEP OSCILLATORS

Single-frequency measurements performed at many points are not satisfactory when making measurements on wide-frequency-band devices because it takes so long to make them and, even then, not all the frequencies in question are covered. However, voltage-controlled oscillators are capable of providing swept-frequency signal sources. One such device, the backward-wave oscillator, discussed in this chapter, makes a reasonable swept-frequency oscillator.

Basic requirements of such sweep oscillators include variable sweep width, sweep speed, modulation capability, reasonable cleanliness, stability, and repeatability of the signal. Present-day sweep oscillators even have built-in leveling capabilities. The ALC amplifier is usually an integral part of the electronics of such instruments. Figure 4.11-1 shows a simplified block diagram of such a sweep oscillator's electronics section.

The frequency control section determines the sweep oscillator output frequency. It generates a ramp that sweeps the RF output, or a dc voltage that produces single-frequency output, or a combination ramp and dc voltage for narrow-band sweeps, or a combination of a dc voltage and an external signal to give external frequency modulation. The section also produces automatically repetitive or triggered sweeps and permits manual sweeping as well as individual tuning of the frequency markers.

The frequency control section consists of a tuning voltage generator and a helix voltage generator. The tuning voltage generator consists of a ramp generator that generates a linear, negative-going ramp voltage, a

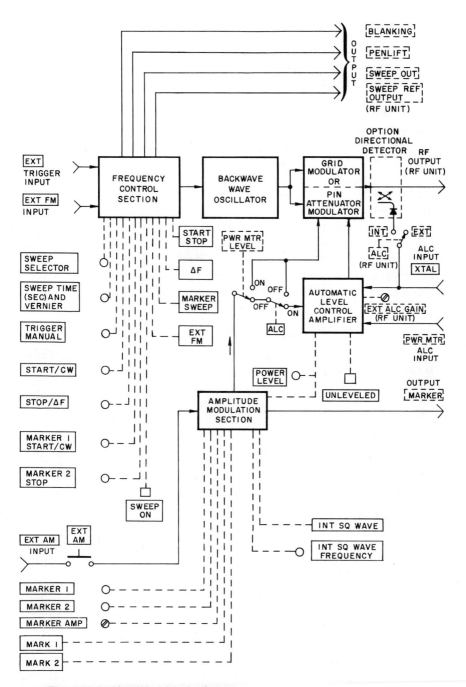

Fig. 4.11-1. Simplified block diagram.

reciprocal amplifier that produces a mirror-image positive-going ramp voltage, and a ramp-combining circuit that combines the two ramps and produces either a positive- or negative-going ramp continuously adjustable in amplitude that is supplied to the helix voltage generator for application to the BWO helix. This adjustable ramp controls the output frequency of the backward-wave oscillator.

When CW operation is selected, the adjustable ramp is replaced by an adjustable dc voltage. When ΔF is selected, a small, adjustable ramp is superimposed on the CW dc voltage. When external FM is selected, the external signal is superimposed on the CW dc voltage.

The frequency control section also supplies a positive-going sweep voltage for operating the X-system of graphic recorders and oscilloscopes, a penlift contact for lifting the pen of a recorder between sweeps, and a blanking pulse that cuts off RF output between sweeps.

The ramp generator produces a linear, negative-going sawtooth of fixed amplitude and adjustable period which is applied through switching circuits to the cathode follower and ramp-limiting (clamp) diodes. These diodes fix the voltage limits at the start and stop ends of the ramp. The low-voltage end is clamped to near 0 volts by diodes. The high-voltage end is clamped to the regulated supply. The clamped ramp is then applied to the START/CW side of the ramp-combining circuit and to the reciprocal amplifier.

The ramp generator also supplies two output pulses during its flyback period: (1) a short negative pulse to the blanking switch to turn off the RF output during flyback, and (2) a longer negative pulse to the penlift circuit that opens relay contacts to lift the pen of an external graphic recorder between sweeps. The ramp generator operates in a free-run mode for automatically repetitive sweeps or in a trigger mode to start a ramp from a trigger received. It can also be switched off to permit manual frequency control.

The sweepers generate swept RF signals in 1.5:1 and 2:1 frequency bands from 500 MHz to 40 GHz using a backward-wave oscillator (BWO) and its associated circuits. Sweepers in bands from 1 to 12.4 GHz contain PIN diode attenuators for amplitude-modulating and/or leveling the RF output. The PIN line absorbs RF power out of the BWO proportional to an applied dc bias voltage or modulating signal. Since the BWO grid-cathode voltage is not varied for modulation or leveling, excellent amplitude and frequency stability are achieved with freedom from incidental FM.

Reflections from connectors, cables, etc., within the leveling loop are compressed by the ALC system so that, ideally, source mismatch is canceled and system accuracy improved. Detection for leveling and readout of data is accomplished by either crystal detectors or power meters. When fast sweeps with oscilloscope readout are desired, coax or waveguide crystal detectors are used for ALC and readout detection. Where good square-law characteristics are required over a large dynamic range, these detectors are used with their matching square-law load resistor. Improved accuracy and convenience made possible by the modern crystal detectors include:

1. flat frequency response.
2. low-reflection coefficient over the full band.
3. high sensitivity.
4. good square-law response.

When the optional load is used with the detectors, sensitivity is 0.1 mv/μw of applied RF. Deviation from square law is less than ±0.5 dB from low level up to 50 mv out of the detector/load combination (approximately −3 dBm of RF input). This configuration gives maximum square-law dynamic range and is used extensively for readout detection in the swept systems. Without the optional load, the sensitivity increases to 0.4 mv/μw of applied RF. Good square-law performance without the optional load can be expected up to about −20 dBm. In certain applications where low source power is a problem, the unloaded crystal may be preferred for its higher sensitivity and good square-law range below −20 dBm. ALC detectors need not be loaded, since they operate at a constant power level within the feedback loop and since square-law dynamic range is not required. Good frequency response and low reflection, however, remain important considerations for the ALC detector. Figure 4.11-2 shows the square-law characteristics and sensitivity of detectors for loaded and unloaded conditions.

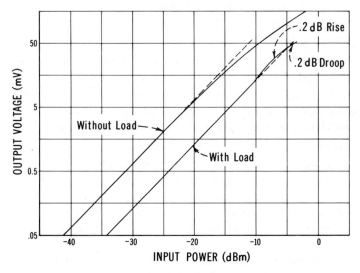

Fig. 4.11-2. Square-law response of typical detector with and without matched square-law load resistor.

Where knowledge of absolute power level is required, power meter and coax waveguide thermistor mounts are used for ALC or readout detection. Longer sweep time is required with power meter detection, so an X-Y recorder readout is used instead of an oscilloscope. A dc current proportional to the power meter deflection is used for ALC or recording purposes. The thermistor mount/power meter combination is a good square-law detector and has good broadband impedance characteristics.

Figure 4.11-3 shows the equipment setup for obtaining leveled swept-

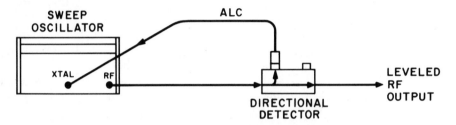

Fig. 4.11-3. System for obtaining crystal leveled sweep oscillator output in coaxial (type N) transmission line, 500 MHz to 12.4 GHz. All systems from 1 to 12.4 GHz use directional detectors comprising a directional coupler and crystal detector matched for flat frequency response.

frequency signals in coax from 500 MHz to 12.4 GHz. The ALC feedback voltage is derived from the directional detector, which consists of a directional coupler with a crystal detector mounted directly on the secondary arm. Although a separate directional coupler and crystal detector may be used, tighter correlation between output power and detector voltage is achieved by combining the two in a single integrated unit.[18]

POWER-METER LEVELING. The design of the temperature-compensated power meter and associated thermistor mounts provides the required stability and metering configuration for a recorder output current proportional to meter deflection. This current (0–1 mA dc into 1 kΩ) is suitable for either recording purposes or driving the ALC circuit in any sweeper for leveling power output.

In addition to a constant output across the band, power-meter leveling provides a continuous indication of the absolute power output that is useful in setting up specific test conditions. Because of excellent zero carry-over,

[18] Prickett, R. Y. "New Coaxial Couplers for Reflectometers, Detection, and Monitoring," *Hewlett-Packard Journal*, Vol. 17, No. 6 (February 1965).

the range switch and zero controls make effective variable attenuators since they control the ALC loop in a precisely known manner. This feature can be used accurately to control output power from the sweeper or calibrate an X-Y recorder in dB or power.

The dc substitution feature on the power meter allows very accurate dc power level changes to be made in the thermistor using an external supply. When this is used in the ALC loop, a precise dc power change in the thermistor causes a precise RF output change from the sweeper. A basic system for power-meter leveling is shown in Fig. 4.11-4.

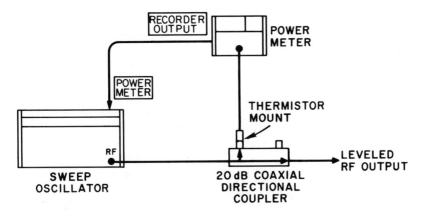

Fig. 4.11-4. Power meter leveling of sweep oscillators
in coaxial systems.

QUESTIONS

1. What are the basic requirements a signal source must satisfy to become a sweep oscillator?
2. What are the two benefits of using automatic level control (ALC) with sweep oscillators?
3. What are the advantages and disadvantages of power-meter leveling?

4.12 PHASE-LOCKING OF SIGNAL SOURCES

Microwave signal sources are quite stable devices for most practical purposes in everyday measurements, but there are applications when frequency stability is of utmost importance. Signal source frequency can drift due to changes in ambient temperature. Tube-type sources may vary considerably because of their actual warmup and other temperature effects.

Furthermore, the finite filtering and imperfect regulation of the power supplies used also can contribute to small frequency instability and inherent frequency modulation of the signal source. If measurement errors due to frequency instability or inherent FM of standard signal sources are intolerable, phase-locking of the source is a good practice.

For example, nearly absolute control of reflex klystron oscillator frequencies is made possible with an oscillator synchronizer. Applications requiring extremely stable signals include Doppler systems, radio astronomy receivers, microwave spectroscopy measurements, microwave frequency standards, parametric amplifier pumps, and radar cross-section studies. In all such applications the oscillator synchronizer provides an oscillator stability previously obtainable only by direct harmonic multiplication from a quartz crystal oscillator. Moreover, the oscillator synchronizer is not subject to the lower power output and incomplete harmonic rejection inherent in harmonic multiplication.

Automatic frequency control (AFC) systems are available that lock klystron oscillators to crystal reference harmonics to provide increased power output and eliminate adjacent harmonics, but they must introduce a frequency error to provide a means for correction. No frequency error of this nature is necessary for the operation of the oscillator synchronizer. The automatic phase control system used completely eliminates long-term drift due to klystron frequency pulling and also minimizes all incidental FM due to klystron and power supply noise. Sideband noise power is typically down 70 dB below the carrier, measured in any 1-kHz band from 3 kHz to beyond 100 kHz on either side of the carrier. An oscillator synchronizer can be applied to a klystron oscillator for frequency modulation. The synchronizer may be used to monitor a specified signal frequency and provide a deviation frequency output for measurement by an electronic counter or frequency meter. A strip chart recorder can be added to the counter or meter for permanent records of signal drift.

The oscillator synchronizer is essentially a crystal-controlled superheterodyne receiver terminating in a phase comparator (see Fig. 4.12-1). An external oscillator sample is mixed with harmonics of the internal RF reference by the harmonic generator-mixer to produce an intermediate frequency of 30 MHz. The output of the IF amplifier is compared in phase with a 30-MHz reference derived from a 10-MHz crystal oscillator. The resultant phase-error voltage is added in series with the klystron reflector power supply voltage. The phase comparator output terminals are floated and insulated for this purpose. The IF oscillator can also be switched to operate as a variable-frequency oscillator (VFO) or as an amplifier for application of an external IF reference. The frequency control loop is stabilized by an adjustable phase-lag network which permits operation with klystrons having modulation sensitivities as high as 4 MHz/v and provides a loop bandwidth (equivalent FM) in excess of 100 kHz.

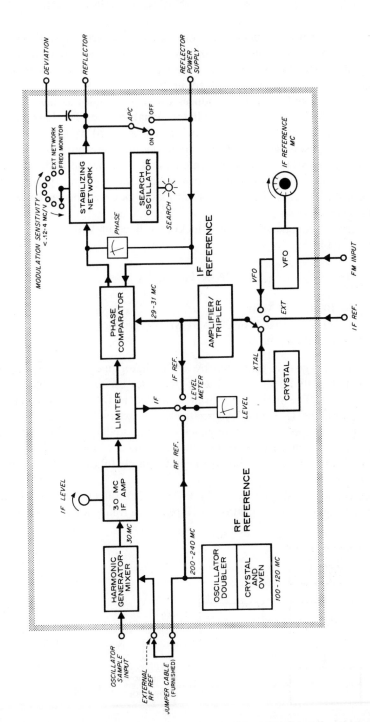

Fig. 4.12-1. Oscillator synchronizer.

The RF reference frequency is controlled by a quartz crystal that is oven mounted for temperature stabilization. The oven and circuit accommodate a fifth-overtone crystal ground for a frequency between 100 and 110 MHz. The oscillator output is doubled in frequency and applied to the harmonic generator-mixer, which produces useful harmonics as high as 18 GHz for mixing with the external oscillator signal sample. The harmonics of the internal reference are spaced between 200 and 220 MHz apart, depending on the crystal selected. For each harmonic there are two "lock" frequencies, one 30 MHz above the harmonic and the other 30 MHz below. A number of lock points are therefore available for a given crystal. For example, a 100-MHz crystal produces 42 available lock frequencies between 8.2 and 12.4 GHz (X-band).

The signal frequencies at which locking will occur with a particular crystal are given by the formula

$$F_{\text{signal}} = 2N \times F_{\text{xtal}} \pm F_{\text{IF}}$$

where $F_{\text{xtal}} = 100$ to 110 MHz,
$\quad F_{\text{IF}} = 30$ MHz (fixed) or 29 to 31 MHz (variable),
$\quad N = $ harmonic number.

The IF reference signal is generated by an oscillator operating either as a 10-MHz crystal oscillator or as a variable-frequency oscillator tunable from 9.7 to 10.3 MHz. The oscillator output is tripled in frequency for phase comparison with the IF signal. Alternatively, an external reference of 30, 15, or 10 MHz may be employed, in which case the oscillator circuit will act as an amplifier or multiplier, according to the frequency applied. The oscillator output is also made available, through a cathode follower, for external reference or measurement.

Since the locked frequency of the klystron is a direct function of the IF reference frequency, any variation in IF reference frequency is transferred

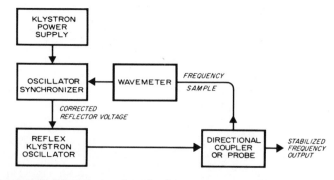

Fig. 4.12-2. Klystron synchronization.

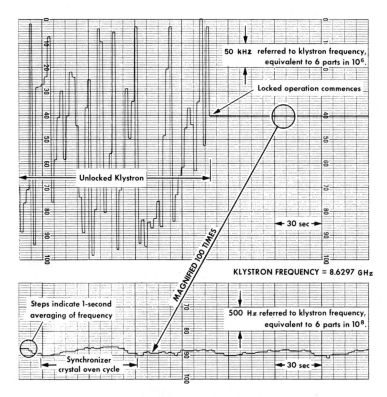

Fig. 4.12-3. Strip chart recording showing reduction of
klystron frequency jitter when phase-lock
is applied.

via the phase comparator to the klystron. This permits manual tuning (over a
range of ±1 MHz) or frequency modulation of the klystron either in the
internal VFO mode or with an external reference. When used as a crystal
oscillator, the frequency stability of the IF reference is sufficiently high
(being an additive term) that the overall stability is primarily that of the RF
reference.

A typical phase-locked klystron system is shown in Fig. 4.12-2. To
close the frequency control loop, a sample of the klystron output is fed to
the oscillator synchronizer input. The sample may be obtained with a directional coupler or a power divider or signal sampler probe, depending on the
klystron and application. The effect of the oscillator synchronizer is illustrated in a typical strip chart recording in Fig. 4.12-3. Such a synchronizer
will hold a klystron in synchronization over a frequency range dependent on
the klystron modulation sensitivity and its electronic tuning range. In the
particular oscillator synchronizer shown in Fig. 4.12-1, a total-control voltage
swing of 40 v is available from the phase comparator. The lock range for a

given klystron will therefore approach 40 times its modulation sensitivity unless limited by its electronic tuning range. Synchronization of the klystron is made very simple by a unique search oscillator feature of the oscillator synchronizer. The search oscillator automatically sweeps the frequency of the free-running klystron such that the klystron need only be tuned to enter the lock range of the synchronizer for capture to occur. The search oscillator is automatically suppressed when the klystron is captured and locked. The search oscillator also ensures that the klystron is automatically recaptured if its output is removed from and then reapplied to the synchronizer as long as the frequency remains within the synchronizer lock range. Center the phase meter by adjusting the cavity tuning and the reflector power supply voltage, and maximize the IF level meter reading; the klystron will now be centered in the synchronizer lock range and tuned for maximum power output. Any physical or electrical change that tends to shift the frequency within the lock range will then be exactly compensated for by the synchronizer. Small changes in phase will be indicated by a phase meter, showing that the reflector voltage is being corrected to compensate for the attempted changes. Deviations from zero indicate only a phase difference between the signal and reference. No frequency change can occur while the klystron remains within the lock range. Oscillator frequency drift and jitter are virtually eliminated by phase-lock synchronization. If, for example, a reflex klystron having a modulation sensitivity of 1 MHz per volt exhibited a jitter rate of 1 part in 10^3 per second, the synchronizer would reduce this to an offset error of less than 1 part in 10^{11}, which is much smaller than jitter of the reference frequency itself. This effect is shown in the typical strip chart recording in Fig. 4.12-3. Similarly, sinusoidal frequency disturbances are reduced by the synchronizer loop gain at the disturbance rate. For example, if 1 millivolt ripple in reflector voltage at 120 Hz would cause 1 kHz deviation of the unstabilized klystron, ripple deviation on the locked klystron would be reduced by 100 dB to about 0.01 Hz.

The oscillator synchronizer can also be used as a frequency monitor for microwave signals. (See Fig. 4.12-4.) In this role it acts as a fixed-frequency, double-conversion, superheterodyne receiver whose output frequency represents the difference between the signal frequency and the end

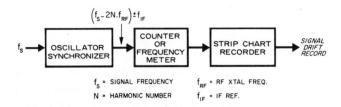

Fig. 4.12-4. Frequency monitoring.

harmonic of the RF reference, ± the IF reference. The output error frequency can be monitored with a counter or with a frequency meter; both these instruments will drive a strip chart recorder to provide a permanent record. Drift or error measurements can be made in this manner over a range of ±1 MHz about the nominal signal frequency. With the use of an electronic frequency meter it is particularly useful in providing low-frequency FM information as well.

To monitor frequency modulation on a microwave signal, the oscillator synchronizer is operated in the VFO mode and the phase comparator output is connected to the VFO modulation input. This locks the IF reference oscillator to the IF signal. The phase comparator output therefore reflects the FM on the IF signal. This can be examined with an oscilloscope. The maximum deviation that can be measured with such an instrument is ±500 kHz.

Figure 4.12-5 shows a setup making such measurements. An oscillator

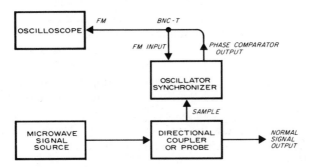

Fig. 4.12-5. FM monitoring.

synchronizer may be used directly with signal generators where the instruments provide connectors for frequency sampling and control input. Advantages of the synchronizer–signal generator combination are multiple output frequencies stabilized to 1 part in 10^8 short term, 1 part in 10^6 long term, and automatic power leveling in up to 20 dB of pulse and amplitude modulation of the synchronized signal. If special modulators are used, a sophisticated, high-speed, low-jitter modulation system is required.

Not only signal generators can be phase-locked; sweepers using tube-type oscillators, such as backward-wave oscillators, need even more concern about cleanliness of signal and stability. A stabilized sweep oscillator system can provide the stable signals required by today's more sophisticated microwave applications. In these systems, a sweep oscillator with appropriate interchangeable RF unit is phase-locked to a 240–400 MHz reference oscillator. The reference oscillator stability is thereby transferred to the sweep oscillator. The reference oscillator is continuously tunable, so the sweep oscillators can be stabilized at any frequency in their respective ranges quickly and easily; there are no crystals to change. In addition, stabilized systems providing swept as well as CW operation are available. Frequency indication is unambiguous; it can be read directly from the sweeper dial. A counter can be added for more accurate frequency indication. Furthermore, the phase-lock IF in this system is chosen to be 20 MHz. This choice of frequency eliminates IF feedthroughs and related problems which arise in applications such as sensitive receiver testing when receiver IF is the same as the phase-lock IF. The output of the sweep oscillator is mixed with the harmonics of the reference oscillator, which is continuously tunable from 240 to 400 MHz. Then the mixing products are applied to a synchronizer which has a 20-MHz IF amplifier terminated in a phase comparator. This circuit compares the phase of the IF signal to the output of an internal 20-MHz crystal oscillator. The synchronizer then supplies the sweeper RF unit with a dc error signal proportional to the phase difference between the IF and crystal oscillator signals. The error signal corrects the output frequency of the sweep oscillator completing the phase-lock loop. For a system like this, the lock range is about 50 MHz. The synchronizer also includes a search circuit which sweeps the output-error voltage and thereby the sweeper frequency to facilitate the initial locking process. The search or capture range is about 20 MHz. The search circuit is automatically disabled when lock is achieved. Although the phase-lock IF amplifier passes either 20-MHz mixing-product reference oscillator 20 MHz above or below the sweeper frequency, the search oscillator is disabled only when the reference oscillator harmonic is 20 MHz below the sweeper frequency. Therefore adjacent lock points are spaced by the reference oscillator frequency rather than twice the IF (40 MHz).

Since the accuracy and resolution of the frequency dial on the sweeper are adequate to distinguish between lock points, it is easy to determine to which harmonic the system is locked. Phase-locking the system is simple; the desired frequency is set on the sweeper dial and the reference oscillator is then tuned for lock as indicated by a front panel light on the synchronizer.

For systems which include a counter, the reference oscillator is tuned to obtain the desired sweeper frequency on the counter. Although the counter indicates sweeper frequency, it actually counts the reference oscillator. The sweeper is then tuned to the appropriate lock point. The wide spacing of the lock points makes picking the right lock point easy. For higher stability the reference oscillator can be replaced by a more stable source, such as a frequency synthesizer. The stability is essentially that of the reference source. When a substitute oscillator is used, an amplifier may be required to provide enough drive to the mixer or to the multipliers.

The system operation in swept mode is somewhat similar to the operation in CW mode. However, the reference oscillator and the sweeper must track throughout the sweep. The method of tracking depends upon the sweep width. For broadband sweeps, a mechanical servo control loop tunes

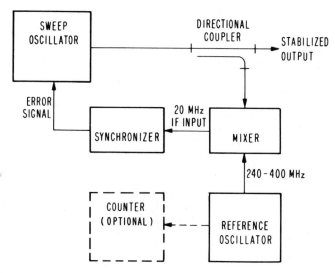

Fig. 4.12-6. Coaxial systems (1 to 12.4 GHz)
stabilized in CW only.

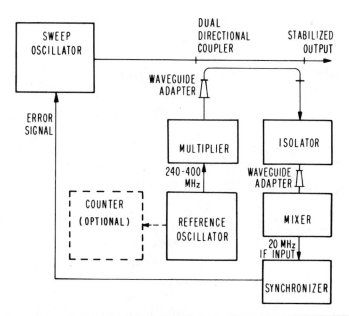

Fig. 4.12-7. Waveguide systems (12.4 to 40 GHz)
stabilized in CW only.

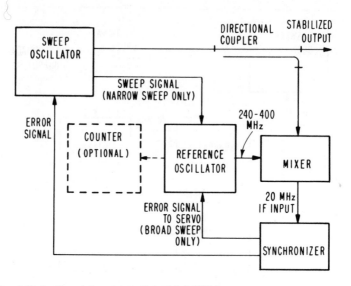

Fig. 4.12-8. Coaxial systems (1 to 12.4 GHz) stabilized in swept or CW mode.

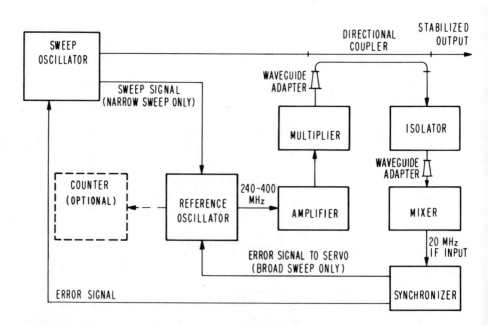

Fig. 4.12-9. Waveguide systems (12.4 to 40 GHz) stabilized in swept or CW mode.

the reference oscillator. The error signal for the servo loop is obtained from a low-level output from the phase comparator in the synchronizer. Although the reference oscillator determines short-term stability, long-term stability is that of the unstabilized sweep oscillator. For the narrow or ΔF sweep mode, the mechanical servo loop is opened. The reference oscillator is swept electrically by the sweep voltage from the sweep oscillator. This voltage is applied to the Varactor in the oscillator circuit as the reference oscillator sweeps. An error signal developed in the phase comparator causes the sweep oscillator frequency to track the reference oscillator. In this case, then, both the long- and short-term stability are that of the reference.

Figure 4.12-6 shows a coaxial system used from 1 to 12.4 GHz stabilized in CW only. Figure 4.12-7 shows a waveguide system from 12.4 to 40 GHz stabilized in CW only. Figure 4.12-8 shows a coaxial system from 1 to 12.4 GHz stabilized in swept or CW mode. Finally, Fig. 4.12-9 shows a waveguide system from 12.4 to 40 GHz stabilized in swept or CW mode.

QUESTION

1. What does phase-locking provide to a signal source?

4.13 FREQUENCY STANDARDS[19]

The demand continues for ever greater precision and accuracy in frequency control. Basic to such control is the atomic standard with unprecedented frequency stability. The cesium and hydrogen atomic standards are truly primary frequency standards and require no other reference for calibration.

Atomic resonance standards use quantum mechanical effects in the energy states of matter, particularly transitions between states separated by energies corresponding to microwave frequencies. Transitions having properties well suited to standard use occur in atoms of cesium, rubidium, thallium, and hydrogen. Considerable attention has been directed to three devices: the cesium atomic beam, the rubidium buffer gas cell, and the hydrogen maser. The cesium and rubidium devices utilize passive atomic resonators to steer conventional oscillators, usually of the quartz crystal type, via feedback control circuits. The hydrogen maser, an active device, derives its signal from stimulated emission of microwave energy amplified by electronic means to a useful power level.

Three other devices of interest as frequency standards are the thallium beam, the ammonia maser, and the rubidium maser. The ammonia maser

[19] Hewlett-Packard Application Note 52: Frequency and Time Standards.

is attractive because of the high spectral purity and excellent short-term stability it offers, but its development has not reached the stage of widespread use in general frequency-control applications. The thallium beam, similar to the cesium beam standard, seems capable of even higher precision than cesium and is being investigated at the present time. The rubidium maser is a recent development which offers the prospect of spectral purity exceeding that of any existing atomic frequency standard.

Secondary frequency standards are those which must be referenced to an accepted source, such as a primary standard. Quartz crystal oscillators are widely used as high-quality secondary standards. Both the cesium beam and the rubidium cell devices make use of slaved quartz crystal oscillators. For applications not requiring the precision and accuracy available in a quartz crystal oscillator referred directly to a primary atomic standard, quartz oscillators referred to the U.S. Frequency Standard by means of phase comparisons with low-frequency radio signals from WWVB or WWVL offer high accuracy at moderate cost.

Common to all atomic frequency standards are means for (1) selecting atoms in certain energy state, (2) enabling long lifetimes in that state, (3) exposing these atoms to (microwave) energy, and (4) detecting the results. Two primary atomic standards which have reached a high state of development are the hydrogen maser and the cesium beam.

The hydrogen maser is a primary frequency standard in that it provides a frequency which is well defined without reference to any external standard. A system with the high Q needed to generate such a frequency results when arrangements are made to allow a relatively long interaction time between atoms in a selected energy state and a microwave field in a manner leading to maser action.

Hydrogen atoms are formed into a beam and passed through a magnetic field of high gradient to select those atoms in the higher quantum mechanical energy states, among which are the useful atoms in the state $F = 1$, $m_F = 0$. The selected atoms pass into a quartz cell having a protective coating that neither adsorbs hydrogen nor causes unwanted energy transitions to occur. A microwave field surrounds the cell. Within it, the atoms make random transits, reflecting from the walls. Their zigzag path lengthens possible interaction time, and they interact with the microwave field until they relax either by giving their energy usefully to the field or through some (unwanted) collision event. The long interaction time permits coherent stimulation of the hydrogen atoms and sustained maser oscillations.

The hydrogen maser is potentially capable of extremely high stability, and existing units have reached stabilities to few parts in 10^{13} over months.[20]

Cesium beam standards are in use wherever high precision and accuracy

[20] *Frequency*, 2, 4, July–August 1964, p. 33.

in frequency and time standards are the goals; in fact, cesium beam units are the present basis for the U.S. Frequency Standard.

For the cesium beam standard, the quantum effects of interest arise in the nuclear magnetic hyperfine ground state of the atoms. A particularly appropriate transition occurs between the $(F = 4, m_F = 0) \leftrightarrow (F = 3, m_F = 0)$ hyperfine levels in the cesium-133 atom, arising from electron-spin nuclear-spin interaction. This transition is appropriate for frequency control by reason of its relative insensitivity to external influences, such as electric and magnetic fields, and of its convenient frequency (in the microwave range, 9192+ MHz).

A typical atomic beam device which takes advantage of this invariant transition is so arranged that cesium atoms (in all sublevels of states $F = 3$ and $F = 4$) leave an oven, are formed into a beam, and are deflected in a nonuniform magnetic field ("A" field) by a vector component dependent upon m_F and the field. For the two $m_F = 0$ states, the deflection is equal and opposite. The atoms then pass through a low and uniform magnetic field space ("C" field) and are subject to excitation by microwave energy. At resonance, change of state occurs by absorption or by stimulated emission, depending on the initial state, $F = 3$, or $F = 4$. On crossing a second magnetic field ("B" field) identical to the first one, those atoms and only those atoms which have undergone transitions are focused on the first element of a detector, a hot wire. The cesium atoms, ionized by the hot wire, are attracted to the first dynode of an electron multiplier. The resulting amplified current serves to regulate the frequency of an external crystal oscillator. Oscillator output is multiplied and fed back into the cesium beam through a waveguide, closing the loop.

In the Hewlett-Packard cesium beam standard, the microwave field, derived from a precision quartz oscillator by frequency multiplication and synthesis, is phase-modulated at a low audio rate. When the microwave frequency deviates from the center of atomic resonance, the current from the electron multiplier contains a component alternating at the modulation rate with amplitude proportional to the frequency deviation and with phase information which indicates the direction, that is, whether the frequency lies above or below center frequency. This component is then filtered, amplified, and synchronously detected to provide a dc voltage used automatically to tune the quartz oscillator to zero error.

The control circuit provides a continuous monitoring of the output signal. Automatic logic circuitry is arranged to present an indication of correct operation. The new, compact cesium beam tubes exhibit frequency perturbations so small that independently constructed tubes compare within a few parts in 10^{12}. Outstanding reliability is obtained from these tubes with a presently guaranteed life of 10,000 hours.

The quartz crystal oscillator used exhibits superior characteristics even without control by the atomic resonator. Drift rate is less than 5×10^{-10} per 24 hours, and short-term stability is better than $\pm 1.5 \times 10^{-11}$ for a one-second averaging time. The 5-MHz quartz crystal is housed in a two-stage, proportionally controlled oven. Output variation due to temperature is less than $\pm 1 \times 10^{-10}$ from 0° to 40°C.

The striking properties of the quartz crystal oscillator give it such an advantage over all earlier frequency stable systems, such as those relying on tuning-fork resonators, that its use for exacting measurements of frequency and time quickly became almost universal in national and industrial laboratories around the world.

Crystalline quartz has great mechanical and chemical stability and a small elastic hysteresis (which means that just a small amount of energy is required to sustain oscillation; hence frequency is only very slightly affected by variable external conditions). These are most useful in a frequency standard. The piezoelectric properties of quartz make it convenient for use in an oscillator circuit. When quartz and certain other crystals are stressed, an electric potential is induced in nearby conductors; conversely, when such crystals are placed in an electric field, they are deformed a small amount proportional to field strength and polarity. This property by which mechanical and electrical effects are linked in a crystal is known as the piezoelectric effect.

An inherent characteristic of crystal oscillators is their resonant frequency changes (usually increases) as they age. This "aging rate" of a well-behaved oscillator is almost constant. After the initial aging period (a few days to a month), the rate can be taken to be constant with but slight error. Once the rate is measured, it is usually easy to apply corrections to remove its effect from data. Over a long period the accumulated error drift could amount to a serious error. (For example, a unit with drift rate of 1 part in 10^{10} per day could accumulate in a year an error of several parts in 10^8.) Thus periodic frequency checks and corrections are needed to maintain a quartz crystal frequency standard.

The rubidium vapor standard, as is the case for the cesium beam, uses a passive resonator to stabilize a quartz oscillator. The Rb standard offers excellent short-term stability in a relatively small apparatus easily made portable. It is a secondary standard because it must be calibrated against a primary standard such as the cesium beam during construction; it is not self-calibrating.

Operation of the rubidium standard is based on a hyperfine transition in Rb-87. The rubidium vapor and an inert buffer gas (to reduce Doppler broadening, among other purposes) are contained in a cell illuminated by a beam of filtered light. A photo detector observes changes near resonance in the amount of light absorbed as a function of applied microwave frequencies. The microwave signal is derived by multiplication of the oscillator frequency.

A servo loop connects the detector output and oscillator so that the oscillator is locked to the center of the resonance line.

By an optical pumping technique, an excess population is built up in one of the Rb-87 ground-state hyperfine levels within the cell: the population of the $F = 2$ level is increased at the expense of that of the $F = 1$ level. The illuminated Rb-87 atoms are optically excited into upper-energy states, from which they decay into both the $F = 2$ and $F = 1$ levels. The exciting light has been filtered to remove components linking the $F = 2$ level to the upper-energy states. Since the light therefore excites atoms out of the $F = 1$ level only, while they decay into both, an excess population builds up in the $F = 2$ level. The optical absorption coefficient is reduced because fewer atoms are in the state where they can absorb the light.

Application of microwave energy, corresponding to that which separates the two ground-state hyperfine levels $F = 2$ and $F = 1$, induces transitions from the $F = 2$ to $F = 1$ level with the result that more light is absorbed. In a typical system arrangement, photo-detector output reaches a minimum when the microwave frequency corresponds to the Rb-87 hyperfine transition frequency (approximately 6835 MHz).

Resonance frequency is influenced by the buffer gas pressure and, to a lesser degree, by other effects. For this reason, an Rb vapor standard must be calibrated against a primary standard. Once the cell is adjusted and sealed, the frequency remains highly stable.

Frequency- and time-standard broadcast stations make possible worldwide comparisons of local standards to national standards. In the United States, the National Bureau of Standards operates standards stations, and the U.S. Navy operates transmitters that are frequency stabilized.

Fast and precise frequency calibrations traceable to the U.S. Frequency Standard (USFS) are possible through comparisons of a local standard against phase-stable low-frequency standard signals now being transmitted by the National Bureau of Standards (NBS). These low-frequency broadcasts from NBS transmitters at Fort Collins, Colorado, WWVB (60 kHz) and WWVL (20 kHz) are capable of yielding comparison precisions as high as a few parts in 10^{12} against the U.S. Frequency Standard (in the continental United States under good propagation conditions). The USFS is located at the Boulder, Colorado, laboratories of the NBS.

A critical requirement for useful modern standards is that they be rigorously consistent along the entire chain of measurements, tracing back to international prototype standards representing the fundamental units of mass, length, and time. To provide the nation with the central basis for a self-consistent and uniform system of physical measurement is the primary mission of the NBS. Boulder, Colorado, is the NBS center for precision measurements of frequency and time.

At the NBS Electronic Calibration Center, which occupies one wing of

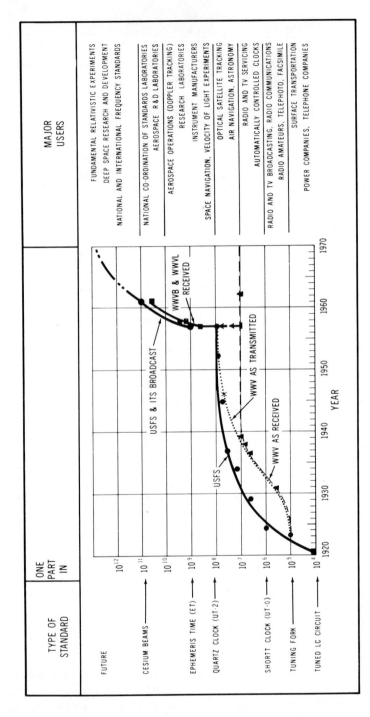

Fig. 4.13-1. Improvements in the precision of the US frequency standard (USFS) and its dissemination.

the Radio Standards Laboratory at Boulder, NBS makes available to government, industry, and the military services access to the nation's primary electronic standards. The Radio Standards Laboratory is part of the Institute for Basic Standards, one of the three institutes which presently constitute the NBS. The NBS shifted its radio standards work to Boulder in 1954. Earlier, radio standards were the responsibility of the Central Radio Propagation Laboratory, originally located in Washington, D.C. Radio standards work was begun in 1911.

The Radio Standards Laboratory maintains the U.S. Frequency Standard at Boulder and disseminates it via low-frequency and very-low-frequency broadcasts from nearby Fort Collins (WWVB and WWVL) and from the older high-frequency stations (WWV and WWVH).

The U.S. Frequency Standard has been maintained since 1920. Improvement in its precision has been steady, and it took a sharp upturn in 1960, when a cesium beam apparatus became the device which provides the Standard.

The NBS states the present standard to be accurate to five parts in 10^{12} (3σ limits), an accuracy higher than that achieved in the measurement of any other quantity. One to two parts in 10^{13} are attainable for measurement times of about one hour. The NBS is now working on a standard to exceed even this high accuracy. Figure 4.13-1, reproduced from a publication of the NBS,[21] shows improvements since 1920 in the precision of the U.S. Frequency Standard and its dissemination. The level of accuracy required by different users is indicated.

Effective January 1, 1965, the NBS low-frequency station WWVB began to broadcast the international unit of time based upon the atomic standard. The atomic definition of the second was authorized in October 1964 by the Twelfth General Conference of Weights and Measures, meeting in Paris. The conference action based the definition on a transition in the cesium atom for the present, in anticipation that an even more exact definition may be possible in the future. The international second has been redefined before, always with the realization of increased exactness. In 1956 a second called the ephemeris second was defined as a certain fraction of the time taken by the earth to orbit the sun during the tropical year 1900. Before 1956 the second was defined as a fraction of the time required for an average rotation of the earth on its axis with respect to the sun.

The atomic definition realizes an accuracy much greater than that achieved by astronomical observations. It results in a time base more uniform and much more convenient. Now determinations can be made in a few minutes to greater accuracy than was possible before in measurements that took

[21] National Bureau of Standards, "Precision of the U.S. Frequency Standard," *NBS Technical News Bulletin*, Vol. 48, No. 2, February 1964, 31.

many years. The exact wording of the action of the Twelfth General Conference is: "The standard to be employed is the transition between two hyperfine levels $F = 4$, $m_F = 0$ and $F = 3$, $m_F = 0$ of the fundamental state $^2S_{1/2}$ of the atom of cesium-133 undisturbed by external fields and the value 9,192,631,770 hertz is assigned."

QUESTIONS

1. What is now the most accurate frequency standard?
2. What mechanism is used in atomic frequency standards to obtain the kind of accuracies quoted?
3. Is the rubidium vapor standard a primary frequency standard? Tell why or why not.

5 SIGNAL ANALYSIS

Microwave signal sources for general laboratory applications were discussed in the previous chapter, and different kinds of signal generation techniques were dealt with. This chapter deals with the analysis of the signals being generated.

Several questions can be asked as one starts to analyze the signal available at a port. The first concern is the available power output, which is probably the most basic information about the signal source. An almost equally important consideration is the frequency of the source.

Then, of course, several other questions can be raised: What is the quality of the signal which is given? Is it modulated or not? Is it stable? What kind of response does it have? Is it distorted? What is the harmonic content? All these questions have to be answered. Therefore signal analysis can be divided into three categories. The first two deal with the basic measurements, such as power and frequency measurements. The third category deals with all the other questions and is called *spectrum analysis*. Consequently, signal analysis includes: power measurement, frequency measurement, and spectrum analysis.

5.1 POWER MEASUREMENT[1]

The term "power" is used to describe the rate at which energy is made available to do work. In the field of microwaves, this "work" is usually the transmission of aural, visual, or coded (radar) intelligence over a given distance. Other examples of work accomplished by microwave power include the excitation of molecules in a medium to produce heat or the acceleration of particles for nuclear studies.

From an economic standpoint we are interested in measuring power to determine the cost of work performed in terms of energy expended. In applied science, power measurement establishes a means for evaluating the work capability and efficiency of a given device. The continuous effort to design more efficient systems and to increase the work capability of microwave devices leads to higher accuracy requirements in the measurement art. Along with this, greater measuring speed is desirable so that advancements will not be limited by impractical test methods.

Voltage and current measurements are basic at low frequencies and dc because they can be made conveniently with high accuracy. These are fundamental electrical standards, whereas power is a derived term of necessarily lower accuracy. At microwave frequencies one is faced with a distributed circuit constant and transmission line lengths that are appreciable fractions of a wavelength or more. Furthermore, impedance at any given measurement

[1] Application Note 64, Hewlett-Packard Co.

point is not easily determined, and any impedance mismatch between source and load sets up standing waves along the transmission line such that voltage measurement becomes arbitrary. Since power remains invariant with position in a lossless line, the practical method of determining energy available in a microwave circuit becomes one of power measurement.

5.1.1 BASIC DEFINITIONS

There are two types of power measurements in microwave work. Most applications require a knowledge of average power in a device; the others require peak power over periodic pulse. By definition, the power in a load resistor at any instant in time is the product of the voltage and current, where the voltage and current figures are instantaneous voltage and current values. From this it can be seen that instantaneous power varies at a rate that is twice the frequency of the generator. Instantaneous power does not relate the amount of work done over any interval in time, which is of primary importance in most cases. The average power over a given time period is the sum of all instantaneous values of power divided by the period. It can be shown that

$$P = \frac{1}{T} \int_0^T ei \, dt \qquad (5.1\text{-}1)$$

where P = average power, T = time, and e and i = instantaneous values of voltage and current.

Figure 5.1-1 shows that the instantaneous power in load resistor R_L varies at a rate twice the applied generator frequency. With a CW source, the shortest time of interest is generally the period of one cycle of output signal.

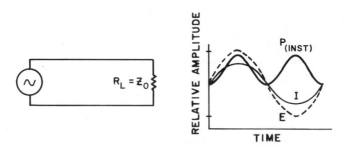

Fig. 5.1-1. Instantaneous power in load resistor R_L varies at a rate twice the applied generator frequency. Average power is the integral of instantaneous power averaged over one or more cycles.

The average power in this period is the same as for any integral number of cycles, provided a steady-state condition exists. The average power for an ac circuit can be given mathematically by

$$P = E_{\mathrm{rms}} \times I_{\mathrm{rms}} \cos \phi \qquad (5.1\text{-}2)$$

Peak pulse power may be defined in one way as the average power during the time the pulse is on. Figure 5.1-2 shows the envelope of a periodic rectangular

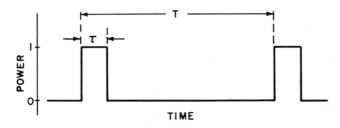

Fig. 5.1-2. Peak power of periodic rectangular pulse is readily defined by average power-duty cycle relationship. Pulse width (τ) and repetition rate ($\gamma\tau$) are customarily measured with reference to 50% amplitude points.

pulse of RF energy of width τ and period T. It should be noted at this point that any variations in the pulse, such as overshoot or ringing, are averaged out by this definition, and their instantaneous peak values cannot be considered as peak pulse power. For a rectangular microwave pulse the peak power expression is

$$P_{\mathrm{PK}} = \frac{P_{\mathrm{ave}}}{\text{duty cycle}} \qquad (5.1\text{-}3)$$

where the duty cycle is τ/T.

At more complicated pulses where the pulse is intentionally not rectangular or contains aberrations which cannot be calculated accurately enough, the absolute determination of τ is not possible. Peak power at the pulse's maximum amplitude could differ considerably from peak power as defined in Eq. (5.1-3). A constant of correction can be calculated or determined with a planimeter for many nonrectangular pulses; however, this is a time-consuming task. In applications involving nonrectangular pulse shapes, peak pulse power may be defined as

$$P_{\mathrm{PK}} = \frac{e_p^2}{R} \qquad (5.1\text{-}4)$$

where $R =$ a constant load resistance absorbing the power, and $e_p =$ peak rms voltage across a constant resistance R. This expression does not rely on pulse shape for an accurate determination of the pulse's maximum amplitude point. This fact makes it especially useful for voltage breakdown checks. Since peak pulse power may have different meanings, depending upon the application, the method of measurement should be flexible enough to allow a choice as to what is considered the peak power of a pulse.

Average Power Measurements

Fundamental standards of microwave power lie in a dc or low-frequency ac power which may be used for comparison or substitution when accurately measured. Instrumentation used for such comparison or substitution must be efficient in design and properly employed for minimum loss of time and accuracy in the transfer. As might be expected, these criteria often impose stringent requirements, so the approach must be one of compromise. There are two basic average power measurement techniques: the calormetric and the bolometric power measurements.

5.1.2 BOLOMETRIC POWER MEASUREMENTS

Microwave power up to about 10 mw average is usually measured by bolometric means. The basis for bolometric power measurement is that, when power is dissipated in a resistive element, a corresponding change occurs in the element's resistance. By proper construction of the element, the heating effect of dc and microwave power will be nearly the same, and the resistance

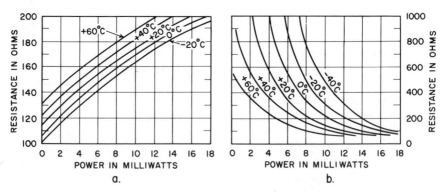

Fig. 5.1-3. Resistance versus dissipated power in (a) typical wire barretter, and (b) typical bead thermistor at various ambient temperatures.

change may be sensed by associated circuitry to indicate power. The resistive elements are termed "bolometers" and fall into two classifications: (1) barretters, which consist of either a short length of fine wire suitably encapsulated or a strip of thin metallic film deposited on a base of glass or mica, and (2) thermistors, which are made of a compound of metallic oxides. Barretters exhibit a positive temperature coefficient (i.e., as more power is dissipated in the element, its resistance becomes greater, and vice versa). Thermistors have a negative temperature coefficient and are physically and electrically more rugged than barretters. Fig. 5.1-3 shows the resistance-versus-power characteristics of typical wire barretters and thermistors.

Power Meter Bridge

One of the simplest methods for bolometric power measurements is to place a bolometer (usually a thermistor) in one leg of a Wheatstone bridge, as in Fig. 5.1-4. The bridge is excited by a regulated dc supply whose am-

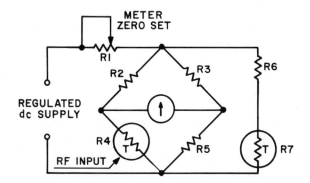

Fig. 5.1-4. Unbalanced thermistor bridge provides simple power monitor for limited range and accuracy requirements.

plitude may be adjusted with R1. Since R4 is a thermistor, its resistance may be controlled by the amount of current allowed to pass through it. In operation, R1 is adjusted until just enough current passes through the bridge to make the thermistor's resistance equal to R5, bringing the bridge into balance and causing the meter to read zero. Microwave power is then applied to the thermistor, and the heating effect causes the thermistor's resistance to decrease, unbalancing the bridge in proportion to the power applied. The unbalance current is indicated on the meter, which is calibrated directly in milliwatts. This method is called the unbalanced-bridge technique.

The bolometer resistance curves shown in Fig. 5.1-3 vary with temperature and require some form of temperature compensation to minimize meter drift. Therefore a disc-shaped thermistor, R7, is placed in close thermal proximity with R4 but isolated from the microwave power. This is usually done by placing the disc against the outside wall of the thermistor mount for maximum heat transfer. The temperature characteristics of R7 are chosen to be as nearly equal to those of R4 as possible so that, as ambient temperature increases, for example, resistance decreases proportionately in both thermistors. The reduction in R4's resistance is seen as an unbalance in the bridge, as if more power were being applied to R4. The reduction in R7's resistance results in more current shunted away from the bridge, thus lowering the dc power in R4. This action causes R4's resistance to increase, compensating for the temperature change and restoring bridge balance. This scheme works well over limited temperature variations but depends upon the R-T curves of both thermistors being the same, which is seldom the case. Also, the system has slow response because of the time required for the thermistor mount and compensating disc to reach equilibrium with the measuring thermistor's temperature. With a full-scale range of 2 mw, meter drift over an extended period of time might be as high as half scale in a typical unbalanced bridge with this form of temperature compensation.

Note in Fig. 5.1-3(b) that the thermistor's resistance is linear with applied power over only a very small range. This range extends for about a 2-mw increment of power in the knee of the R-P curves, thus limiting dynamic range. At an operating temperature of 40°C, a bias power of approximately 6 mw is required to place the thermistor in the linear portion of its R-P curve. Thermistor resistance at this point is about 200 ohms, so this value is chosen for the other resistors in the bridge. As power in excess of 8 mw (6 mw dc bias + 2 mw microwave) is applied, thermistor resistance change for a given power change becomes less and resolution is drastically reduced.

One consideration in the accuracy of the unbalanced bridge is the fact that bolometer resistance varies with the level of microwave power applied. This changes the impedance of the bolometer mount, resulting in mismatch error, since part of the microwave power is reflected rather than absorbed in the bolometer element. Mismatch error due to a 2:1 resistance change in the bolometer can be as high as 0.5 dB. This technique leaves quite a bit to be desired.

In the automatically balanced bridge, ac substitution and readout are done automatically, eliminating all operations required in the manually balanced bridge except the zero adjustment for initial balance. The technique combines much higher accuracy and wider range than the unbalanced bridge with far greater convenience and speed than the manual bridge. These features have led to industry-wide use of the automatic bolometer bridge. Such

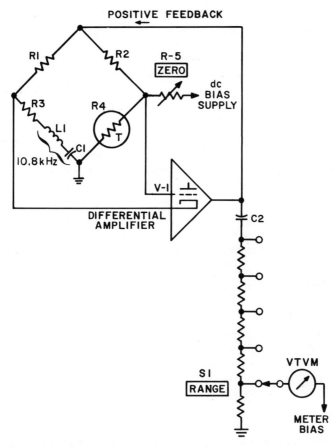

Fig. 5.1-5. Simplified schematic of an automatically balanced power meter bridge.

power meter bridges operate with thermistors or barretters that require up to 16 milliamps of bias current to reach an operating resistance of 100 or 200 ohms. Full-scale power ranges are from 0.1 mw to 10 mw in a 1,3,10 sequence. A simplified schematic of such a power meter bridge is shown in Fig. 5.1-5. The accuracy of substituted power on these types of power meter bridges is ±5%.

To illustrate the operation of such a power meter, it is assumed that the thermistor mount being used requires a total of 28 mw at its ambient temperature to reach the operating resistance set by the circuit. This power may be dc, audio frequency, microwave, or any combination thereof. In most of these power meters a combination of audio and dc is initially applied to the mount. Whatever dc power is applied, the circuit automatically supplies the balance in audio power up to a total of, in this example, 28 mw. Any microwave power applied then automatically reduces the audio power by a like amount to maintain the total constant. A VTVM then measures this reduction and indicates it as applied power. Referring to Fig. 5.1-5, it may be seen that feedback from the amplifier to the bridge is positive for one side of the bridge and negative for the other. The positive feedback is temperature sensitive, depending on the resistance of the thermistor. Any increase in feedback power, for example, decreases the feedback factor. The negative feedback on the other side of the bridge is frequency sensitive and reaches a minimum value at the resonant frequency of the tuned circuit L_1 and C_1.

When the circuit is first turned on, the thermistor will be cold, its resistance high, and the positive feedback very large. Oscillations will immediately start up, putting power into the thermistor and reducing its resistance until the positive feedback exceeds the negative by a very small margin. This condition occurs at the resonant frequency where the negative feedback is a minimum. The degree by which the positive feedback exceeds the negative is set by the gain of the circuit and is made very small by making the gain high. If dc or microwave power is added to the thermistor, the thermistor resistance tends to decrease. However, this tendency automatically decreases the net positive feedback, causing the oscillations to decrease by an amount exactly equal to the applied dc or microwave power. The small degree of bridge unbalance is always preserved, and the thermistor resistance does not change.

To indicate power on the VTVM meter scale, it is necessary to start from a known reference condition. If there were 20 mw of audio power in the bolometer and 5 mw of RF were added, a certain scale would be required to operate between audio power of 20 and 15 mw. If the audio decreased instead from 10 to 5 mw, a different scale would be required. A simple gain adjustment in the VTVM would not suffice because of the square-law relation between the power and the rectified voltage actually applied to the meter. Hence a dc-bias supply is incorporated into the instrument, and enough dc power is added to reduce the audio power to a convenient reference value. Usually, in these types of power meter bridges, this is arranged to be 1.2 times the full-scale value on each range. Thus, on the 10-mw range, the dc is adjusted so there is exactly 12 mw audio in the bolometer and the meter circuit is calibrated to read zero. The audio drops to 2 mw when 10 mw RF is applied, and the circuit is calibrated to read full scale. This 1.2:0.2 relation holds on all ranges, and the meter scale is accurate at all points. Table 5.1-1 illustrates how combinations of power are used to balance a typical bolometer in such a power meter bridge for zero and full-scale indications on the 10 mw range. Since it is desirable to have the meter read upscale as RF power is applied, a residual dc current is applied to the meter movement in the forward direction and rectified audio is applied in the reverse direction.

Table 5.1-1.

	Power Meter at:	
Power to Bolometer	Zero	Full Scale
Microwave	0 mw	10 mw
dc	16 mw	16 mw
Audio	12 mw	2 mw
Total	28 mw	28 mw

Range switching is accomplished by changing the sensitivity of the VTVM to the audio level. As lower ranges are selected, VTVM sensitivity is increased and the dc bias increased to reduce the audio level. Eventually, drift becomes a problem, since instability in the dc supply becomes magnified. In addition, any change in the ambient temperature of the bolometer changes its total power requirement, causing the audio level to change. If this occurs during the course of a measurement, an error results, since the temperature change is indicated along with the RF power. Dynamic range is normally limited to about 20 dB.

Bolometer Mounts

So far, only the instrumentation portion of the bolometer bridge has been discussed in detail. Before introducing temperature-compensated automatic bridges, it would be good to complete the discussion on uncompensated systems with more detail on bolometers and how they are mounted for practical use.

Bolometer elements are mounted in either coaxial or waveguide structures so they are compatible with the common transmission line systems used at microwave and RF frequencies. The bolometer-mount combination must be designed to satisfy four important requirements so the bolometer element absorbs as much of the power incident to the mount as possible. In this regard the mount must: (1) present a good impedance match to the transmission line over the frequencies of interest; (2) keep I^2R and dielectric losses within the structure minimized so that power is not dissipated in electrical contacts, waveguide walls, or insulators; (3) provide isolation from thermal and physical shock; and (4) keep leakage small so that microwave power does not escape from the mount in a shunt path around the bolometer. Shielding is also necessary to keep extraneous power from entering the mount. One other consideration should not be overlooked, although it is usually accounted for by other mount construction requirements: the mount should have enough mass to minimize the effects of ambient temperature variation.

Bolometers generally can be divided into two basic groups: one is the coaxial thermistor mount or barretter mount structure; the other type is waveguide mount. Figure 5.1-6 shows a variety of coaxial and waveguide bolometer mounts which may be used with manually and automatically balanced bridges. Figure 5.1-7 shows a cutaway view of a typical waveguide thermistor mount. Bolometers may be subdivided into tunable, fixed tuned, and broadband untuned types. In Fig. 5.1-6, types A, B, and D are broadband units which do not need any tuning. Type C shows a tunable bolometer mount which takes either a barretter or a crystal detector but has to be tuned

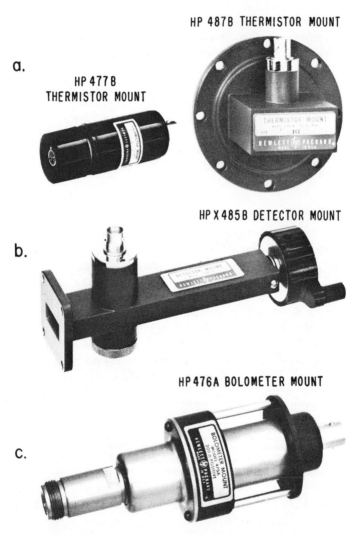

Fig. 5.1-6. Variety of coaxial and waveguide bolometer mounts.

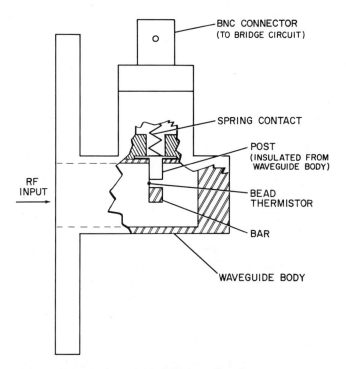

Fig. 5.1-7. Cutaway view of HP 487B broadband waveguide thermistor mount shows post and
bar arrangement for mounting thermistor.
Configuration allows good impedance match
over waveguide band without tuning.

at every frequency in its broad range. It is not a broadband device by its
nature, but it can be tuned to any frequency within its waveguide band.

A Typical Broadband Coaxial Thermistor Mount

A broadband coaxial thermistor mount such as the one shown in Fig.
5.1-6(a) has two thermistor beads which are connected in the coaxial mount,
as shown in Fig. 5.1-8. Connection to the bolometer bridge is made through
J2. The bridge provides approximately 30 mw of bias power to the two
thermistors in series, reducing their resistance to 100 ohms each. Total
resistance presented to the bridge is thus 200 ohms, since reactance of C1 and
C2 is high at the dc and 10.8 kHz bias frequency of the bridge. When microwave power is applied at J1, it passes unattenuated through the dc blocking

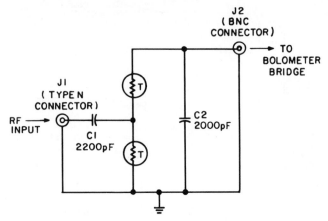

Fig. 5.1-8. Dual-element coaxial thermistor mount operates over 10-MHz to 10-GHz frequency
range. SWR is 1.3 or less over most of the
band. Circuit arrangement results in 50-
ohm impedance match to RF input, and
200-ohm resistance to bolometer bridge.

capacitor C1 to the junction of the two thermistors. Since C2 is a low impedance at microwave frequencies, the two 100-ohm thermistors appear to be
parallel, presenting a combined resistance of 50 ohms necessary for impedance matching to the coaxial transmission line. This design was chosen
primarily because the dc return to the bridge would be difficult over broad-
bands using a single thermistor. All components, including the parts beyond
the transmission line, are shielded to prevent extraneous pickup and radiation.
This type of thermistor mount is capable of measuring up to 20 mw of power
in manual bridges.

Waveguide Bolometer Mounts

Untuned waveguide bolometer mounts are used for power measurements in waveguide systems to avoid losses and mismatch errors caused by
waveguide-to-coax adapters. Figure 5.1-7 shows a cutaway sideview of an
untuned broadband waveguide thermistor mount. It consists of a shorted
section of fabricated brass waveguide with the thermistor bead mounted on a
post and bar arrangement in the center of and parallel to the E field. The
thermistor is biased to an operating resistance of 100 ohms by applying dc
and/or audio frequency bias current through the BNC jack shown in the
figure. The size and location of the post and the bar are designed to present
the conjugate impedance match in the shorted waveguide section over a wide

range of frequencies—typically, across an entire waveguide band. This arrangement provides broadband impedance matching of the mount for low-voltage, standing-wave ratio. Average power (up to 10 mw) may be measured directly with such mounts connected to a suitable bridge. If excess power is applied, the bridge can no longer remove enough bias power to remain in balance. The thermistor therefore decreases in resistance rapidly, causing a large impedance mismatch to reflect a portion of the applied power. This prevents all the applied power from being dissipated in the thermistor bead. Accidental overloads of 25 to 30 mw will not harm the element. Although it is not recommended, these thermistors will withstand considerable overloads without burning out.

Tunable Waveguide Bolometer Mounts

These types of bolometer mounts, as shown in Fig. 5.1-6(c), are usually designed to cover the entire waveguide band. The mounts may also be used with suitable crystal diodes for detection of modulated RF signals. Figure 5.1-9 is a cutaway sideview of a tunable detector mount with a barretter installed. The barretter is positioned in the center of and parallel to the E field

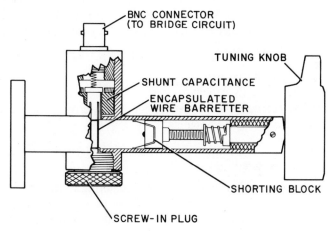

Fig. 5.1-9. Detector mount used encapsulated wire barretter for power measurements in waveguide bands from 2.6 to 12.4 GHz. Movable shorting block is adjusted for best impedance match at test frequency. Screw-in plug provides easy access for changing barretter or installing crystal detector for other applications.

in the guide, the same as the untuned mount previously described. The barretter's lower contact is made to ground with a screw-in plug which allows easy removal of the barretter element for replacement. The upper contact is insulated from the guide and fastens to a shunt capacitance to minimize RF leakage and allow connection to the bias in the bridge circuit. The movable shorting block is highly conductive to ensure a low-loss short across the guide. The block is positioned at any odd multiple of a quarter-wavelength behind the barretter element at the operating frequency. It is best to use the position closest to the element that still allows peaking of an indicator power, since this minimizes I^2R losses in the waveguide. When the plunger is properly tuned, maximum power is absorbed by the barretter and the mount presents a good impedance match to the line. Usually these mounts are specified to have quite low standing-wave ratios in the order of 1.5:1. The small reduction of standing-wave ratio offered by tunable mounts over the broadband mounts is generally not significant enough to warrant the tuning operation required for each frequency change or measurement location change. When the best possible impedance match is required for maximum accuracy, a slidescrew or an EH tuner is used ahead of the bolometer mount. The chief advantage of using such a device is that it employs a barretter rather than a thermistor. From the curve of Fig. 5.1-3 we see that ambient temperature variations have much less effect on a barretter than on a thermistor in nontemperature-compensated systems. For this reason, bridge zeroing is less frequent using the barretter type of bolometer instead of the thermistor type. Disadvantages of using the barretter are:

1. It is easy to burn out;
2. It is mechanically delicate;
3. Its fast thermal time constant can cause error in average power measurements of AM signals.

Temperature-Compensated Power Meters

In the preceding balanced bridge power-metering systems there has been no way to distinguish between ambient temperature variations and applied power changes at the bolometer. The temperature compensation described for an unbalanced bridge is satisfactory for slow-temperature changes in relatively insensitive bridges. However, the problem becomes more serious with increased sensitivity until ultimately the null drift prevents making the measurement. If power is to be measured in the microwatt region, some effective means of temperature compensation must be employed to make highly sensitive bridges practical.

A dual bridge can be designed to operate the temperature-compensated thermistor mounts. Power may be measured with these mounts in a 50-ohm coaxial system or in any waveguide system typically with less zero drift than with the other types (on the order of 100th of that in the compensated systems). It allows measurements to be made as small as 10 μw full scale. Many other advantages that result from temperature compensation will be discussed later.

Figure 5.1-10 shows the block diagram in two halves to facilitate description of circuit operation. On the left is the RF substitution bridge with the RF detection thermistor R_d connected. On the right is the compensation and a metering bridge with the compensating thermistor R_c connected. Both R_d and R_c are contained in the same thermistor mount, with R_d being mounted to intercept applied microwave power. R_c is electrically isolated from R_d and the microwave power. However, the two elements are physically and thermally in very close proximity. Thus any change in mount temperature effects both thermistors identically, but only R_d is exposed to RF.

Each thermistor forms one leg of two separate bridges that are excited by a common 10-kHz oscillator amplifier through transformers T101 and T102. The action of the RF bridge is such that it is self-balancing (similar to the uncompensated power bridge discussed earlier), except that in normal operation only ac bias is applied to R_d. Initially, the zero control is adjusted until balance is also reached in the metering bridge. At balance, a minimum 10-kHz error signal is applied to the three-stage amplifier so the synchronous detector and differential amplifier outputs are minimum and the meter indicates zero. The circuit is arranged to avoid perfect balance, as this would result in zero dc feedback to the metering bridge; with zero feedback the metering loop could appear open. With an open metering loop the circuit would be unstable and zero reference would not be accurately established. The bridge is therefore operated at a point very near balance, and compensation is made in the current-squared generator for zero indication on the meter.

When microwave power is applied to R_d, the RF-substitution bridge tends to unbalance and the 10-kHz oscillator backs off an equal amount of bias power to maintain bridge balance. Since the primaries of T101 and T102 are in series, the metering bridge feels an equal reduction in bias power. R_c is no longer biased for bridge balance, and an error signal is fed to the three-stage amplifier and synchronous detector. The pulsating dc from the detector is filtered and amplified by the differential amplifier and fed back to the metering bridge to rebalance it. Part of the dc feedback current is squared and applied to the meter to indicate the equivalent microwave power applied. The squaring circuit allows a linear calibration of power on the meter scale, since the power is proportional to bridge current squared.

Now that we have described how microwave power is sensed and measured by the system, let us change the ambient temperature around the thermistor mount and see how it is compensated. After the meter is zeroed,

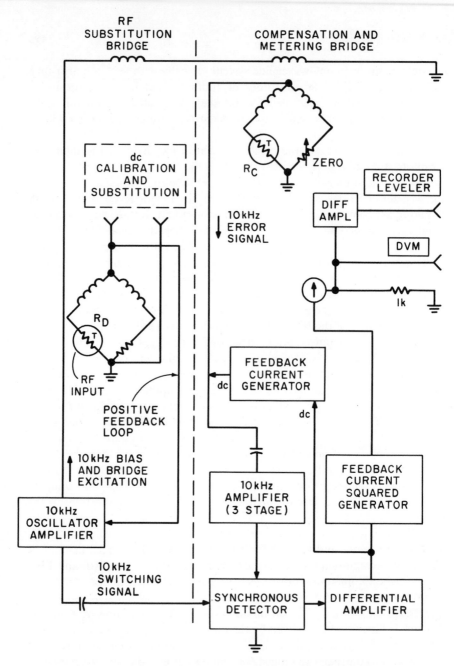

Fig. 5.1-10. Block diagram of HP 431C temperature-compensated power meter. Dual bridge design differentiates between changes in applied power and unwanted temperature variations at the termistor mount. Used with the HP 478A or 486A compensated thermistor mounts, this automatic power meter provides useful ranges from 10 μw to 10mw with a single zero and null arrangement.

suppose the temperature increases. Since both thermistors are in equal thermal environments and of identical types, their resistance will decrease by an equal amount. The RF bridge attempts to unbalance, causing a reduction in 10-kHz bias power applied to T101 and T102, which brings the RF bridge and R_d back into balance. The reduction in bias necessary for R_d to rebalance in the RF bridge is sensed by R_c in the compensating bridge as being just enough reduction to compensate for the heating effect of the ambient temperature increase. Thus the metering bridge remains in balance and no error signal is fed to the three-stage amplifier. Consequently no change in dc feedback occurs and the meter remains at zero. The same compensation would take place if the meter were measuring power and an ambient temperature change occurred, since the system is able to distinguish between RF and mount temperature variation.

The synchronous detector is a solid-state circuit used to minimize common-mode signals and noise entering the metering circuit. Substituted power equivalent to RF power sensed by the thermistors is automatically measured to an accuracy of 3% of full scale.

Temperature-Compensated Thermistor Mounts

Temperature-compensated power meters depend heavily on thermistor mount design and construction for drift-free, accurate measurements. In addition to the general requirements listed for uncompensated bolometer mounts, the compensated mount must: (1) provide as closely as possible identical temperature-resistance characteristics of R_d and R_c; (2) maintain electrical isolation between R_d and R_c; and (3) keep both elements in thermal proximity.

The coaxial thermistor mount contains four identical thermistors electrically connected, as in Fig. 5.1-11(a). The mount is similar to the non-compensated one described earlier in that two 100-ohm thermistors appear in *series* to the 10-kHz bias and bridge excitation signal from the bridge and in *parallel* to applied RF. These two thermistors are the detection thermistors R_d which are mounted within a comparatively massive thermal block, as shown in Fig. 5.1-11(b). The coaxial center conductor from the RF input jack connects to the junction of the two thermistors at point A through a dc and audio-blocking capacitor (not shown).

The compensating thermistors R_c are mounted on small blocks that are completely enclosed in the cavity B when the unit is assembled. With this arrangement, electrical isolation and thermal proximity of R_b and R_c are accomplished with the thermal block massive enough to prevent sudden temperature gradients across the thermistors. Capacitors C2 through C5 are mounted on a circuit board on the back of the thermal block, which is shielded by the metal case. Capacitors C2 through C4 provide the RF bypass

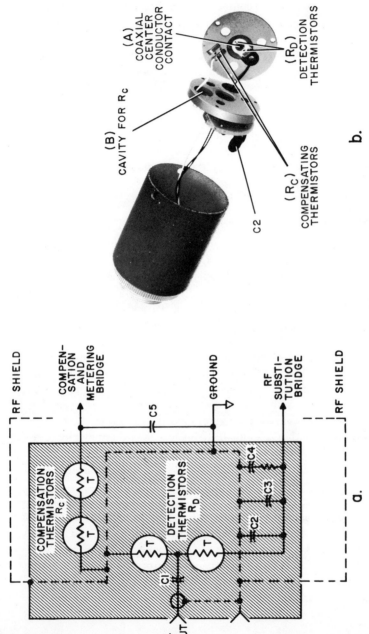

Fig. 5.1-11. Temperature variations around HP 478A coaxial thermistor mount are compensated by (a) aricuit arrangement and (b) construction shown above. Operating with HP 431B power meter, RF power is sensed by dual-element arrangement of detection thermistors R_d. Unwanted temperature variations are sensed by identical compensation thermistors R_c to reduce meter zero-drift by 100:1.

required for a parallel thermistor configuration for RF. Phase shift of the 10-kHz signal in the metering bridge must be canceled to achieve bridge balance. Capacitor C5 is used to swamp the bridge-to-mount connecting cable capacitance so variations in cable capacity will have little effect on 10-kHz phase shift.

Waveguide thermistor mounts usually cover frequencies from 2.6 to 40 GHz. Those mounts in the 2.6- to 18-GHz bands utilize a post-and-bar mounting arrangement for the detection thermistor similar to that described for the noncompensated mounts. The detection thermistor, R_d, is a single-bead thermistor mounted with its axis parallel to and centered in the E field within a shorted section of waveguide. Figure 5.1-12(a) shows the electrical configuration of R_d and the compensation thermistor R_c. Figure 5.1-12(b) shows the post and bar which are thermally isolated from the waveguide structure by a circular section of glass epoxy. Electrical continuity across the epoxy is obtained by a thin gold plating on the epoxy surface inside the guide. This thin plate exchanges a minimum of heat from the waveguide to the bar and thermistor element, isolating them from ambient temperature changes while providing a good electrical path. The detection thermistor is usually further isolated from ambient thermal changes by a block of polystyrene foam inserted into the waveguide opening. This foam prevents convective temperature changes at R_d and also keeps out foreign objects which could enter the waveguide input. The foam has very little effect on the mount SWR and does not change mount efficiency. Compensating thermistor R_c is also a single bead which is thermally strapped to the bar, as shown in Fig. 5.1-12(b). Thermistor R_c and swamping capacitor C1 are shielded by a metal case which also serves to prevent convective thermal changes from reaching the compensating thermistor.

Those mounts covering K- and R-bands (18 to 26.5 GHz and 26.5 to 40 GHz) utilize thermistors that are biased to an operating resistance of 200 ohms rather than the 100 ohms used in lower-frequency units. These mounts differ somewhat from the post-and-bar arrangement but are schematically the same as that shown in Fig. 5.1-12(a).

5.1.3 ACCURACY CONSIDERATIONS OF BOLOMETRIC POWER MEASUREMENTS

A number of factors are responsible for the overall accuracy attained in a microwave power measurement. By knowing the cause and quantitative effects of errors, we can correct or account for them, thereby improving the measurement accuracy. Sources of errors in making power measurements can be categorized as follows:

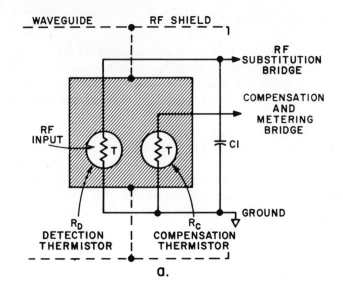

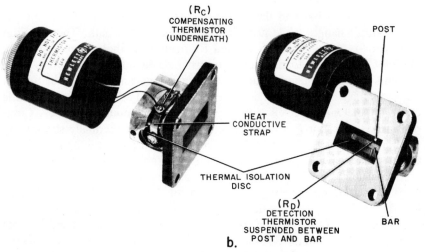

b.

Fig. 5.1-12. HP 486A waveguide thermistor mount (a) circuit and (b) construction show details of detection and temperature-compensation design. Single thermistor R_d is mounted in a post-and-bar arrangement to sense RF power. Post and bar are thermally isolated from the waveguide body by epoxy discs with thin gold plating to maintain electrical continuity. Compensating thermistor R_c senses equal temperature variations as R_d because of a large head conductive strap connected to bar.

1. mismatch errors
2. dc-to-AF-to-microwave substitution errors
3. thermoelectric effect error
4. instrumentation error
5. dual-element bolometer mount error

Each of these errors will now be analyzed.

Mismatch Errors

Consider a dc source with an internal resistance R_G and an external load R_L. Obviously, no power can be delivered to R_L when it is either zero or infinite, so there must be some in-between value at which the power delivered is maximum. It is easily shown that this occurs when R_L equals R_G. Any other value of R_L results in the delivery of less than the maximum available power, or a "mismatch loss."

Consider next the extension to the general case, where the source has an internal impedance which can be represented at any frequency by a resistance and a reactance. Whether series or parallel equivalent circuits are used is immaterial, but the reactances must be of equal magnitude and opposite sign so that they will be in resonance and therefore have no effect on the power delivered. Since the two resistances should be equal, the load impedance should be the complex conjugate of the source impedance to get maximum available power from the source. When the two actual impedances are known, the power delivered can be calculated and compared with the maximum available power to determine the mismatch loss.

At microwave frequencies, a complication arises. The length of transmission line used to connect the load and source can be long enough electrically to transform the load impedance to some other value at the source terminals. What the source "sees" is determined by the actual load impedance (Z_0) of the line. In the optimum situation all elements in a system have the characteristic impedance of the line, and there is a maximum transfer of power. In general, however, neither source nor load has Z_0 impedance. Furthermore, the actual impedances are almost never known completely. They are given only in the form of SWR's, which lack phase information. As a result, the power delivered to the load, and hence the mismatch loss, can be described only as lying somewhere between two limits. This uncertainty increases with SWR, which is one of the fundamental reasons why manufacturers strive to reduce the SWR's of microwave components.

In the special case where either source or load has unity SWR, the mismatch loss is unique and calculable from the other SWR. The accuracy specifications on some commercial power measurement systems are based on the assumption of a Z_0 source. Practically speaking, however, almost no sources have exactly Z_0 impedance, and such an accuracy specification is unrealistic.

To analyze a particular case of mismatch, a convenient basis of calculation must be chosen. On a conjugate basis, the power actually delivered to the load is compared with the maximum available from the source. On a Z_0 basis, comparison is made with the power the source will deliver to a Z_0 load. These are two of a number of possibilities.

Considerable confusion has arisen over the use of terms such as "match" and "mismatch," since it is not always clear just what basis is intended. R. W. Beatty[2,3] of NBS proposed a complete set of specific terms and definitions which would eliminate this confusion if adopted. His terms and definitions pertaining to mismatch analysis in power measurement are as follows:

CONJUGATE MATCH: The condition for maximum power absorption by a load in which the impedance seen looking toward the load, at a point in a transmission line, is the complex conjugate of that seen looking toward the source.

CONJUGATE MISMATCH: The condition in the situation above in which the load impedance is not the conjugate of the source impedance.

CONJUGATE MISMATCH LOSS: The loss resulting from conjugate mismatch.

Z_0 MATCH: The condition in which the impedance seen looking into a transmission line is equal to the characteristic impedance of the line.

Z_0 MISMATCH: The condition in which the impedance seen looking into a transmission line is not equal to the transmission line characteristic impedance Z_0 (in general, conjugate match is a case of Z_0 mismatch).

Z_0 MISMATCH LOSS: The loss resulting from a Z_0 mismatch.

CONJUGATE AVAILABLE POWER: Maximum available power.

Z_0 AVAILABLE POWER: The power a source will deliver to a Z_0 load.

The general expression for power transfer between a source and a load of reflection coefficients Γ_G and Γ_L is:

$$\frac{(1 - |\Gamma_G|^2)(1 - |\Gamma_L|^2)}{|1 - \Gamma_G\Gamma_L|^2}$$

[2] Beatty, R. W., "Intrinsic Attenuation," *IEEE Transactions on Microwave Theory and Techniques*, Vol. MTT-1, No. 3, May 1963, p. 179.

[3] Beatty, R. W., "Insertion Loss Concepts," *Proceedings IEEE*, Vol. 52, No. 6, June 1964, 663.

where $|\Gamma_G|$ and $|\Gamma_L|$ can be obtained from the SWR's σ_G and σ_L by the simple relation:

$$|\Gamma| = \frac{\sigma - 1}{\sigma + 1}$$

The expression is the fraction of the maximum available power actually absorbed by the load. The Z_0 mismatch loss associated with the source is expressed by the term $(1 - |\Gamma_G|^2)$. The Z_0 mismatch loss resulting from the load is expressed by the term $(1 - |\Gamma_L|^2)$. The *uncertainty* in the power transfer is expressed by $(1 - \Gamma_G\Gamma_L)^2$, since Γ_G and Γ_L are complex quantities. The limits of uncertainty are obtained by evaluating $(1 \pm |\Gamma_G|\,|\Gamma_L|)^2$. (It can be seen that the rather vague term "mismatch error" applies in general to a combination of calculable mismatch losses and uncertainties.)

If a conjugate basis is to be used, the effects of all three terms in the expression are included, and the entire expression lies between two limits never exceeding unity. Now consider the case if a Z_0 basis is to be used; since the first term in the numerator gives the fractional power delivered to a Z_0 load by the source, only the remaining two terms need be evaluated to determine the Z_0 mismatch loss. Note, however, that Γ_G must still be known to determine the uncertainty in the loss calculation. This fact must be recognized somehow in any statement of power measurement accuracy. The expression for Z_0 mismatch can have limits above and below unity, as well as both below unity.

When the conjugate basis is used, mismatch is always expressed as a loss. Figures 5.1-13 and 5.1-14 are charts giving conjugate mismatch loss limits for different ranges of SWR. The diagonal lines running upward to the right give the minimum possible loss for any combination of source and load SWR's; the lines running upward to the left give the *maximum* possible loss.

When the Z_0 basis is used, the Z_0 mismatch *loss* is obtained from the bottom scale in Fig. 5.1-15 and the *uncertainty* from the chart above. There is only one set of diagonal lines in Fig. 5.1-15; but note that the upper-left half of the chart gives the upper limit of uncertainty, the lower-right half gives the lower limit, and that these begin to differ appreciably in the upper-right corner.

An example may help clarify the use of these charts in power measurement analysis. Suppose that the power output of a signal generator having an SWR not greater than 1.80 is measured with a power meter and a bolometer having an SWR not greater than 1.35. If it is inconvenient to measure the actual SWR's, these values may be taken as a worst possible case. Figure 5.1-13 (point A) shows a lower *conjugate mismatch loss* limit of -0.090 dB and an upper limit of -0.83 dB, producing an uncertainty range of 0.74 dB. The loss limits may also be found in percentage of the conjugate available power by interchanging source and SWR data to enter Fig. 5.1-13 as shown by point A'. Now suppose the actual SWR's are measured and found to be

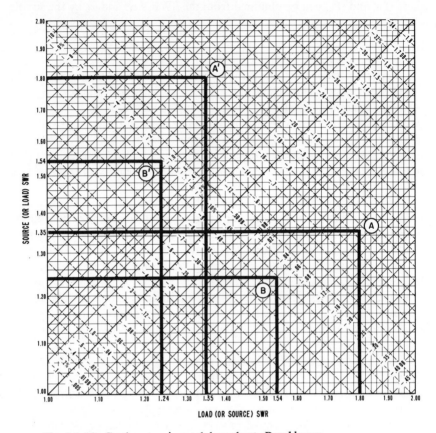

Fig. 5.1-13. Conjugate mismatch loss chart. Read losses
at either intersection of SWRs in question
(load and source SWR scales are interchangeable). Diagonal lines running up
to right indicate minimum loss; diagonal
lines running up to left show maximum
loss. Upper left half of chart shows losses
in percentages, while lower right half shows
losses in decibels.

1.54 and 1.24, respectively. Points B (and B′) in Fig. 5.1-13 now give limits
of -0.050 dB (1.2%) and -0.445 dB (9.8%), an uncertainty range of 0.395
dB.

Signal generators are customarily rated in terms of the power they will
deliver to a Z_0 load ("Z_0 available power"). When testing such generators,
use the Z_0 mismatch loss chart to determine if the generator meets output
power specifications. For example, suppose the generator SWR is 1.54 and

bolometer mount SWR is 1.24; using the bolometer mount SWR of 1.24, we enter the lower scale in Fig. 5.1-15 (point A) and find the Z_0 mismatch loss of the mount is -0.050 dB. There is an uncertainty of $+0.200$, -0.195 dB (point B) on this loss. Adding the loss and uncertainty data results in a total Z_0 mismatch "loss" of $+0.15$ dB to -0.245 dB. Thus the power available to a Z_0 load would be between 0.150 dB below and 0.245 dB above the power absorbed by the bolometer mount.

In case a mismatch loss chart is not available, the maximum and minimum conjugate mismatch losses corresponding to two SWR's, σ_1 (the larger) and σ_2, may readily be determined as follows. The maximum loss corresponds to that which would occur if one SWR were equal to the product

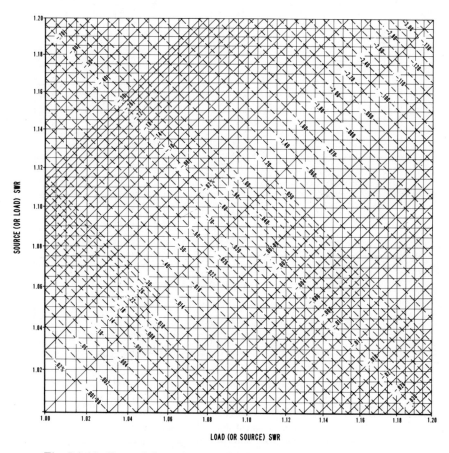

Fig. 5.1-14. Expanded conjugate mismatch loss chart covering SWRs from 1.00 to 1.20. Chart is read in the same manner as Fig. 5.1-13.

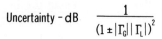

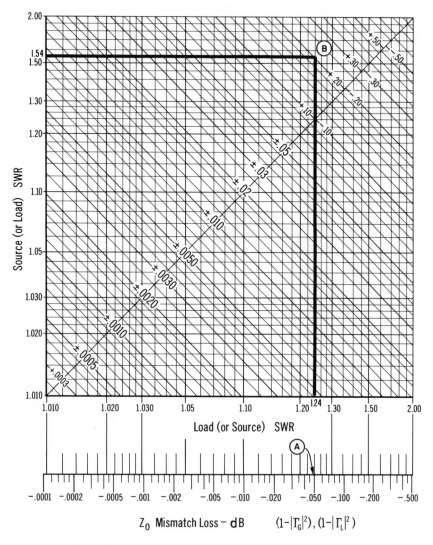

Fig. 5.1-15a. Z_0 mismatch loss and uncertainty chart. Z_0 mismatch loss (in decibels) is found on bottom scale by entering the appropriate load SWR above the scale and reading the loss immediately below. Z_0 mismatch uncertainty (in decibels) is found at the intersection of source and load SWRs in the chart.

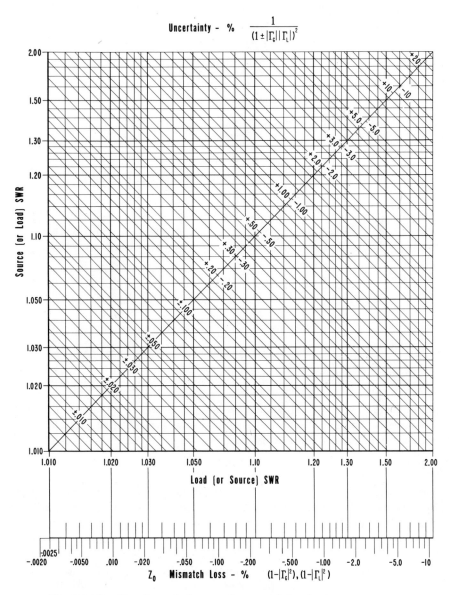

Fig. 5.1-15b. Z_0 mismatch loss and uncertainty chart calibrated in per cent of measured power.

$(\sigma_1 \cdot \sigma_2)$ and the other to unity; the minimum loss corresponds to that which would occur if one SWR were equal to the quotient (σ_1/σ_2) and the other to unity. Using the relation between SWR and the reflection coefficient, the fractional expression for minimum power transfer (maximum loss) is

$$\frac{4\sigma_1\sigma_2}{(\sigma_1\sigma_2 + 1)^2}$$

and the fractional expression for maximum power transfer (minimum loss) is

$$\frac{4\sigma_1\sigma_2}{(\sigma_1 + \sigma_2)^2}$$

The loss in dB is ten times the log of either expression (or, conveniently, its reciprocal, for a positive number of dB).

Another example will clearly show how mismatch errors can really be considered the largest source of error in microwave power measurement if care is not taken to minimize their effects. Figure 5.1-16 shows a simple

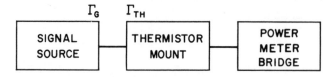

Fig. **5.1-16.** Simple power metering setup.

power measuring setup. Γ_G stands for the generator mismatch and Γ_{TH} for the thermistor mount mismatch. Assuming some typical values for reflection coefficients of such equipment, let:

$$\Gamma_G = 0.33 \text{ max } (\sigma_G = 2.0\!:\!1)$$

$$\Gamma_{TH} = 0.20 \text{ max } (\sigma_{TH} = 1.5\!:\!1)$$

Multiple mismatch errors will become uncertainties if the length of transmission line between the signal source and the thermistor mount is unknown, as is true in most cases. Since the actual values of reflections are also not known exactly, and only their maximum values are given, simple mismatch figures will become only ambiguity limits. For instance, if the power meter bridge measures exactly 1 milliwatt, the question is (disregarding all other sources of errors and taking only into consideration the errors due to mismatches): How much power is actually available in the transmission line? Another question that can be raised is: How much power can be taken out from the signal source? The first question can be answered by the available Z_0 power; the second question can be answered by the available conjugate power. The first case, the available Z_0 power, will not take into consideration the simple mismatch of the signal source $(1 - \Gamma_G)^2$ term; it will include only the simple mismatch of the thermistor mount and the multiple mismatches of the thermistor mount and the signal source. Mathematically,

$$\text{Error} = \frac{1 - \Gamma_{TH}^2}{(1 \pm \Gamma_G \Gamma_{TH})^2}$$

Substituting the actual values of mismatches:

$$\text{Error} = \frac{0.96}{(1 \pm 0.066)^2}$$

In terms of dB: $-0.18 \pm 0.6 = \begin{array}{l} +0.42 \text{ dB} \\ -0.78 \text{ dB} \end{array}$

In percentages: $\begin{array}{l} +10\% \\ -16\% \end{array}$

In terms of power, if one milliwatt was read on the power meter, not taking into consideration any other errors except the mismatch errors, the power actually could be as high as 1.1 milliwatts or as low as 0.84 milliwatt or any value in between. The error can be quite large. If one is interested in finding out, as the second question asks, what the error due to mismatch is if conjugate available power is the interest, both the generator and the thermistor mount simple mismatch losses, and their multiple mismatch errors have to be taken into account. Mathematically,

$$\text{Error} = \frac{(1 - \Gamma_G^2)(1 - \Gamma_{TH}^2)}{(1 \pm \Gamma_G \Gamma_{TH})^2}$$

Substituting the actual values of mismatches,

$$\text{Error} = \frac{(0.89)(0.96)}{(1 \pm 0.066)^2}$$

In terms of dB: $-0.5 - 0.18 \pm 0.6 = \begin{array}{l} -0.08 \text{ dB} \\ -1.28 \text{ dB} \end{array}$

In percentages: $\begin{array}{l} -1\% \\ -25\% \end{array}$

Thus with 1 milliwatt measured, the power can be as high as 0.98 milliwatt and as low as 0.75 milliwatt, not even considering any other sources of errors that could also contribute quite a bit more. This shows clearly that, if due care is not taken, power measurements can be in error by 25% or more.

How can this error be reduced? Figure 5.1-17 shows a setup in which an attenuator with low mismatches is connected between a thermistor mount and a signal generator. The attenuator has to have sufficiently high attenuation to mask the reflections of each device so they will become insignificant. In this particular case, a 15-dB attenuator would be sufficient to lower the mismatches to insignificant values, under 1%. If this is the case, then two

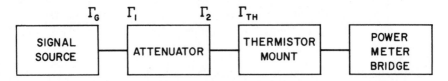

$\Gamma_G \qquad \Gamma_1 \qquad\qquad \Gamma_2 \qquad \Gamma_{TH}$

| SIGNAL SOURCE | ATTENUATOR | THERMISTOR MOUNT | POWER METER BRIDGE |

Fig. 5.1-17. Attenuator (pad) between signal source
and thermistor mount.

sets of multiple mismatch errors have to be taken into account. In our previous examples, the error of ± 0.6 dB would be reduced to a total of 0.12 dB uncertainty. Of course, the attenuator being used has to have a sufficient calibration accuracy so that what one gains by decreasing the uncertainty due to the multiple mismatches is not lost by the uncertainty of the attenuator calibration. Assuming that Γ_1 and Γ_2 absolute values are smaller than 0.05, the multiple mismatch terms become ± 0.13 and ± 0.08—all together ± 0.21 dB, which is much smaller than 0.6 dB. As was mentioned above, the attenuator's calibration must also be quite accurate to make this move worthwhile.

Another way to measure and to eliminate mismatch errors is to place a slidescrew tuner or an EH tuner, which really is a susceptance transformer, into the line to achieve Z_0 match. In the conjugate match case, one has to tune the susceptance transformer to the complex conjugate of the thermistor mount. The thermistor mount has to be tuned with the help of the susceptance transformer to become the complex conjugate of the source mismatch at that point. This can be achieved by tuning the susceptance transformer for maximum power transfer. Here another error comes into the picture—the loss introduced by the susceptance transformer. Three susceptance transformers can be used to calibrate the actual loss of the one susceptance transformer being used. This is quite cumbersome, and many times it is not worthwhile to do unless much accuracy is needed. If the mismatches are small enough to tune out, most susceptance transformers will not introduce enough extra loss to be considered.

RF Losses and dc-to-Microwave Substitution Error

RF losses account for power that enters the thermistor mount but is not dissipated in the detection thermistor and not reflected by mount mismatch. [Such losses may be in the walls of a waveguide mount or in the center conductor of a coaxial mount, in the capacitor's dielectric or in poor connections before the mount, radiation, etc.] From the previous discussion on the operation of bolometer power meters, we know that the power meter can sense only the power dissipated in the detector (thermistor(s) or barretter).

Dc-to-microwave substitution error is caused by the difference in heating effects of the dc or audio bias power substituted in a bolometer and the microwave power being measured. This difference results from the fact that the spatial distributions of current, power, and resistance within the bolometer element are different for dc and RF power (skin effect).

Dc-to-microwave substitution error and RF losses in thermistor mounts are virtually impossible to separate in a quantitative measurement. For this reason, the total effect of these two errors is measured by the National Bureau of Standards (NBS) and some commercial standards laboratories and presented as a figure of merit called the *effective efficiency* (η_e) of a mount. Effective efficiency is defined as the ratio of substituted dc power in the bolometer element to the microwave power dissipated within the bolometer mount. This may be stated symbolically as

$$\eta_e = \frac{P_{(\text{dc substituted})}}{P_{(\text{MW dissipated})}}$$

a ratio less than unity in actual practice,[4] generally expressed in percentage. Note that this expression does not include the effects of mismatch error. Effective efficiency is largely independent of the level of input power; therefore efficiency test results at 10 mw are valid at 10 μw in well-designed mounts.

Another calibration service available from NBS and some commercial standards laboratories is the *calibration factor* (K_b) of a bolometer mount. Calibration factor is defined as the ratio of the substituted dc power in the bolometer element to the microwave power incident upon the bolometer mount. This may be stated symbolically as

$$K_b = \frac{P_{(\text{dc substituted})}}{P_{(\text{MW incident})}}$$

Note that this expression *includes the effects of mismatch loss*, since $P_{(\text{dc substituted})}$ decreases when RF power is reflected. If a tuner is not used to cancel mismatch effects in a power measurement, the calibration factor can be directly applied as a correction to the meter reading to improve overall accuracy. Since a tuner is not often used, K_b is a highly practical term and is therefore measured and included with the efficiency data on temperature-compensated thermistor mounts.

Thermoelectric Effect Error

Thermoelectric error is peculiar to the temperature-compensated power meter. This error is found in power meters where nearly the total bias power

[4] Dc-microwave substitution error could be positive or negative, so in theory η_e could exceed unity. In actual practice, however, RF losses are always large enough to cause the combined effect to be less than unity.

is supplied by ac. Although this thermoelectric phenomenon has not been fully investigated, it is felt that it can be explained on the basis of the two thermocouples that are formed by the contacts of the thermistor leads to the thermistor oxide. The bead thermistor used in thermistor mounts operates more than 100°C above room temperature. It is likely that both contacts to the bead will not be at the same temperature; thus a thermocouple will result. The ac bias power heats the thermistor, and, if there is not perfect cancellation of these thermocouples, the result will be a dc current flow. This dc current flow will either add to or partially cancel any dc currents that are being applied by the power meter. This error is significant only on the more sensitive ranges of the power meter. However, this effect also becomes significant in the RF thermistor when a dc calibration or dc substitution measurement is being made and can amount to a 0.3-μw error. Where accurate substitution or calibration is required at low power levels, this thermoelectric effect can be eliminated by using a dc supply whose polarity is easily reversed to supply the substitution or calibration power to the RF thermistor. By reversing polarity, the sum and difference of substitution power and thermoelectric power can be determined.

Instrumentation Error

Instrumentation error is the inability of the power meter to measure exactly the dc or ac substituted power in the bolometer element. This type of error is somewhat analogous to the error found in VTVM's. Examples of the causes of instrumentation error are: (1) range resistor tolerance, (2) non-linearity in feedback amplifiers, (3) matching of bridge transformers, and (4) meter tracking error. In specifying the accuracy of a power meter, instrumentation error does not exceed 3% of full scale in most temperature-compensated power bridges.

Dual-Element Bolometer Mount Error

A detailed analysis of dual-element error has been made by G. F. Engen[5] of NBS and by I. A. Harris.[6] This error does not occur in the waveguide thermistor mounts but must be considered in nearly all coaxial mount designs.

Figure 5.1-18(a) shows the basic circuit of a coaxial mount. R_{T1} and R_{T2} are the detection thermistors. Figure 5.1-18(b) shows the equivalent cir-

[5] Engen, G. F., "A DC-RF Substitution Error in Dual Element Bolometer Mounts," *NBS Report* 7934.

[6] Harris, I. A., "A Coaxial Film Bolometer for the Measurement of Power in the UHF Band," *Proc. IEE* (British), Vol. 107, Part B, No. 31, January 1960.

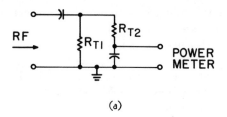

(a)

$$P_{BIAS} = \left[(I_{BIAS})^2 R_{TI}\right] + \left[(I_{BIAS})^2 R_{T2}\right]$$

$$P_{RF} = \left[\frac{(E_{RF})^2}{R_{TI}}\right] + \left[\frac{(E_{RF})^2}{R_{T2}}\right]$$

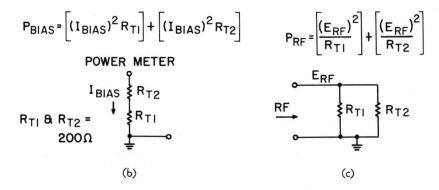

(b)

(c)

Fig. 5.1-18. Actual circuit of dual-element coaxial ther-mistor mount is shown in (a). The dc and low-frequency equivalent circuit (b) ap-pears as two series thermistors to power meter bridge. The equivalent circuit for RF input power (c) appears as two parallel thermistors. If resistance of thermistors is unequal, dc and RF power will have unequal effects on the circuit, causing "dual-element error."

cuit for dc and 10 kHz. The thermistors are in series, and, if the resistance division is unequal, the greatest power will be dissipated in the thermistor with the largest resistance. However [refer to the equivalent circuit for RF in Fig. 5.1-18(c)], the RF power sees the two thermistors in parallel. In this case, if the resistance division is unequal, the greatest power will be dissipated in the thermistor with the least resistance. Unfortunately, the situation at microwave frequencies is not quite this simple inasmuch as circuit reactance may cause a different power split. In any case, Engen has shown that the resulting error is equal to

$$E = \left(\frac{1}{\gamma_2} - \frac{1}{\gamma_1}\right) \Delta r$$

where γ_1 and γ_2 are the "ohms-per-milliwatt" coefficients of the thermistors and Δr is the shift in resistance division.

Dual-element error may be as large as 1% for conventional thermistor mounts at 10-mw RF levels, increasing sharply as power increases. Conversely, as the measured RF power decreases, this error decreases rather rapidly. This, of course, represents one of the problems involved with dual-element error, inasmuch as it is not a constant error but rather is a function of the RF power being measured. Since automatic bridges have 10-mw upper-range limits, this error is generally quite small when using mounts of good quality.

5.1.4 CALORIMETRIC POWER MEASUREMENT

The most fundamental method of measuring microwave power is to dissipate it as heat and measure the resulting temperature rise. True calorimetry involves dissipating a certain amount of energy, containing the resulting heat, and measuring the temperature rise. This principle has been used for years in chemical and physical measurements. The calorimetric power meter dissipates power (a continuous flow of energy), controls the resulting flow of heat through some thermal path, and measures the temperature rise set up by this flow.

Calorimetric power meters used for microwave work may be divided into two categories: (1) flow (liquid) and (2) static (dry). Although there are several designs of both types, an example of each will acquaint the reader with the principles involved. The dry calorimetric power meter was an outgrowth of the basic calorimeter and is still used in some applications.

Flow Calorimeters

Figure 5.1-19(a) shows the basic circuitry of a 10-watt calorimetric power meter. The circuit consists of a self-balancing bridge which has identical temperature-sensitive resistor gauges (one in each leg), a high-gain amplifier system, an indicating meter, and two load resistors, one for unknown input power and one for comparison power. The input load resistor and one gauge are in thermal proximity so that heat generated in the input load resistor is carried to its gauge by the oil stream, tending to unbalance the bridge. The unbalance signal is amplified and applied to the comparison load resistor that is in thermal proximity to the other gauge. The heat generated in the comparison load resistor is carried to its gauge by the oil stream and automatically rebalances the bridge. Power supplied to the comparison load to rebalance the bridge is equivalent to the RF power applied to the input load

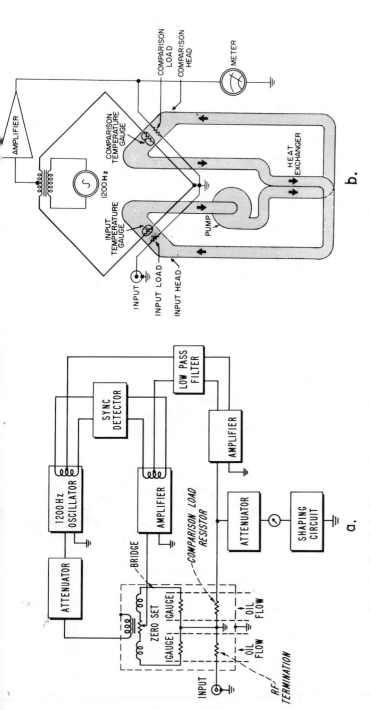

Fig. 5.1-19. Block diagram (a) and oil flow diagram (b) of HP 434A calorimetric power meter. Moving oil stream and bridge-feedback arrangement result in fast response time to input power changes, enabling dynamic tuning of units in test. Series oil flow arrangement eliminates flow rate errors usually encountered in flow calorimeters.

and is measured by the meter. Because of this feedback arrangement, total response time is less than 5 seconds for full-scale deflection of the meter.

Resistance characteristics of the gauges are the same, as are the heat-transfer characteristics of the loads, so the meter may be calibrated directly in input power. The calorimetric power meter measurement is accurate because volumetric oil flow through the two heads is made identical by placing all elements of the oil system in series, as shown in Fig. 5.1-19(b). Differential temperature between oil entering the two heads is also zero because the oil to each head is brought to the same temperature by passing it through a parallel-flow heat exchanger. Microwave power dissipated in the RF head resistor is matched by dc power in the comparison head to give a meter reading.

CALORIMETRIC POWER METER ACCURACY. The calorimetric power meter is inherently accurate due to its circuit stability, controlled thermal flow, and rigid mechanical design. Overall measurement accuracy is better than 5% of full scale, excluding mismatch loss. Accuracy can be made even higher by being aware of, and correcting for, various sources of error such as mismatch, head efficiency, and instrumentation error.

MISMATCH ERROR. As in the case of the bolometric power meter, mismatch error is a very important factor in making accurate calorimetric power meter measurements. The same laws of impedance and phase apply to the calorimetric head as to bolometer mounts. When neither source nor load is Z_0, there is an ambiguity as to how much power is being delivered. On a conjugate basis, there is a mismatch loss when the calorimetric input impedance is not the conjugate of the source.

CALORIMETRIC HEAD EFFICIENCY. The calorimetric power meter does not depend upon the validity of the substitution process in the comparison of microwave power and dc or low-frequency power. This is because all power delivered to the terminating resistor raises the temperature of the oil stream, and skin effect or thermal distribution within the sensor is not of primary importance, as it is in the bolometer mount.

Efficiency in a calorimeter head is defined as the microwave power dissipated in the head resistor divided by the microwave power dissipated in the entire RF circuit. Calibration factor is the microwave power in the head resistor divided by the microwave power incident upon the RF input circuit.

It may be seen that these definitions are essentially equivalent to those for bolometers except that the sensors are different. Loss in the RF circuit occurs in the transmission line between the front panel connector and the head load resistor that heats the oil stream. Typical efficiencies are between 97% and 99% in the X-band region and approach 100% at lower frequencies down to dc.

INSTRUMENTATION ERROR. Causes of instrumentation error in the calorimeter are largely the same as in bolometer power meters in that measurement of the feedback power cannot be exact. Instrumentation error accounts for 1%–2% in the overall specification of 5%. The calibration accuracy (which includes head efficiency) is usually more accurate than 3% to 10 GHz and is within 4% through 12.4 GHz.

Dry Calorimeters

The basic dry calorimeter measures the thermal EMF generated in a thermopile placed between a reference load and an active load dissipating input power. There are a number of dry calorimeter designs, one of which is of particular interest. This is the microcalorimeter, described by G. F. Engen[7] of NBS, which is the U.S. national reference standard for effective efficiency measurements at microwave frequencies. NBS Working Standard bolometer mounts are calibrated in the microcalorimeter and then used to calibrate suitable bolometer mounts submitted for efficiency and calibration factor measurements.

Microcalorimeter Design

Figure 5.1-20 is a cutaway view of the microcalorimeter showing a double-walled thermal shield housing two identical bolometer mounts of special design. RF entry is made through a length of waveguide thermally isolated from the bolometer mounts to prevent conduction of heat away from the mounts. Applied RF power passes through a waveguide bend and is dissipated in the lower (active) bolometer mount. A physical junction is made at the waveguide bend with an isometric bend connecting to the upper (reference) bolometer mount. RF is prevented from entering the upper bend by a partition at the guide junction; thus the reference mount remains constant in temperature with power applied to the entry point. This arrangement provides equal mass in the two arms so they are isothermal with no power applied and remain so with ambient temperature variation. A number of series-connected thermojunctions are made alternately between the waveguide flanges facing each bolometer mount. This thermopile senses the temperature gradient between the bolometer mounts that results from the application of power to the system. The thermopile output can be accurately calibrated by applying known amounts of dc power to the active bolometer mount.

[7] Engen, G. F., "A Refined X-Band Microwave Microcalorimeter," *Journal of Research of the National Bureau of Standards*, 63C:77 (1959).

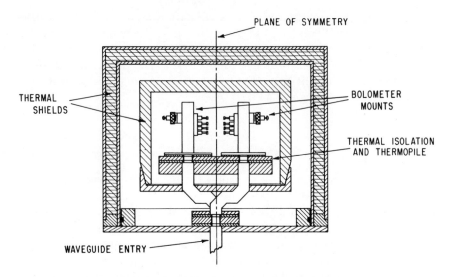

Fig. 5.1-20. Cross section of NBS microcalorimeter with special bolometer mounts installed. Efficiency calibrations in the 8.2- to 18.0-GHz region by NBS based upon instruments of this design.

5.1.5 THERMOCOUPLE POWER METERS

The use of high-frequency thermocouples for measuring power has the same advantage as diodes in that only a simple dc millivolt meter is needed to indicate the detector output. Most high-frequency thermocouples are of the direct-heating variety that have a resistive wire-heating element situated between two thermocouple wires forming a series circuit. As RF current is passed through the circuit, power is dissipated in the heating element, causing a thermal voltage to be developed by the thermocouple wire. The millivolt output may be calibrated by applying known low-frequency power levels to the heater element.

A number of disadvantages have precluded significant use of this technique at microwave frequencies. One is the problem of impedance-matching the thermocouple to the transmission line. As in the unbalanced bolometer bridge, element resistance varies with applied power and results in a mismatch error partially dependent on power level. This makes accurate measurements difficult over any appreciable dynamic range. Another problem is that the thermocouple heater must have a very small diameter to minimize skin effects so that the output is not frequency sensitive. With small heater diameter, the element operates very near burnout and will not withstand even short duration or transient overloads of more than about 50%.

The use of thin-film techniques is a recent approach to microwave power measurements with thermocouples. This has led to a power meter, comparable in range to bolometric meters, that uses a number of thin-film thermoelectric heads for various power ranges in certain frequency bands. Each head uses a thin-film metallic load which absorbs the input power. A number of thermojunctions are formed by thin films of bismuth and antimony which are heated by the metallic load, causing a thermal EMF to be developed. The output is then applied to a dc amplifier and meter circuit to indicate power. Unlike a bolometer bridge, the system is an open loop and subject to undetected calibration changes caused by aging or usage of the thermoelectric heads.

With the advent of integrated circuit technology, the thermocouple type of power measurement is gaining increased attention. Furthermore, certain prebiasing techniques now being developed promise new approaches for more sensitivity and probably higher accuracy in the future for average power measurements.

5.1.6 PEAK POWER MEASUREMENT

Measurement of peak pulse power has been a frequent requirement in microwave work since the early development of pulse radar. Various approaches to peak pulse power measurement include the following techniques: (1) average power-duty cycle, (2) notch wattmeter, (3) direct pulse, (4) dc-pulse power comparison, and (5) barretter integration-differentiation.

The first three methods simply involve the interconnection and use of a number of standard instruments, such as crystal detectors, bolometers, and oscilloscopes. For this reason we prefer to describe them as "techniques" of peak power measurement.

Average Power-Duty Cycle Method Number 1

Peak power of a rectangular RF pulse envelope may be determined by measuring average power, pulse width, and repetition frequency. The product of pulse width and repetition frequency is defined as the duty cycle of the pulsed source and relates the time the pulse is on to the period of the pulse rate. Figure 5.1-21 shows a setup for determining peak power in a waveguide system using the average power-duty cycle method. It is important that the maximum peak power and maximum energy-per-pulse ratings of the thermistor mount not be exceeded. At pulse-repetition frequencies *less than* 1 *kHz*, the maximum energy per pulse is 2.5 *watt-microseconds*, i.e., any com-

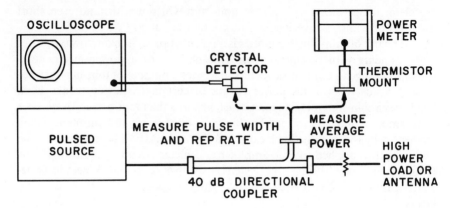

Fig. 5.1-21. On-line test of peak power sources may be
conveniently made using system shown.
Power meter reads average power, which
is then divided by system duty cycle and
power ratio of coupler.

bination of peak power and pulse width which when multiplied together produces 2.5 watt-microseconds. For pulse-repetition frequencies *above* 1 *kHz*,
the maximum is 5 *watt-microseconds*. The maximum peak power in *any* case
should not exceed 100 watts at the thermistor mount. Another highly important factor in peak power systems is that all waveguide flanges must be
very carefully aligned and the waveguide must be clean and dry to avoid
arc-over.

The procedure for the test (setup shown in Fig. 5.1-21) is as follows:

1. Connect the crystal detector and its square-law load to the auxiliary
   arm of the directional coupler as shown.
2. Measure and note pulse-repetition frequency (PRF) and pulse width
   (τ) using the oscilloscope's calibrated sweep.
3. Replace the detector with the thermistor mount and power meter.
4. Note the average power reading, correcting for coupling factor, thermistor mount calibration factor, and mismatch uncertainty between
   mount and the coupler auxiliary arm.
5. Calculate peak pulse power using the following equation:

$$P_{pk} = \frac{P_{average}}{(PRF \times \tau)\left[\log_{10}^{-1}\left(-\frac{dB}{10}\right)\right]}$$

where P_{pk} = peak power of source,
$\quad P_{average}$ = average power indicated on power meter, and
$\quad\quad$ dB = coupling factor of directional coupler.

Example.

Suppose the following readings are made from a pulsed source with the setup shown in Fig. 5-1-21:

$$PRF = 1\ kHz$$

$$\tau = 1.5 \times 10^{-6}\ seconds$$

$$P_{average} = 7.5 \times 10^{-3}\ watts$$

$$coupling\ factor = 40\ dB$$

Peak pulse power is calculated as follows:

$$P_{pk} = \frac{7.5 \times 10^{-3}\ watts}{(1 \times 10^3)(1.5 \times 10^{-6})\left[\log_{10}^{-1}\left(\frac{-40}{10}\right)\right]} = 50\ kw$$

Sources of error and examples of the resulting uncertainties are summarized in the table in Fig. 5.1-21. Since mismatch cannot be corrected with a tuner in pulse power measurements, it is important to calculate the possible limits of mismatch loss using the coupler auxiliary arm and thermistor mount SWR,[8] as indicated by step 4 of the foregoing procedure. The accuracy of this method of pulse power measurement depends upon the pulse being rectangular. As the RF pulse envelope departs from a true rectangle, error increases rapidly. A better method for peak power measurements of non-rectangular pulses is described under "Direct Pulse Power Technique," which does not depend on pulse shape. Peak power range of the system shown is about 100 kw maximum at X-band, limited by the maximum power ratings of the termination, thermistor mount, detector, and coupler. This power range decreases at waveguide sizes smaller than WR-90.

Average Power-Duty Cycle Method Number 2

One method of improving overall accuracy is to improve the method of average power measurement and use the duty cycle calculation for peak power. A more accurate average power measurement is possible in situations in which all the power may be terminated and measured in a water load, thus avoiding directional coupler calibration error. The system shown in Fig. 5.1-22 is suitable for permanent laboratory installations that will accommodate the required calorimeter and water circulation equipment. Using the system shown in Fig. 5.1-22, average power is calculated from the temperature difference of water entering and leaving the load, flow rate, and specific

---

[8] Thermistor mount ρ may be calculated from efficiency and calibration factor data included with new thermistor mounts by using the relations

$$\rho = 1 - K_{b}/\eta_{e} \quad and \quad SWR = \frac{1 + \rho}{1 - \rho}$$

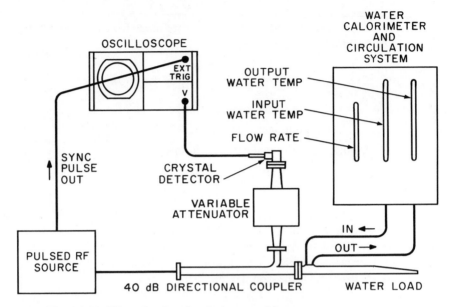

Fig. 5.1-22. Water load and calorimeter of best accuracy in high power measurements (over 10 watts average) largely because directional coupler uncertainty is not involved. Coupler shown merely samples pulsed power for determination of PRF and pulse width on oscilloscope. System is useful when all the source power can be terminated with the water load during the measurement, and water circulation equipment can be readily set up.

heat of the water. Duty cycle is determined from the oscilloscope measurements of pulse width and repetition rate.

The calorimeter system shown in Fig. 5.1-22, if a well-matched water-load is used, can have a total possible uncertainty of $\pm 7\%$. Common-water-load systems handle average power levels from 5 kw to 30 kw in X-band, with capabilities increasing at the lower microwave bands to about 100 kw. Larger systems utilizing power splitters handle average power in the megawatt region.

Direct Pulse Power Technique

An improved waveguide setup for peak power measurement of non-rectangular pulses or multipulse systems is shown in Fig. 5.1-23. In this

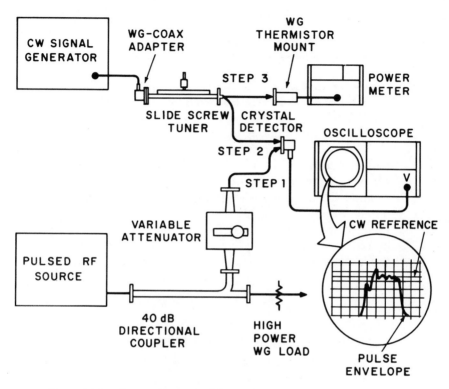

Fig. 5.1-23. Direct pulse method for peak power measurement in waveguide system. This tech-
nique is useful for measuring peak powers
of nonrectangular pulse envelopes.

arrangement the pulse envelope is first detected by a fast-response crystal
detector. The detected pulse amplitude is noted on a dc-20 MHz oscilloscope
and then accurately duplicated with a known CW power. The directional
coupler and attenuator should limit the power into the detector to 1 to 2 mw
peak, since this is about the maximum CW power available for comparison.
The procedure is as follows:

1. Connect the equipment as shown in Fig. 5.1-23, step 1. Set the
   attenuator to a convenient dB reference that will limit the expected
   peak power to the detector to about 1 mw.
2. Adjust the oscilloscope trigger and vertical sensitivity controls for a
   stable pulse display. *Use dc coupling* at the oscilloscope vertical input, and make no further adjustments to the oscilloscope sensitivity
   once the reference pulse amplitude is set.

3. Move the detector/square-law load combination to the output of the slidescrew tuner and signal generator, as shown in step 2.

4. Adjust the signal generator for a CW output frequency equal to the pulsed source's RF frequency.

5. Adjust the slidescrew tuner for maximum deflection of the oscilloscope trace. *DO NOT ADJUST THE OSCILLOSCOPE VERTICAL SENSITIVITY.* Use the generator's output attenuator to keep the trace on the screen during the tuner adjustment.

6. Carefully set the generator output attenuator for a trace deflection on the oscilloscope equal to the pulse amplitude noted in step 1. Make no further adjustments to the output attenuator.

7. Connect the thermistor mount and power meter to the generator output, as shown in step 3.

8. Readjust the slidescrew tuner for a maximum reading and note the average power, correcting for mount efficiency and tuner loss.

9. Calculate peak power at the source output from the equation

$$P_{pk} \text{ (source)} = \frac{P_{average}}{\left[\log_{10}^{-1} \left(-\frac{dB}{10} \right) \right]}$$

where dB = coupling factor of directional coupler in dB, and
$P_{average}$ = average power of generator measured on the power meter.

Example.
Suppose $P_{average}$ is 0.5 mw and the combined attenuation of the directional coupler and attenuator used is 80 dB at the test frequency. Peak power is then:

$$P_{pk} \text{ (source)} = \frac{5 \times 10^{-4} \text{ watts}}{\left[\log_{10}^{-1} \left(-\frac{80}{10} \right) \right]}$$

The chart in Fig. 5.1-23 shows the sources of error and examples of the resulting uncertainties using the direct pulse technique. Comparing the uncertainties of Figs. 5.1-21 and 5.1-23 for equal values of peak power indicates the direct pulse technique to be less desirable. Greater ambiguity exists because of the greater decoupling required between source and detector. However, the advantages of the system in Fig. 5.1-21 are quickly lost if the pulse envelope departs considerably from a rectangular shape or if multiple pulses are being measured.

In addition to the usual limitations of peak and average power ratings of the components, this system must provide a CW power level equal to the peak power detected at the variable-attenuator output. Using the generators shown, this power level is on the order of 1 to 2 mw. At X-band, the system shown would be limited to about 100 kw peak power and 500 w average with convection cooling of the load.

Notch Wattmeter

The notch wattmeter gets its name from a technique that involves gating a reference CW power source off during the interval that the pulsed source in test is on. Thus the reference output is "notched" at the same rate, duration, and amplitude as the pulse output of the device in test. The resulting CW power is then measured by conventional means. This system can be useful in situations where the average-to-peak-power ratio of a pulse is too great to be measured directly by a power meter. An example of this would be a radar of very low-duty cycle.

Figure 5.1-24 shows a notch wattmeter setup. The technique requires that the pulsed source in test have a sync pulse output of sufficient amplitude and polarity to operate an external pulse generator.

The measurement procedure is as follows:

1. Connect the equipment as shown, presetting the pulse generator *pulse amplitude* to about 50 volts positive when operating with the signal generators.
2. Null and zero the power meter and turn on the pulsed source in test.
3. Adjust the signal generator to the pulsed source's RF frequency.
4. Set the pulse generator as follows:
   trigger mode to *ext,*
   slope for same polarity as sync pulse out of source in test,
   int. rep. rate to approximately the same frequency as the source PRF,
   pulse width to approximately that of the pulsed source,
   pulse position to 0–1 *μs, pulse delay.*
5. With the generator output attenuator set for minimum output power, adjust the oscilloscope vertical sensitivity, sweep time, and trigger controls for a stable display of the source's detected output pulse. Use dc coupling at the oscilloscope vertical input.
6. Now increase the generator output until the oscilloscope base line rises to an amplitude equal to the detector pulse top.
7. Adjust the *int. rep. rate, pulse width,* and *pulse position* as required to obtain a straight and nearly unbroken trace on the oscilloscope. The signal generator output is now being pulsed *off* for the precise interval that the source is pulsed on. The generator's output amplitude has also been made equal to the peak pulse power at the output of the 70-dB directional coupler, resulting in a CW power which is easily measured with the power meter.

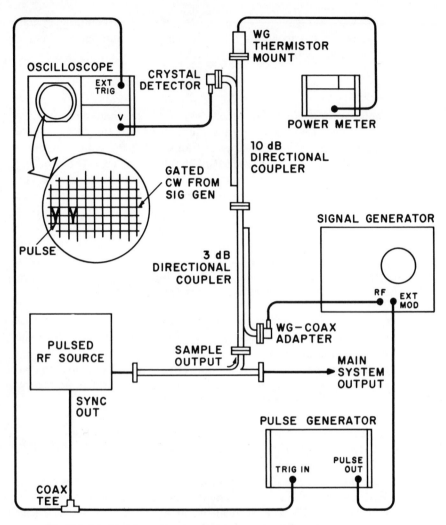

Fig. 5.1-24. Notch wattmeter requires more elaborate
setup than other methods but is useful with
very low duty cycles and nonrectangular
pulse shapes.

8. Read the power level indicated by the power meter and calculate
peak power at the pulsed source output by the following equation:

$$P_{pk} = \frac{P_1}{\left[\log_{10}^{-1} \left(-\frac{dB}{10} \right) \right]}$$

where P_1 = power read on the power meter,

dB = total attenuation of directional couplers between pulsed source and thermistor mount.

Example.

Suppose P_1 is 1 mw and the total attenuation between source and thermistor mount is 73 dB at the test frequency. Peak power out of the source may then be calculated as follows:

$$P_{pk} = \frac{1 \times 10^{-3} \text{ watts}}{\left[\log_{10}^{-1}\left(\frac{-73}{10}\right) \right]} = 20 \text{ kw}$$

Overall uncertainty of the notch wattmeter depends on factors listed in the error analysis of Fig. 5.1-24. The values shown are examples of X-band measurements which result in an overall uncertainty of about 28%. The notch wattmeter does not depend upon pulse shape for its accuracy if the duty cycle is low. The waveform shown in Fig. 5.1-24 illustrates how a pulse of poor rise and decay time might appear in the notch display. At low duty cycles, the "dead" time between the generator output and pulsed source power is so small that the thermistor averages out the error. Conventional barretters are not satisfactory for this application because their comparatively fast thermal time constant tends to follow the instantaneous power variation, introducing pulse-shape error into the measurement.

QUESTIONS

1. What are the two kinds of power measurements?
2. What is the value of the peak power delivered into a load having a resistance of 200 ohms and a peak voltage of 20 volts?
3. What are the two kinds of bolometers? What is the basic difference between them?
4. In what kind of circuit arrangement is a bolometer power-measuring device connected?
5. What kind of accuracy can be achieved with a simple, automatically balanced bridge power meter?
6. What are the four basic requirements in the construction of a bolometer mount?
7. What parameter is taken into consideration and corrected for when using temperature-compensated power meters? Why?
8. What is the mismatch in terms of dB for Z_0 available power if the source has a VSWR of 1.6:1 and the thermistor has a VSWR of 1.35:1?
9. What are the units of the actual power absorbed by the thermistor mount

if the source is capable of delivering 5 milliwatts into a Z_0 line and the source VSWR is 2.64:1 and the thermistor VSWR is 1.5:1?

10. What does calibration factor include?

11. How can thermoelectric effect error be eliminated?

12. What are the contributing factors to instrumentation error in power bridges?

13. What is the basic sensing element of a flow calorimeter?

14. How is power measured in a calorimeter?

15. What are the sources of errors in calorimetric power meters?

16. What is the basic power-sensing element in a microcalorimeter?

17. What is the basic difference between conventional bolometers and thermocouples as power-measuring devices?

18. What are the five different approaches to making peak pulse power measurements?

19. How many parameters must be known to measure rectangular RF peak pulse power? What are they?

20. Calculate the peak pulse power using average power-duty cycle method number 1. PRF = 2 kHz; average power measured with the power meter = 3 mw; a 20-dB directional coupler is used; and measured pulse width is 2 μs.

21. How does a notch wattmeter work?

5.2 FREQUENCY MEASUREMENTS

In signal analysis the second basic type of information about a signal is the frequency. Frequency measurement can be done with the greatest accuracy, a fact which can be contributed to the available standards developed. Since frequency and time do relate to each other, $F = 1/T$. These types of measurements have quite a history. Accurate clocks have been made for quite some time. Accurate frequency control became of utmost importance by the ever-increasing number of radio-transmitting stations.

Even with the simplest techniques, in microwaves, with a wavelength, measurement information about frequency can be obtained in the order of 1%

Table 5.2-1.

Signal generator dial accuracy	<1%
Coaxial wavemeters (over octave BW)	<0.2%
Waveguide type TE_{111} wavemeters (WG bandwidth, $\frac{1}{2}$ octave)	<0.1%
Echo boxes TE_{01N} types (10% BW)	<0.005%
Heterodyne techniques (transfer oscillators)	< 10^{-7}

accuracy, since $F = v/\lambda$, where $v = $ the velocity of propagation in the transmission line.

Highly accurate frequency measurements can be achieved with electronic counters. In fact, with present-day atomic frequency standards, accuracies in the order of 10^{-12} are possible. Table 5.2-1 lists some frequency measurements with these accuracies.

There are two basic techniques for obtaining frequency information. One is electromechanical and measures wavelength; the other is electronic, a frequency-beating technique in which a known frequency or its harmonic is compared with the unknown through a "zero-beating" procedure.

5.2.1 ELECTROMECHANICAL MEASUREMENT TECHNIQUES

Electromechanical devices commonly use such elements as resonant cavities and slotted lines, both of which depend upon physical dimensions for their operation and accuracy. Resonant devices range all the way from lumped constant L-C circuits calibrated to be resonant at various frequencies in the band to microwave cavities that are also resonant at specific frequencies in their band.

Resonant Circuits

In general, resonant devices are coupled into the circuit under test as the frequency passes through the resonant points, and a reaction is evident in the circuit. A typical parallel resonant circuit is shown in Fig. 5.2-1. The

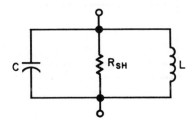

Fig. 5.2-1. Parallel resonant circuit.

resonant frequency of this parallel resonant circuit is expressed by the Thompson formula:

$$F_0 = \frac{1}{2\pi\sqrt{LC}} \qquad (5.2\text{-}1)$$

An ideal parallel resonant circuit without the shunt resistance, at resonance, will provide infinite impedance. The input admittance at the frequency different from the resonant frequency is given in Eq. (5.2-2).

$$Y_{\text{input}} = \frac{1}{R_{\text{SH}}} + j \left(\frac{F}{F_0} - \frac{F_0}{F} \right) \sqrt{\frac{C}{L}} \qquad (5.2\text{-}2)$$

where F_0 = resonant frequency,
F = frequency.

It can be clearly seen from this equation that the admittance at resonance will be equal to the reciprocal of shunt resistance. The parallel resonant circuit Q is defined as

$$Q = \frac{R_{\text{SH}}}{2\pi F L} \qquad (5.2\text{-}3)$$

Quarter-Wave and Half-Wave Resonators

Lumped-circuit elements cannot be used at microwave frequencies because their physical size would be comparable with an appreciable portion of the operating wavelength. Therefore resonant circuits for these frequencies are made of sections of coaxial lines or waveguides that are resonant at known frequencies. A transmission line can be made resonant by terminating both ends in extreme impedances—either with a short or an open or both. Two kinds of such resonant lines can be imagined: the quarter-wave resonator in which one end is terminated in a short and the other in an open (Fig. 5.2-2); and the half-wave resonator in which both ends are either opened or shorted

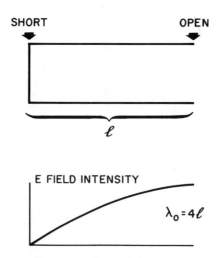

Fig. 5.2-2. Quarter-wave resonator.

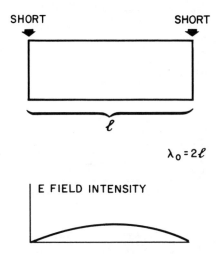

SHORT   SHORT

ℓ

$\lambda_0 = 2\ell$

E FIELD INTENSITY

Fig. 5.2-3. Half-wave resonator.

(Fig. 5.2-3). The resonant wavelength λ_0 of the quarter-wavelength resonators is equal to four times the length of the line L. The resonant wavelength λ_0 of the half-wave resonator is equal to twice the length of resonator L. If a transmission line in which the actual length of the resonant cavity (L) can be varied, it becomes a variable-tuned circuit and can be used for frequency measurements.

Let us consider the usable bandwidth of the quarter-wave and the half-wave resonator. "Usable bandwidth" means the width of the frequency band where no spurious resonances appear other than the main resonance (the quarter-wave or the half-wave resonance). What other resonance can appear on a quarter-wave resonator? The boundary conditions for carrying a resonance on a quarter-wave cavity are that a very low and a very high impedance end have to terminate the transmission line. Resonances can occur on a quarter-wave resonator at each odd quarter-wavelength, the 3-quarter, 5-quarter, 7-quarter, and so forth. This fact shows that a quarter-wave resonator is able to maintain a 3:1 spurious-resonance-free bandwidth. The half-wave resonator can resonate every multiple of half-wavelengths, such as half, full, $1\frac{1}{2}$, 2, $2\frac{1}{2}$, and so on. This means that a half-wave resonator can carry only a 2:1 bandwidth without any spurious resonances.

Coaxial Transmission Line Resonators

The most commonly used microwave frequency meter is built of a co-axial transmission line. Figure 5.2-4 shows a quarter-wave and a half-wave resonator basic cross section. To vary the resonant frequency of a quarter-

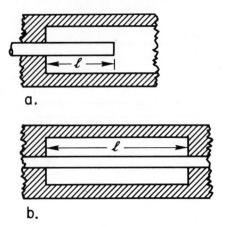

Fig. 5.2-4. Coaxial resonant cavity: (a) quarter wave
(b) half wave.

wavelength cavity, the length of the center conductor has to be varied. The resonant frequency of a half-wavelength cavity can be varied by moving one of the end plates of the cavity. This could be quite troublesome, since the outer and center conductor have to be kept shorted, whereas only the center conductor in the quarter-wavelength cavity has to make good contact with the end plate. The mechanical problems of keeping good contact and providing very low contact resistance with the moving parts can be clearly seen. Furthermore, as was seen above, a quarter-wave resonator has by itself a wide spurious-free bandwidth. Theoretically, a half-wavelength coaxial cavity should have a higher Q, since

$$Q = \frac{R_{\mathrm{SH}}}{Z_0}$$

But possible problems due to contact resistance usually make it easier to design and to gain higher Q on a quarter-wave-type cavity in coaxial resonators, to make a broader bandwidth wavemeter without spurious resonators.

To make a broader bandwidth[9] wavemeter without spurious resonances, a loading capacitance has to be added to the open circuit end of a quarter-wavelength cavity, as shown in Fig. 5.2-5. At the high-frequency end of the band this type of cavity appears to be a pure quarter-wave cavity, but as it approaches the low-frequency end, the step in the outer conductor acts as a capacitance and the cavity becomes a quarter-wave cavity loaded with a capacitance. As the center conductor approaches the lower impedance line, the capacitive loading will have a continuously greater effect. This technique

[9] Anthony S. Badger and Stephen F. Adam, "Wideband Cavity-Type Frequency Meters," *Hewlett-Packard Journal*, Vol. 18, No. 4 (December 1966).

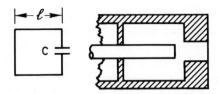

Fig. 5.2-5. Capacitively loaded quarter-wave resonant cavity.

will move the spurious resonance further out on the band, and wider spurious-free bandwidth can be achieved. Bandwidths of 10:1 have been obtained, but the tradeoff in cavity Q is quite significant over 4:1 or 5:1 bandwidth.

To obtain the highest possible Q, the main concern in cavity design is to achieve low losses. The larger the cavity, the smaller the shunt losses. Of course, there is a limit to how large a cavity can be made, and this limit will be determined by the first high-order mode which would come into the frequency band. In a coaxial transmission line, the first higher-order mode is the TE_{11} mode, and the cutoff frequency of that mode has to be set above the top frequency end to be covered.

Another aspect of coaxial transmission lines to be taken into consideration is the minimum loss. This can be obtained when the characteristic impedance of the line is 77 ohms. This was explained in detail in Chap. 2. A characteristic impedance of 77 ohms in an air-insulated line can be achieved with an outer and inner conductor ratio of 3.6:1.

Coupling Mechanisms

To put energy into a resonator and to get an indication that it is resonating, a coupling mechanism is needed. Many types can be employed, depending on what kind of transmission line is to be used. Basically, these coupling mechanisms are such that they excite either the electric or magnetic fields inside the cavity. Generally, either an antenna terminated with an open-center conductor to excite the electric field or a loop configuration to excite the magnetic field is employed for coaxial transmission lines. Figure 5.2-6

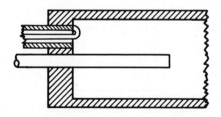

Fig. 5.2-6. Coupling loop in a quarter-wave cavity.

shows a coupling-loop arrangement in a quarter-wavelength-type cavity. The coupling loop excites the magnetic field of the TEM mode prevailing in the cavity and is oriented to couple into the magnetic field of the TEM wave. Furthermore, it has to be placed where the magnetic field concentration is the highest. This is exactly at the shorting plate, since the magnetic fields are directly proportional to currents in the wall. The highest current point in a transmission line is at the short.

Two kinds of coupling are used in frequency meters (often called wave-meters)—the absorption type of coupling and the transmission type. The absorption type of coupling to the cavity is analogous to a series resonant network connected parallel to a transmission line (Fig. 5.2-7). Such a circuit

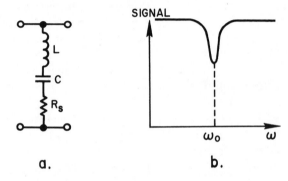

a.   b.

Fig. 5.2-7. (a) Equivalent circuit and (b) frequency response of an absorption-type coupled-cavity resonator.

will react as a dip in the signal level at resonant frequency. A transmission-type coupling mechanism in a cavity resonator will react as a bandpass filter. For instance, an equivalent circuit with lumped-circuit elements will show a resonant network coupled inductively to another resonant network through mutual inductance (Fig. 5.2-8). The frequency response of such a cavity is shown in Fig. 5.2-8(b). How actual coupling loops are located in a quarter-wavelength-type cavity for reaction- and transmission-coupled cases is shown in Fig. 5.2-9. Figure 5.2-9(a) shows a reaction-type coupling where a coaxial transmission line is transformed in such a manner that two of these lines join together with a short piece of center conductor exposed. This is also devised to have 50 ohms of characteristic impedance to maintain the least amount of reflection. If there is no resonance in the cavity, the trans-mission line has the least amount of reflection, and energy will be drawn from the transmission line when resonance occurs. In the transmission-coupled coaxial cavity resonator, the transmission line is terminated in a loop. Often

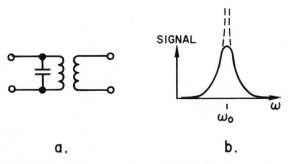

a.   b.

Fig. 5.2-8. (a) Equivalent circuit and (b) frequency
response of a transmission-type coupled-
cavity resonator.

a series 50-ohm resistor is inserted in the center conductor to achieve acceptable impedance match. These types of cavities have a Q in the order of 1,500
to 4,000, depending on the frequency range. The Q of the cavity determines
the maximum accuracy achievable with such wavemeters.

Wavemeters

Moving the center conductor is usually done mechanically with a precision thread and a precision nut, using a spring-loaded device to eliminate
backlash errors. Calibration dials or micrometer readout devices can be calibrated to great accuracies and maintained in extreme environments. If tem-

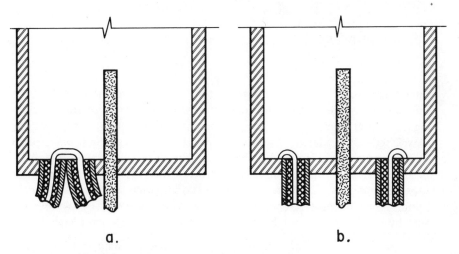

a.   b.

Fig. 5.2-9. (a) Reaction and (b) transmission coupled-
coaxial-cavity resonator.

perature compensation is assured by good selection of material, several tenths of a percent accuracy or better can be achieved. To achieve better accuracies with cavity resonators, higher-order-mode types of transmission lines can be used.

WAVEGUIDE WAVEMETERS. A cylindrical type of cavity resonating in the first high-order mode can be used. This type of wavemeter has only about 50% of the bandwidth of the coaxial types because the TE_{11} mode is dominant; however, it is much more accurate because the Q of the cavities is on the order of 8,000 to 12,000. Since waveguide open circuits are not so reflective as coaxial line open circuits (waveguide open acts as a reasonably good antenna), only short circuits can provide acceptable terminations of extreme values of impedances. This is why waveguide frequency meters are the half-wavelength type only. Figure 5.2-10 shows the arrangement of a waveguide wavemeter

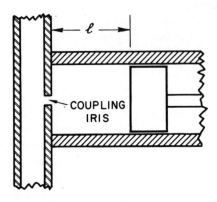

Fig. 5.2-10. Waveguide TE_{111}-type cavity resonator
wavemeter.

operating in the TE_{111} mode. On the bottom of the wavemeter, an iris couples into the waveguide as shown in the figure. Coupling holes or irises are designed so that either the electric or the magnetic field will be coupled in. The wavelength is calculated, of course, in the particular waveguide wavelength in that mode where the cavity resonates. The third subscript on the mode description stands for the number of half-wavelength resonances present. The wavemeter in Fig. 5.2-10 shows the typical absorption-type wavemeter coupling. Transmission-type coupling can be achieved by putting two waveguides side by side—one waveguide coupling in and the other one coupling out.

EFFECTS OF TEMPERATURE AND HUMIDITY. The resonant frequency of a cavity wavemeter is determined primarily by its physical configuration and the dielectric constant of the medium inside the cavity. Consequently, most wavemeters are affected by temperature changes, which cause differential expansion in the cavity. Humidity causes a slight change of the dielectric constant inside the cavity if it is used. Both these effects are overcome in more expensive precision wavemeters by the use of hermetically sealed cavities and special temperature-compensating material. Temperature and humidity calibration or correction charts may also be used to improve accuracy.

Slotted Lines

A less frequently used mechanical frequency-measuring technique is done with a slotted line. This technique depends on the fact that the standing-wave ratio set up in the transmission line produces nulls or minimums every half wavelength. If these nulls are detected and the distance between them is measured, the frequency may be determined. It is necessary, however, to make a correction for the guide wavelength when waveguides are used. Accuracies obtainable with this technique are usually limited to approximately 1% because the guide wavelength depends primarily upon very accurate waveguide dimensions.

Frequency measurement in a coaxial system is not dependent upon the physical dimensions of the cross-sectional geometry of the transmission line. A typical frequency-measuring setup using a frequency meter of the cavity type is shown in Fig. 5.2-11. It is a good rule of thumb to use an attenuator

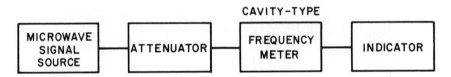

Fig. 5.2-11. Typical frequency measuring setup.

between the microwave signal source and a frequency meter. At least 10 dB of attenuation should be used to eliminate frequency pulling of the microwave signal source while the frequency meter is tuned on and off resonance. Resonant frequency is measured when the indicator shows a dip on the frequency response as the meter is tuned.

5.2.2 ELECTRONIC MEASUREMENT TECHNIQUES

The purely electronic frequency measurements are those in which a known standard frequency is compared with an unknown frequency by a null-beating technique. This technique is used to calibrate wavemeters because of the high accuracy obtained. Typically, the signal from a frequency standard is taken, amplified, and then connected into a harmonic generator to provide a so-called comb of standard frequencies. Ahead of the harmonic generator this signal contains the frequency from the standard, for example 10 MHz. When it is amplified and connected into the harmonic generator, its waveshape will be distorted, resulting in harmonic distortion. Harmonic distortion provides a spectrum coming out of the harmonic generator consisting of harmonics of the original frequency. Good harmonic generators provide harmonics up to the 100th or higher harmonic number. If one takes a mixer and mixes the harmonics with a swept-frequency signal and connects it through a detector to an indicator such as an oscilloscope, beat frequency notes can be seen. Since a mixer will pass the sum and the difference of the frequencies, a low-pass filter is usually connected to the end of the mixer to allow only the difference of the frequencies to be passed. If the signal is connected to the vertical amplifier of an oscilloscope having a few MHz frequency response, indication in the vertical direction on the oscilloscope will occur only when the beat frequency is in the passband of that amplifier. Consequently, when a difference frequency from the sweeper and the particular harmonic of the harmonic generator begins to approach a few MHz, indication will be seen. As the difference frequency decreases, the amplitude of the signal becomes larger and larger until zero beat occurs. Of course,

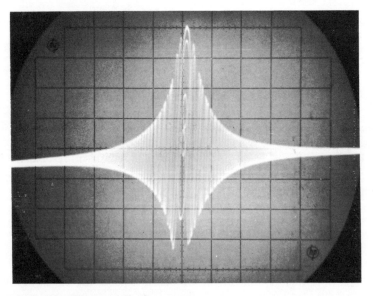

Fig. 5.2-12. "Beat note" of a sweeper.

there is no amplitude at zero beat. As the sweeper frequency passes through zero beat, the mirror image of the former event occurs; from maximum amplitude it decreases to zero (Fig. 5.2-12). These beat notes will appear at every harmonic number on the spectrum being swept. If the fundamental frequency being multiplied is 100 MHz, then a beat note is obtained at every 100 MHz on the spectrum.

A typical wavemeter calibration setup using a 100-MHz frequency standard driven from an atomic primary standard is shown in Fig. 5.2-13.

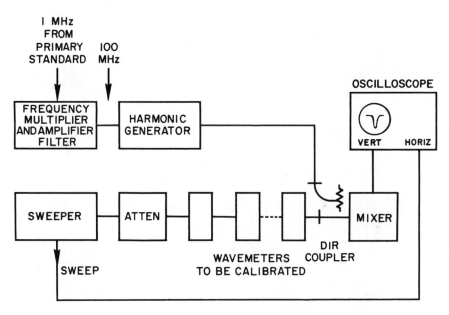

Fig. 5.2-13. Typical wavemeter calibration setup using
the frequency comb generator as a standard.

Figure 5.2-14(a) shows a CRT photo of a wavemeter dip and several beat notes. Figure 5.2-14(b) shows a CRT photo of a wavemeter dip and a single-beat note for high resolution. System accuracy for such a setup is dependent on the accuracy of the beat note markers and the accuracy of the comparison of wavemeter dips to these beat notes. Beat note accuracy depends upon the standard being used. Comparison accuracy is then determined by the wavemeter Q and the overall system resolution. In comparison accuracy, the sweeper resolution with modern sweepers can be spread out so much that

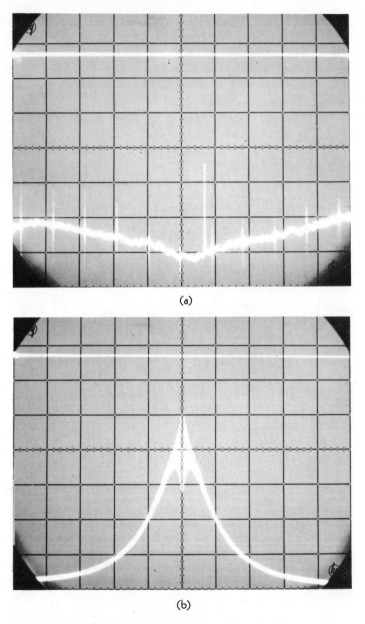

(a)

(b)

Fig. 5.2-14. (a) CRT photo of wavemeter dip and several beat notes for rapid adjustment, (b) CRT photo of wavemeter dip and single beat note for high resolution.

only the cavity Q would come into consideration as far as resolution is concerned. So the usual procedure is to tune the wavemeter dip initially to the beat note at a conveniently wide sweep and then to make the final adjustment on a smaller sweep. This provides the convenience and the resolution necessary for checking the wavemeter accuracy.

5.2.3 CAVITY Q MEASUREMENTS

Cavity Q is defined as the ratio

$$Q = \frac{F_0}{f_1 - f_2} \qquad (5.2\text{-}4)$$

where F_0 is the resonant frequency of the cavity.

$$F_0 = \frac{f_1 + f_2}{2} \qquad (5.2\text{-}5)$$

where f_1 and f_2 are frequencies below and above resonance, respectively, corresponding to the half-power points of the cavity response (Fig. 5.2-15).

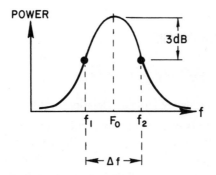

Fig. 5.2-15. Resonance curve of a cavity.

Using this relationship, cavity Q may be easily measured with a system where the sweep width of the oscilloscope is calibrated. Most present-day sweepers have quite accurately calibrated sweep widths. Since only differential frequency readings are of interest, this gives adequate accuracy. Cavity Q is then calculated directly from the true frequency readings at the 3-dB points (half-power points). If a square-law detector is being used, a half-amplitude point on the oscilloscope will correspond to a half-power point.

QUESTIONS

1. What are the three lowest resonant frequencies of a quarter-wavelength coaxial cavity in which the effective length is 1.800 inches?
2. What is the Q of a coaxial cavity in which the inner diameter of the outer conductor is 1.800 inches, the inner conductor diameter is 0.500 inch, and the cavity shunt resistance is 18 kilohms?
3. Which kind of coupling technique is to be used if one chooses to use a cavity resonator as a narrow-band tunable bandpass filter?
4. Which higher-order mode resonator must be used in cylindrical construction to cover approximately half an octave bandwidth, free of spurious responses?
5. To what accuracy can one measure wavelength in the low-microwave region (centimeter range) with a slotted line?
6. What is the frequency of a signal if one measures a half wavelength (2.1 cm) on a 0.400 in. $\times$ 0.900 in. waveguide slotted line (dominant mode of operation)?
7. What is the determining factor of accuracy in an electronic frequency measurement using heterodyne technique with electronic counters?
8. What is the Q of a cavity if the 3-dB points on the resonance curve are at 7,862 MHz and 7,868 MHz?

5.3 SPECTRUM ANALYSIS

5.3.1 BASIC CONSIDERATIONS

Time Domain and Frequency Domain Relationships[10]

An oscilloscope is one of the most widely used instruments in modern radio technology. It allows one to make all types of simple and complex measurements. Today, with an oscilloscope, practically all measurements of interest in the frequency spectrum can be made as high in frequency as the oscilloscope will operate (this is as high as 12.4 GHz with modern sampling techniques). Consider the oscilloscope display closely. What is being displayed on the CRT? How is the signal, which enters the vertical input, displayed on the CRT face?

The horizontal scale of an oscilloscope represents time. The time base, internal to the oscilloscope, generates a ramp voltage that is connected to the horizontal plates of the oscilloscope. This ramp voltage moves the trace horizontally across the base of the CRT at a rate determined by the horizontal time base. (This time-versus-distance relationship is often expressed in terms

---

[10] Application Note 63, Hewlett-Packard Company.

of seconds or milliseconds or microseconds per centimeter of horizontal move-ment.) The CRT display synchronizes the time base and the amplitude of signal entering the vertical input.

We can say that this type of analysis of a signal is in the *time domain*. (*Note:* For a continuous display, a conventional oscilloscope is dependent upon the fact that the signal is a repetitive one.) However, it is often necessary to analyze signals from another point of view. Many parameters of waves can be expressed rather explicitly in the *frequency domain*. Time and fre-quency domain display complement each other, since both provide informa-tion vital for the designer of circuits.

Any periodic wave is comprised of a series of various amplitudes of sine waves at various frequencies. In other words, an arbitrary periodic wave, like the one shown in Fig. 5.3-1 in a time-domain display, can be broken down to

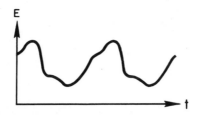

Fig. 5.3-1. Arbitrary periodic wave.

a series of discrete frequencies at multiples of the fundamental, but of differing amplitudes.

Spectrum Analyzer

If a signal is plotted in the frequency domain, such information as how many harmonics are included in the wave and their levels will become apparent. The instrument which presents a display of a wave in the frequency domain is called a spectrum analyzer. Spectrum analyzers were first built during World War II when designers of radars were seeking more information about oscillators and associated modulation circuits. They needed fast rise and decay times of pulses to ensure good resolution of time between trans-mitted and received signals. The pulse duration had to be as short as possible for the best resolution and longest "listening" time. The pulse shape was very important, since the fidelity of the pulse determined the accuracy of the esti-mated target distance. To achieve such goals, scientists at the M.I.T. Radi-ation Laboratory developed the basic spectrum analyzer. With this tool they could view the frequency spectra of their transmitters and be aware of prob-lems as they occurred. This simple spectrum analyzer soon became a basic

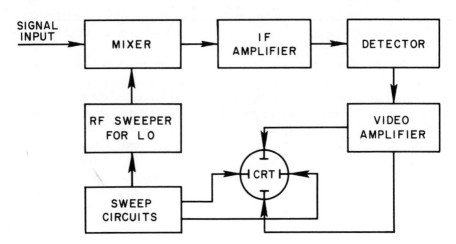

Fig. 5.3-2. Simple block diagram of a basic spectrum
analyzer.

tool for RF pulse circuit designers. Figure 5.3-2 shows the basic block diagram of such an instrument.

A spectrum analyzer is basically a heterodyne receiver in which the local
oscillator (LO) is swept across a certain band. The output of the detector is
applied to the vertical plates of a CRT display, and a voltage proportional to
that applied to the sweep oscillator is applied to the horizontal plates. When
the input signal beats with the local oscillator within the IF amplifier bandwidth, a signal will appear in the video amplifier. This signal deflects the beam
on the CRT. The horizontal deflection on the CRT is derived from the sweep
circuit of the local oscillator, the frequency of which determines the location
of the vertical deflection. This gives a frequency-related amplitude display on
the CRT. Thus spectrum analyzers sweep frequency, just as oscilloscopes
sweep time.

Figure 5.3-3 shows a block diagram of an early type of spectrum analyzer using three superheterodyne conversions. The first mixer's local oscillator is a CW signal tunable to cover various input signals. The wide bandwidth
of the following IF amplifier enables the second local oscillator to be sweepable. This allows a spectrum width as wide as the bandwidth of the first IF
amplifier.

Later spectrum analyzers were designed and built for different frequency
ranges, but the same basic technique was used until 1963, when the first wide-
sweeping, calibrated spectrum analyzer was introduced by Hewlett-Packard.
Because of its calibration, this analyzer is capable of quantitative analysis, in
contrast to the earlier design, which allowed essentially only qualitative analysis. This analyzer has calibrated controls throughout and a 2-GHz sweep

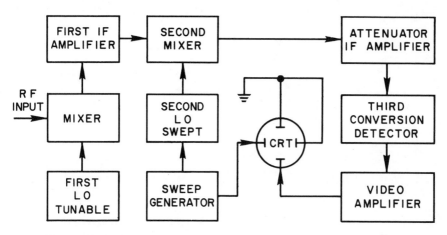

Fig. 5.3-3. Early spectrum analyzer using three
conversions.

width. (Earlier analyzers used a maximum sweep width of about 100 MHz, achieved by sweeping the second local oscillator. The wide sweep of the later analyzers is achieved by sweeping the first local oscillator.)

Fourier Analysis

Any signal which is a function of time can be expressed as a sum of sinusoidal waves. This is the familiar Fourier analysis. Making the transformation of a wave from time domain to frequency domain results in a Fourier series, where discrete frequency components are shown as sinusoidal waves with information about their amplitude and frequency. The periodic function $F(t)$ is a series of sinusoids consisting of the fundamental and its harmonics:

$$F(t) = a_0 + a_1 \cos \omega t + a_2 \cos 2\omega t + a_3 \cos 3\omega t$$
$$\ldots + b_1 \sin \omega t + b_2 \sin 2\omega t + b_3 \sin 3\omega t + \ldots$$

The argument of the trigonometric expressions clearly shows the harmonic nature of the series. The amplitude part of the expression is also closely related to trigonometric functions through mathematical manipulations.[11] Displaying the frequency spectrum of a signal provides very important information of complicated signals. Fourier transformation is the mathematical procedure for turning a *time function* into a *frequency function*.

The Fourier spectrum of a square wave contains all the odd harmonics of the basic repetition rate of the wave with amplitude decreasing as harmonic number increases. In fact, most of you have at one time or another tried to

[11] Whinnery and Ramo, *Fields and Waves in Modern Radio* (New York: John Wiley & Sons, Inc., 2nd ed., 1953), pp. 19–20.

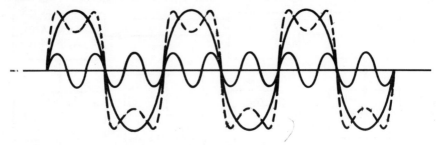

Fig. 5.3-4. Square wave formation of odd harmonics
of the fundamental wave.

reconstruct a square wave out of the odd harmonics of the basic sinusoidal
wave having the square-wave repetition rate (Fig. 5.3-4). It is apparent that
the sum of the two signals starts to take the shape of a square wave. Adding
the fifth harmonic will emphasize the shape of the wave. A perfect square
wave with perfectly rectangular components in the time domain includes all
the odd harmonics up to infinity.

A few examples of different waves will show how Fourier analysis turns
a time-domain display of a wave into its frequency-domain display.

A *CW signal* is a sinusoidal wave in the time domain, and a single
frequency or a spike will represent it in the frequency domain. This is illustrated in Figs. 5.3-5(a) and 5.3-5(b).

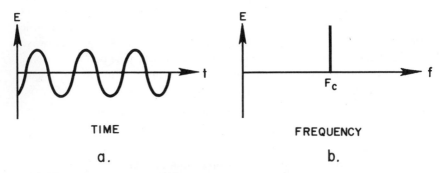

Fig. 5.3-5. (a) Time domain and (b) frequency representation of a signal.

If a CW signal, F_c, is *amplitude-modulated* with a single frequency, f_a,
sidebands will be generated at $F_c \pm f_a$. The Fourier spectrum of this signal
will have three components, $F_c - f_a$, F_c, $F_c + f_a$. Figure 5.3-6 shows a CW
signal amplitude-modulated by a single tone and its Fourier spectrum.

Vectorial manipulation helps in understanding the spectrum of a modu-

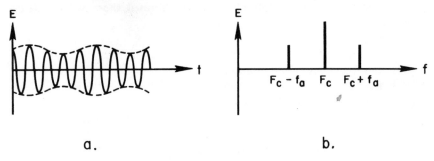

            a.   b.

Fig. 5.3-6. (a) Time domain and (b) frequency representation of a single-tone amplitude-modulated signal.

lated signal. Figure 5.3-7 shows a vectorial diagram of a single-tone amplitude-modulated CW signal. As the carrier vector rotates, all the other
vectors rotate with it, but the modulation vectors are rotating in opposition
to each other. When all the vectors are aligned, amplitudes add up. When
the sideband vectors are in line but opposite to the carrier, they subtract,
giving an understanding of the carrier's behavior in the time domain.

Fig. 5.3-7. Vectorial diagram of amplitude modulation.

It is also interesting to see the Fourier spectrum of a suppressed carrier
signal. Figure 5.3-8 shows such a signal in both the time and frequency
domain; F_c is the carrier and f_a is the modulation frequency.

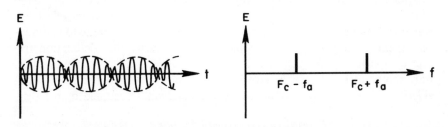

Fig. 5.3-8. Suppressed carrier amplitude modulation.

If the frequency and amplitude axes are calibrated, the viewer can determine:

1. the frequency of the carrier,
2. the frequency of modulation,
3. the percentage of modulation,
4. the nonlinearity of modulation,
5. the presence of residual frequency modulation if spectral lines are jittery, and
6. the spurious signal location and strength, if any.

The spectrum of a complex amplitude-modulated signal will have more components as shown in Fig. 5.3-9. It can be readily seen that the Fourier

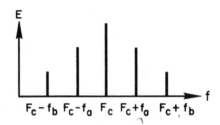

Fig. 5.3-9. Dual-tone amplitude-modulated signal spectrum.

spectrum of any AM signal has a pair of sidebands equally spaced from the carrier by exactly the amount of modulation-frequency component. More complex amplitude-modulated signals can always be identified by their two equally spaced sidebands for each modulation component. Harmonics of the fundamental frequency of the amplitude-modulating signal will show up equally spaced away on both sides from the carrier frequency, since the Fourier components of that wave are harmonically related to each other.

To specify a *frequency-modulated* CW signal, one generally needs to provide information about modulation frequency, index of modulation, deviation of modulation, and spectral width. The Fourier spectrum of such a signal will contain an infinite number of sidebands.[12] For a single-tone modulation, these sidebands will be equally spaced exactly the modulation frequency apart, mathematically, $F_c \pm nf_r$, where $n = 1, 2, 3, \ldots$ In practice, only those sidebands which have significant power will be taken into consideration. A good approximation for the bandwidth occupied by the side-

[12] Cuccia, C. L., *Harmonics Sidebands and Transients in Communication Engineering* (McGraw-Hill Book Company, 1952), p. 226.

bands with significant power is the sum of the carrier deviation and the modulation frequency multiplied by 2. Mathematically this is $BW = 2(\Delta F_c + f)$. Figure 5.3-10 shows the Fourier spectrum of a single-tone frequency-modulated signal.

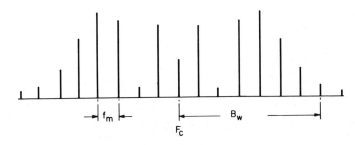

Fig. 5.3-10. Fourier spectrum of a single-tone frequency-
modulated carrier.

Vectorially, a *frequency-modulated* CW signal can be easily understood. In Fig. 5.3-11, as the carrier rotates, a number of sidebands (represented by the other vectors) keep adding vectorially in such a manner that the vectorial

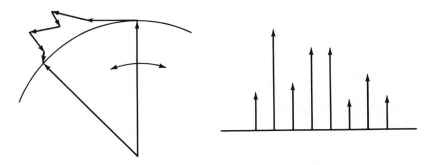

Fig. 5.3-11. Vectorial diagram of frequency modulation.

sum representing amplitude always stays constant. It stays on a circle, but the angular velocity oscillates, representing the frequency modulation.

A *pulse-modulated* CW signal is essentially amplitude-modulated with the carrier turned on and off, or, in more complicated cases, turned partially on and partially off. Let us first analyze the modulating signal, the pulse itself. An earlier part in this chapter showed what is contained in the spectrum of a square wave—a very simple example of a pulse. Another example is shown in Fig. 5.3-12. A periodic, rectangular pulse train is plotted in the

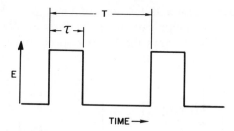

Fig. 5.3-12. Periodic rectangular pulse train.

time domain where perfect rise and decay conditions exist. The original pulse train can be reconstructed by adding the fundamental cosine wave and its harmonics. Figure 5.3-13 shows the first few steps of such an addition procedure.

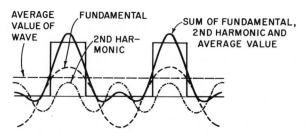

Fig. 5.3-13. Addition of a fundamental cosine wave and its harmonics to form rectangular pulses.

The Fourier spectrum of the perfect pulse train is shown in Fig. 5.3-14. So far we have only taken into consideration the modulation pulse or pulse train. Now let us consider the carrier signal or the relationship of the carrier frequency and the pulse. Basically, pulse modulation is a specialized, very complex form of amplitude modulation. It is complex because there is an infinite number of harmonics involved. The Fourier spectrum of such signals will have an infinite number of sidebands. Each harmonic will produce two lines, the sums and the differences of these harmonics with the RF carrier. Consequently, there will be twice as many spectral lines on the display as there are harmonics contained in the modulating pulse. If we use the pulse train described above to modulate a CW signal, the display will just be shifted from the origin of the frequency axis to the locus of the carrier frequency and, since pulse modulation really is only amplitude modulation, the spectrum will also appear on the other side of the carrier. Figure 5.3-15 shows the

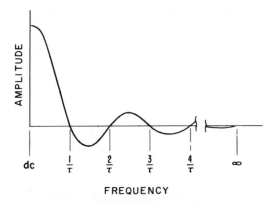

FREQUENCY

Fig. 5.3-14. Spectrum of a perfectly rectangular pulse. Amplitudes and phases of an infinite number of harmonics are plotted, resulting in a smooth envelope as shown.

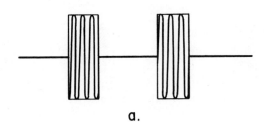

a.

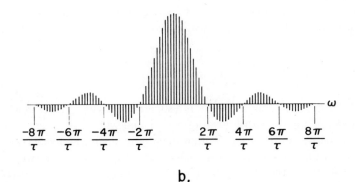

b.

Fig. 5.3-15. Resultant spectrum of a carrier amplitude-modulated with a rectangular pulse.

pulse-modulated carrier in the time and in the frequency domain. The spectral lines are equally spaced by the repetition rate of the modulating pulse

$$F_L = F_c \pm nf_r,$$

where F_L stands for the spectral line frequency, F_c is the carrier frequency, f_r is the repetition frequency of the pulse, and $n = 0, 1, 2, 3$, etc.

The main lobe and the side lobes are the groups of spectral lines above and below the base line. The crossover point between lobes gives us information about the pulse width; the number of side lobes gives important information about the pulse shape (Fig. 5.3-16).

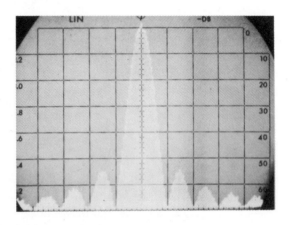

Fig. 5.3-16. Photograph of pulse-modulated carrier
spectrum.

The spectrum analyzer does not present phase information on its CRT display; consequently all the side lobes will appear above the base line. More detailed analysis of pulse displays will be discussed later in this chapter.

5.3.2 ANALYZER THEORY

Basically, a spectrum analyzer is a superheterodyne receiver in which the input signal is mixed with a local oscillator and will be detected only if the difference between the two lies within the intermediate frequency (IF) amplifier bandwidth. As shown in Fig. 5.3-2, the local oscillator is swept. Since the bandwidth of the IF amplifier is not infinitely narrow, a spectral line representing only a single frequency would display the passband curve of the IF amplifier. Figure 5.3-17 shows the analyzer display of a CW signal. If one increases the sweep width of the analyzer until the bandwidth of the IF cannot

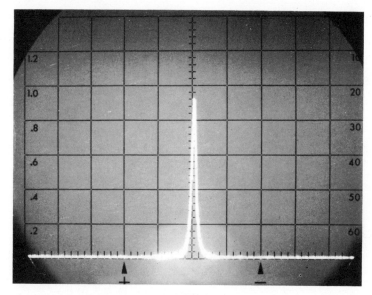

Fig. 5.3-17. Spectrum of a CW signal.

be resolved any more, the display of the analyzer will actually show one spectral line instead of the curve representing the passband of the IF amplifier.

In earlier analyzers, the first conversion was done with a nonsweepable local oscillator that was tunable to cover wide frequency ranges. The IF bandwidth was wide at the first conversion to allow a sweepable second conversion. The second local oscillator was tuned, having a narrow bandwidth IF amplifier. Or, to phrase it differently, the first local oscillator was not swept, and the first IF amplifier had wide bandwidth. The second local oscillator was swept over a narrow range, which was easier to do, since it was at lower frequencies. Triode reactance tube modulator devices were quite advantageous, since stable sweep circuits could be built. Furthermore, sweep calibrations did not change with variation of input frequency, since the first local oscillator was tuned to get response into the bandwidth of the first IF. Usually the IF bandwidth of the analyzers built until the mid-1960's had a bandwidth in the order of only 100 MHz, which allowed a *spectrum width* of the same frequency to be swept.

With this kind of analyzer, the input tuning is done by the first local oscillator, using mechanically tuned klystron oscillators. This technique has mechanical problems, such as backlash and repeller tracking, but it gives a reasonably stable LO signal. For multiband spectrum analyzers, harmonics of the LO are used, but this approach results in loss of sensitivity at higher frequencies. Further problems can occur from spurious responses (discussed later in this chapter). However, these problems can be overcome by careful design of the harmonic generator and mixer. This technique costs a little more, but it extends the tuning range of the spectrum analyzer.

Analyzer Circuits

A spectrum analyzer is a sensitive superheterodyne receiver. After considering some of the faults of earlier analyzers, it is time to get acquainted with a modern one. Figure 5.3-18 is a basic block diagram of such a spectrum analyzer. First, it is apparent that the entire analyzer is split into two parts,

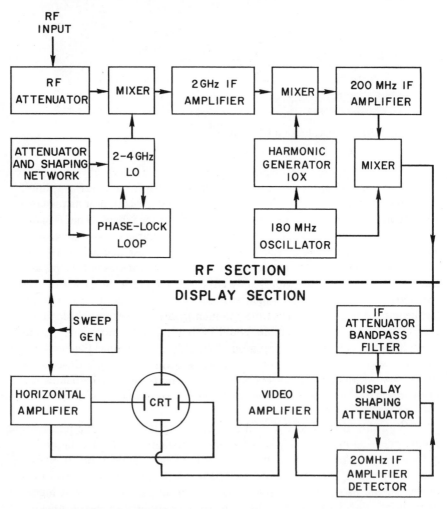

Fig. 5.3-18. Simplified block diagram of a modern spectrum analyzer.

one being the RF section and the other the display section. Triple conversion superheterodyne technique is employed in the RF section with a twice ten-to-one reduction in frequency. 2-GHz IF is converted to 200 MHz and further to 20 MHz, which leaves the RF section and feeds into the display section. The display section houses the basic sweep circuits that feed into the switching network of the RF section that in turn shapes this voltage to suit the helix-voltage-versus-frequency curve of the backward-wave oscillator. The phase-lock loop circuit is also connected to the helix supply. In the display section, the sweep generator output is connected to the horizontal amplifier that drives the horizontal deflection plates of the CRT. The 20-MHz output from the third mixer in the RF section is connected to the display section (housing an IF attenuator, bandpass filter, IF amplifier-detector circuits) and, finally, is connected into the video amplifier that drives the vertical deflection plates of the CRT.

Major Characteristics of Spectrum Analyzers

The widest range of frequencies that can be observed in a single sweep is known as *spectrum width*. The basic limitation to spectrum width is the sweeping capability of the local oscillator.

The maximum sweep width of oscillators is reasonably well fixed by design. Klystrons have long been used as local oscillators because they offer fundamental frequencies in the radar band, but they are capable of only limited electronic sweep. The newest spectrum analyzers can display very wide spectrum widths because they use a backward-wave oscillator (BWO) as the first local oscillator. A 2- to 4-GHz BWO in the analyzer allows it to sweep a 2-GHz band electronically, expanding the spectrum width to 2 GHz; in other words, it is able to display a frequency range of 2 GHz on the CRT. Although using a backward-wave oscillator does away with mechanical problems, such as backlash and repeller tracking, it is not so stable a signal source by itself as the klystron oscillator was in the older analyzers.

This instability is overcome by using phase-locking stabilizing techniques that lower inherent FM in the BWO's from 30 kHz to less than 1 kHz. A mixer and swept local oscillator jointly translate a signal at frequency F_s to a response-producing signal at IF frequency F_{if} whenever

$$mF_s = nF_{lo} \pm F_{if}$$

where $m = 1, 2, 3 \ldots,$
$\quad n = 1, 2, 3 \ldots,$
$\quad F_{lo}$ = local oscillator frequency.

For linear operation of the mixer, amplitude of the F_s signal should be small compared to the amplitude of the local oscillator voltage; then, $m = 1$, which means the operation is in the linear region.

In the new type of spectrum analyzers, IF frequency is chosen to be the low-frequency end (2 GHz) of the local oscillator, extending coverage to low frequencies. Figure 5.3-19 shows a plot of curves of the response locations if

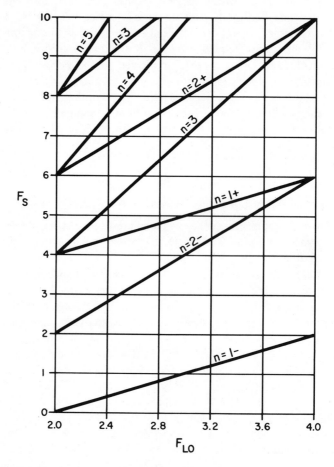

Fig. 5.3-19. Analyzer tuning curves for $F_{if} = F_{lo(min)}$
= 2 GHz.

the IF frequency is 2 GHz. Figure 5.3-20 shows the same kind of plot for a 200-MHz IF.

Two possibilities of signal reception are apparent due to the heterodyning: one response F_{if} above the local oscillator frequency and one response below by the same amount. It is also apparent that these responses are exactly twice the F_{if} frequency apart. On the plot they are denoted as "$n+$" and "$n-$", respectively.

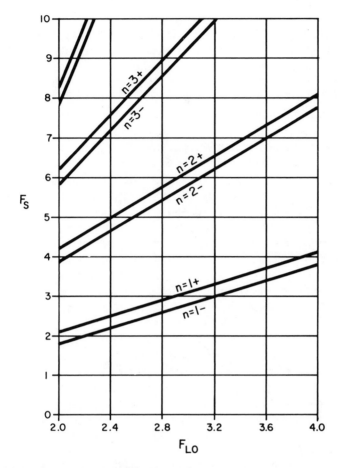

Fig. 5.3-20. Analyzer tuning curves for $F_{if} = F_{lo(min)}$
$/10 = 200$ MHz.

It is often very annoying for the viewer to have these "image" responses present. It makes the usable spectrum width equal to F_{if}, or half the distance of these image responses. In other words, it can be stated that the usable spectrum width is determined by the frequency of the first IF stage. This is the reason why the approach taken in the early 1960's was to increase the IF frequency to 2 GHz. With a sweepable local oscillator of 2.0–4.0 GHz and an IF frequency of 2 GHz, an image-free spectrum display was achieved. Future analyzer IF frequencies can be expected to increase as the state of art allows development of amplifiers at even higher frequencies.

Spurious responses that can be generated in the input mixer are also quite annoying to the user of a spectrum analyzer. The local oscillator's harmonics should be present to allow multiband usage of the spectrum analyzer at higher frequencies, but the harmonics generated by the input signal are undesirable, since they will produce IF signals that would represent responses not present in the spectrum of the applied signal. This effect is caused by excessive signal input, approaching the order of magnitude of the local oscillator voltage level. Harmonics are generated in a mixer if the applied signal amplitude is high enough to vary the conductance of the mixer. These responses are not related linearly to the input signal and can seriously clutter the display. If the signal input is kept extremely low, the conductance of the mixer does not vary with the applied signal and spurious responses will not be seen. But this type of operation would restrict the usefulness of the spectrum analyzer so much that very little useful dynamic range would remain.

By careful design of the mixer and with strong local oscillator drive, the spurious responses can be greatly reduced without imposing extreme, impractical limitations on the dynamic range of an analyzer. For example, Hewlett-Packard's 8551B Spectrum Analyzer employs such mixer design; it suppresses spurious responses to such a degree that the useful dynamic range is not impaired until the input signal level exceeds -30 dBm. Properly designated tuning curves will indicate the location of spurious responses even if they are low in amplitude. Figure 5.3-21 is a plot of the Hewlett-Packard spectrum analyzer tuning curves derived from the equation in Fig. 5.3-20 with the spurious-response-producing frequencies added.

Heavy solid lines represent the best tuning range for each harmonic number n. The light solid-line extensions show the primary responses of these curves extended above and below the recommended tuning range; but they are not useful since the tuning curves become very close together at either end, resulting in a very closely spaced pair of responses of a single signal on the display of the analyzer.

The dashed lines represent input frequencies of -30 dBm amplitude that will produce spurious responses less than 60 dB below the input reference level. The figures in parentheses at the end of each line represent the n, m, and $+$ or $-$ terms, respectively, satisfying the equation $mf_s = nf_{lo} \pm f_{if}$. The dB figure shown along each line is the typical amplitude of the spurious response compared to a -30 dBm signal frequency of mf_s if such a signal were applied to the input. Stronger input signals cause larger and more numerous spurious responses. The analyzer input attenuator should be used to keep signal input to the mixer at -30 dBm or less for minimum spurious response generation and full 60-dB dynamic range (60 dB, because noise level is at approximately -90 dBm and maximum recommended signal input level is -30 dBm).

Notice that some of the dashed lines in the graph do not extend the full range of the local oscillator sweep. This indicates that spurious signals produced above these points are more than 60 dB down referred to the input;

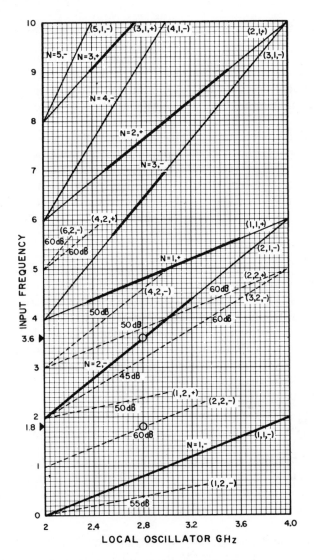

Fig. 5.3-21. HP spectrum analyzer tuning curves. Heavy
solid lines indicate desired response locations for input signals of 10 MHz to 10
GHz. Spurious responses are indicated by
the dashed lines.

thus they are insignificant, since they fall into the noise level on the analyzer display.

There is an easy way to identify spurious responses on a calibrated spectrum analyzer. If the input signal level is decreased 10 dB with the input attenuator, the desired signal on the display will shorten 10 dB, but the magnitude of the spurious responses will decrease a great deal more than that amount.

Bandpass filters are also excellent means of determining whether the spurious response is part of the input or if it is generated at the mixer. If it exists when the filter is connected, it is spurious response. A tunable narrow-band filter or a tunable preselector can easily help clean up and separate cluttered displays by allowing only the desired signal to enter the input. Pre-selectors will be covered later in this chapter.

Resolution

Another very important characteristic of a spectrum analyzer is its *resolution*. As Fig. 5.3-5 shows, a single spectral line of a true CW signal is represented with a finite line width corresponding to the bandwidth of the IF amplifier. The most serious limitation to resolution is the width and shape of the passband of the IF amplifier. It can easily be understood if one displays two CW signals so close in frequency that their frequency spacing is less than the bandwidth of the IF amplifier.

The analyzer display will show only one response. It would seem from this that an extremely narrow passband IF amplifier would solve the problem, but there is another consideration to be taken into account. Spectrum analyzers have swept-frequency displays, and sweep rate can affect resolution. If the local oscillator is swept past the input frequency at too fast a rate, the apparent bandwidth of the IF amplifier will be wider than the actual bandwidth and the amplitude response will be reduced. There is a simple physical explanation. In Fig. 5.3-22 we can see two Gaussian IF amplifier responses superimposed.

If the input frequency is swept slowly through the IF amplifier passband, the output level will have time to reach full amplitude, tracing out curve *A*. Note the 3-dB points of this curve with respect to frequency—this is the true bandwidth. If, however, the local oscillator sweeps through at a high rate, the amplifier output will not have time to reach its full amplitude before the input is gone. This results in the lower amplitude curve *B*. The 3-dB points here are much wider with respect to frequency, giving the impression that IF bandwidth is wider than it actually is, and overall sensitivity has been reduced.

Consider the simple relationship:

$$F_{\text{sweep}} < (\text{BW})^2$$

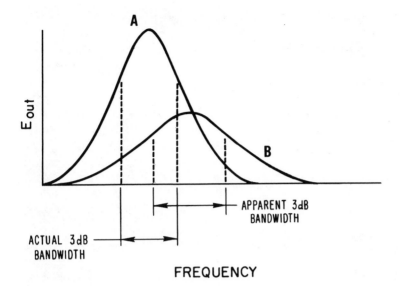

Fig. 5.3-22. Curve *A* shows the actual bandpass for Gaussian IF amplifier. Curve *B* represents output when local oscillator frequency is swept too fast for IF amplifier to respond fully.

That is, the frequency of the sweep rate shall always be less than the square of the bandwidth.

Another factor impairs the resolution of a spectrum analyzer. If the *local oscillator* has *inherent frequency modulation* other than the linear-sweep tuning (which could be quite high with backward-wave oscillators), the effect on the spectral display will be the same as if the input signal were frequency-modulated. This is because the IF stage cannot identify the source of the FM as the signal or as the local oscillator. The degrading effect of local-oscillator FM on a CW signal is shown in Fig. 5.3-23.

Let us consider an FM spectrum for a moment. As a general rule, the approximate bandwidth of an FM spectrum is that of the peak-to-peak deviation of the signal if the frequency deviation is much larger than the modulation rate. Then, if the peak-to-peak deviation of the local oscillator is larger than the IF bandwidth in the analyzer, two CW signals must be separated in frequency by an amount greater than the deviation to be displayed as separate signals.

The local oscillator circuit in the Hewlett-Packard spectrum analyzer is phase-locked, thus reducing the inherent frequency modulation of the BWO

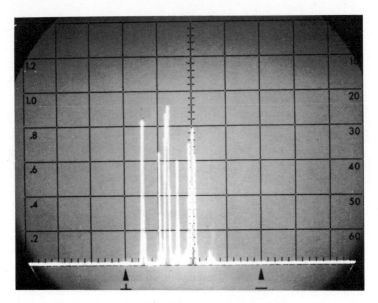

Fig. 5.3-23. Appearance of a CW spectrum when analyzer local oscillator has excessive FM.

below 1 kHz. Full utilization of the 1-kHz IF bandwidth provides resolving
power of about 3 kHz.

Sensitivity

Sensitivity is a measure of the analyzer's ability to detect small signals.
The several common ways to measure sensitivity in receivers depend primarily
on the input stage. In communications, the input stage is usually an RF
amplifier, and the method of verifying sensitivity is to measure signal-to-noise
ratio (SNR) under specific conditions. In radar, the input usually goes directly to a diode mixer whose uncertain noise behavior makes it difficult to
determine SNR accurately. Here the required input power to the receiver for
minimum discernible signal out of the video amplifier is commonly referred
to as sensitivity. This, however, is too arbitrary for consistent measurements.
A more specific definition would be in terms of the system noise figure. In
any case, inherent noise is the ultimate limitation to receiver sensitivity. Since
the spectrum analyzer is a receiver, let us see how sensitivity may be defined
in relation to noise.

The mixer and IF amplifier comprise a system of stages in cascade, each
contributing its own noise. We can therefore express the overall noise figure
of the analyzer in dB as

$$F_r = F_m + L_c(F_{\text{if}} - 1)$$

where F_r = overall noise figure,
$\quad F_m$ = noise figure of mixer,
$\quad L_c$ = conversion loss of mixer $\left(L_c = \dfrac{1}{\text{gain}}\right)$
$\quad F_{if}$ = noise figure of IF amplifier.

Since F_m is not measurable, a noise ratio N_r is generally specified for diode mixers and may be substituted in the above equation. With appropriate factoring, the equation becomes

$$F_r = L_c(N_r + F_{if} - 1) \qquad (5.3\text{-}1)$$

Now, considering the overall noise figure F_r, we can show the equivalent noise power to the input of the analyzer as

$$P_{in} = F_r\,\text{KTB}$$

where K = Boltzmann's constant of $1.37 \times 10^{-23}\ J/°K$,
$\quad T$ = absolute temperature in $°K$,
$\quad B$ = equivalent noise bandwidth of the IF amplifier.

P_{in} as defined in the above equation is generally accepted as receiver sensitivity. A signal of this amplitude would produce an output signal-to-noise ratio of unity if there were no deleterious effects, such as local oscillator FM or excessive sweep rate.

In specifying analyzer sensitivity, it is essential to know the associated IF bandwidth.[13] This can be seen from Eq. (5.3-1) above. Only under these conditions can one predict the SNR for pulse signals or determine an upper limit on sweep rate and the tolerable local oscillator FM. As a rule of thumb, the equivalent noise bandwidth is about the same as the 3-dB bandwidth for a Gaussian IF amplifier, and the value of KT at $290°K$ (room temperature) is -114 dBm/MHz. Thus the above relationships become quite simple for calculation of analyzer sensitivity:

$$\text{Sensitivity (dBm)} = F_r - \frac{114\ \text{dBm}}{\text{MHz}} + 10\log\frac{B}{1\ \text{MHz}}$$

We calculate the sensitivity of a spectrum analyzer where the typical noise figure for fundamental mixing is 29 dB, with an IF bandwidth of 10 kHz, as follows:

$$29 - 114 + 10\log\frac{0.01\ \text{MHz}}{1\ \text{MHz}} \qquad \text{or} \qquad -105\ \text{dBm}$$

There are commonly two measures of a spectrum analyzer's *dynamic range*. The first is the ratio of largest to smallest signals that can be simultaneously displayed on the analyzer screen. This is usually extended by providing the

---

[13] Terman and Pettit, "Electronic Measurements," *Noise Figure of Systems in Cascade* (2nd ed.), McGraw-Hill Book Company, 1952, p. 361.

IF amplifier with an optional logarithmic response. The second measure is the ratio of the largest signal that can be applied at the input without serious amplitude distortion to the smallest signal that can be detected (sensitivity). Distortion occurs when mixer conductance varies with the applied signal. This happens when the peak signal voltage to the mixer approaches the magnitude of the local-oscillator drive voltage. Therefore it is desirable to limit large input signals to the mixer with a suitable input attenuator.

Distortion due to saturation in the IF amplifier stages can also limit dynamic range if it occurs before the mixer is overloaded with large inputs. A well-designed analyzer will have sufficient IF gain control so that the IF amplifier will not be the limiting factor in dynamic range. For most purposes it is preferable to use an envelope (peak) *detector* at the output of the IF amplifier. In some of the earlier spectrum analyzers, square-law detectors were used, which resulted in a response that presented a power spectrum display; however, detector square-law characteristics cannot be relied upon for accurate spectrum measurements. It is generally better to use a linear envelope detector and shape the gain characteristics of the IF amplifier to obtain a response other than linear when required.

It is desirable to have logarithmic as well as linear display when large differences in level are viewed. Logarithmic displays are calibrated in terms of dB. Essentially a logarithmic amplifier compresses the display so the viewer can see the entire spectrum. Square-law display enables the viewer to expand the scale of vertical display if small-amplitude changes that cannot be resolved in linear display are being analyzed.

5.3.3 APPLICATIONS OF SPECTRUM ANALYSIS

Spectrum analyzers were originally invented in World War II to measure pulsed radar's magnetron output to ensure stable oscillation, free of moding and spurious signals. Very shortly after the invention of spectrum analyzers, the users of it found that it had a great number of applications which originally were not thought of. This is even more true with our modern spectrum analyzers, where these features were already well known. Capitalizing on the knowledge of the wide usage, designers have provided wide spectrum width and calibrated controls throughout. Calibration of the controls, plus close attention to human factors, simplify both the evaluation of the display and the operation of such instruments. The accuracy and flexibility of the instrument make it suitable for many applications beyond the capability of other spectrum analyzers.

Distortion analysis, modulation FM linearity, pulse analysis, RFI measurements, spectrum surveillance, semiconductor evaluation, antenna pattern work, and parametric amplifier tuning are some of the applications that the users of spectrum analyzers are making with these instruments today.

DISTORTION AND SIGNAL ANALYSIS functions are easily performed on a spectrum analyzer. Figure 5.3-24 shows the analysis of a signal generator

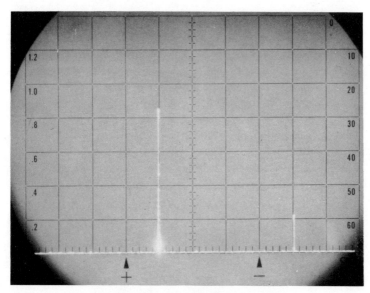

Fig. 5.3-24. 800 MHz signal and 1600 MHz 2nd harmonic shown on one sweep. Horizontal 200 MHz/cm, vertical 10 dB/cm.

output running at 800 MHz and on the same trace showing the second harmonic distortion product at 1600 MHz 32 dB down. In this case the broad sweep characteristics and flat response in fundamental mixing mode allow the signal and its harmonic to be presented on the same screen. The front panel 60-dB RF attenuator should be used to ensure that the second harmonic and other signals are not being internally generated by the analyzer itself. Figure 5.3-25 shows a VHF transistor amplifier operating with a 150 MHz output signal under two input conditions. One shows the specified input and the second harmonic of the output properly 35 dB down (spec 26 dB). The second picture shows overdriving characteristics where many harmonic products are present. Also note that noise appearing in the passband is higher since analyzer gain was turned up.

In the case that the second or other harmonic distortion products occur in another mixing band, it is obvious that they will still appear on the same

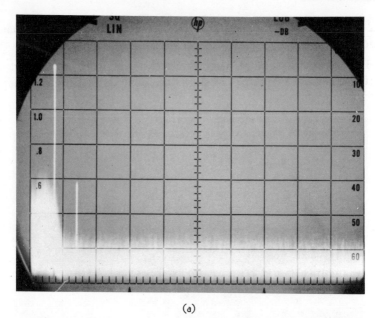

(a)

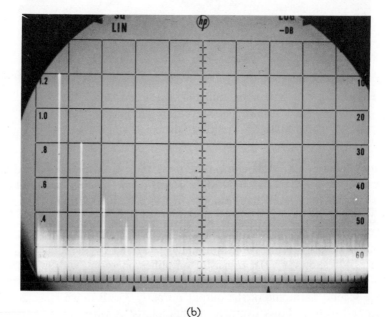

(b)

Fig. 5.3-25. Picture (a) shows normal 150 MHz transistor amplifier output. (b) shows amplifier overloaded with distortion products. Horizontal 200 MHz/cm, vertical 10 dB/cm.

screen presentation, since all harmonics appear (if 2000 MHz sweep is used). However, the actual calibration of the sensitivity of the second (or higher) harmonic mixing must be performed to give the proper comparison between the fundamental and the distortion products. For instance, Fig. 5.3-26 shows

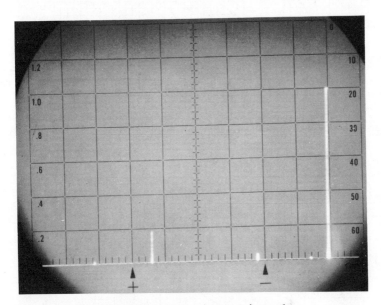

Fig. 5.3-26. The 1800 MHz generator output is on the right in the n = 1⁻ band. 3600 MHz harmonic left of center is in the n = 2⁻ band. Horizontal 200 MHz/cm, vertical 10 dB/cm.

a signal generator input at 1800 MHz with its second harmonic signal appearing on the same screen and identified at 3600 MHz.

The proper use of a multiband spectrum analyzer requires the understanding and application of preselection filter techniques to restrict the input signals at the front panel to a usable minimum.

There are bandpass filters provided in nominal 2-GHz segments from 1 GHz to 10 GHz with modern spectrum analyzers. Bandpass filters should be used with reference to the tuning curves of the spectrum analyzer to eliminate even spurious responses. Figure 5.3-27 shows the Hewlett-Packard spectrum analyzer's tuning curves where the above-mentioned filters should be used.

A 2-GHz notch filter is desirable with the use of this spectrum analyzer. The use of this filter is recommended in any application where there are sig-

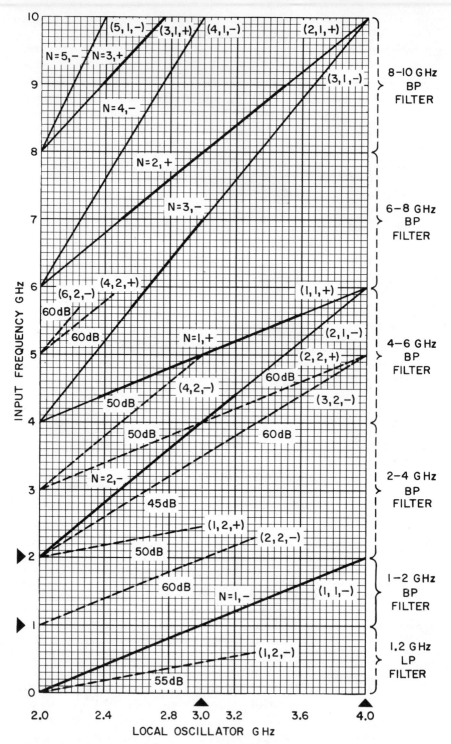

Fig. 5.3-27. RF bandpass filter selection for the HP
851B/8551B spectrum analyzer showing the
tuning curves.

nals at the input jack that involve noise or signal level at precisely 2000 MHz. The reason for this is that the first IF amplifier operating at 2 GHz may be driven directly from the front panel jack. Thus, if broadband noise or other signals at 2 GHz are present, the base line of the entire sweep will shift upward, irrespective of what the local oscillator sweep indicates.

Figure 5.3-28(a) shows the base line being raised by the presence of broadband noise at the input of the analyzer, and Fig. 5.3-28(b) shows its elimination by use of the notch filter.

Measuring Signal Amplitude with the Spectrum Analyzer

Present microwave spectrum analyzers are calibrated to show only relative amplitude. In the logarithmic display mode the vertical display is accurately calibrated in dB/division (usually 10 dB/cm), and in the linear mode it is linear. This makes the spectrum analyzer a good indicator of the relative amplitude differences between signals. However, since it lacks the absolute amplitude calibration of a voltmeter, power meter, or laboratory oscilloscope, the absolute level of a signal in dBm or microvolts cannot be measured except by substitution.

Signal level is measured in this method by first observing the height of the unknown signal on the spectrum analyzer display. Then the output of a calibrated signal generator is substituted for the unknown signal at the spectrum analyzer input. The signal generator output is adjusted until its signal on the display is equal in height to that of the unknown. At this point the signal generator output level is equal to the unknown which can then be read directly off the signal generator. This calibration can be extended to levels above and below this reference level by using the accurate relative amplitude calibration of the analyzer.

An RF spectrum analyzer, covering up to 110 MHz, that has accurate absolute amplitude calibration has been developed.[14] It is calibrated directly in dBm in the logarithmic display mode and in microvolts/cm in the linear mode. This direct-reading amplitude calibration will be expanded into the microwave range, making the microwave spectrum analyzer capable of measuring signal level directly. Because of its great sensitivity, a result of narrow bandwidth capability, the spectrum analyzer can measure signal levels over the very wide range of from approximately -100 dBm to 0 dBm, from milliwatts to picowatts.

Preselector

The new generation of spectrum analyzers is able to display the entire frequency band from 10 MHz to 12.4 GHz with a single sweep and acts like

[14] Hewlett-Packard Model 8552A/8553L.

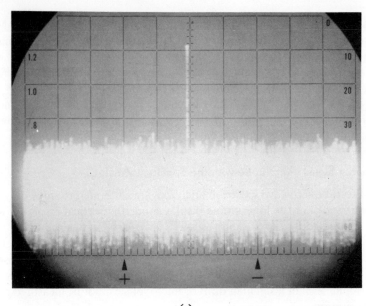

(a)

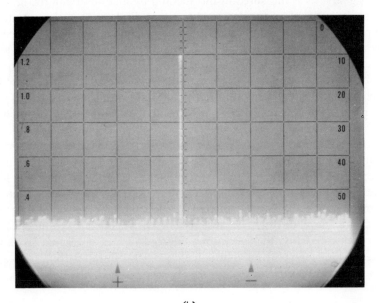

(b)

Fig. 5.3-28. (a) Baseline raised in the presence of broadband noise, (b) same display with 2 GHz
notch filter connected to input.

a wide-open superheterodyne receiver. It accepts all the signals within this frequency band in one coaxial input and mixes them with a local oscillator sweeping from 2 to 4 GHz to produce a 2-GHz IF signal. This signal is further heterodyned down. Finally, a 20-MHz signal is detected and displayed on the CRT. All signals entering the input are able to produce 2-GHz mixing products, consequently being displayed on the CRT. Since some of the signals and their harmonics, with the LO and its harmonics, are capable of producing more mixing products, the display can become fairly cluttered.

The problem blocking the responses to signals outside a desired frequency range was partially solved with the use of interdigital bandpass filters. A device called the *preselector* offers a radical solution to the problems of broadband analysis and fixed interdigital filters. It acts as a voltage-tunable bandpass filter, automatically tracking the desired harmonic of the analyzer's local oscillator. This defines the frequency range displayed on the CRT, simplifying the presentation and making it easier to interpret.

The heart of the preselector is an electronically tunable YIG filter, which was described in Chap. 4. The selectivity characteristic of the YIG filter is at least 6 dB/octave 200 MHz off resonance (see Fig. 5.3-29). There

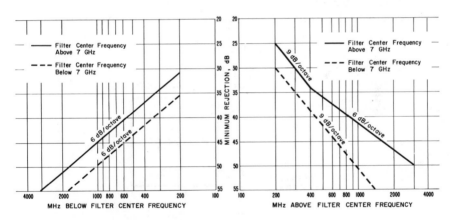

Fig. 5.3-29. YIG filter selectivity for the main
response mode.

are basically three different kinds of operations of such YIG preselectors (such as the HP 8441A):

1. internal sweep mode,
2. external sweep mode, and
3. spectrum analyzer tracking mode.

When the *internal* mode is selected on the function control, a voltage ramp is generated internally to drive the YIG tuning coil at a constant 60-Hz rate. This ramp will sweep the passband across a frequency range selected by the "center frequency-GHz" and the "sweep width-GHz" controls. The former control defines the center frequency of the range to be swept (the midpoint of the ramp), and the latter sets the sweep limits of that center frequency (the endpoints of the ramp). The center frequency is continuously variable from 1.8 to 12.4 GHz; the sweep width can be varied from 0 to 10 GHz. The 1.8- to 12.4-GHz frequency limits are set by the physical characteristics of the YIG spheres and therefore cannot be exceeded. Also, in the internal mode, the sweep width can be set to zero as a fixed tuned bandpass filter. The "center frequency GHz" control will then tune the YIG's passband to any frequency between 1.8 and 12.4 GHz. The nominal filter bandwidth (-3 dB points) is 35 MHz; the exact bandwidth may vary from 20 to 70 MHz over the 1.8- to 12.4-GHz range.

5.3.4 PRESELECTOR

Primarily, the YIG filter was intended to be a *preselector* for the spectrum analyzer. The electronic circuits (see Fig. 5.3-30) match the tuning

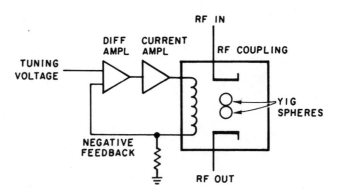

Fig. 5.3-30. Simplified block diagram.

characteristics of the YIG and the spectrum analyzer. The preselector output jack on the spectrum analyzer provides the tuning voltage to the preselector input jack. This voltage is converted to the tuning current needed to drive the YIG electromagnet. In this way the preselector will track a selected mixing mode, rejecting all undesired signals.

Harmonic, multiple, and spurious responses such as intermodulation (IM) distortion can be eliminated by using the preselector. Before discussing how this is accomplished, it is necessary to review the basic operation of the spectrum analyzer. The analyzer responds to input signals according to the following equation:

$$f_{RF} = nf_{LO} \pm 2\,\text{GHz}$$

As the fundamental of the analyzer's local oscillator (LO) tunes from 2 to 4 GHz, the analyzer responds to signal frequencies 2 GHz above and 2 GHz below the local oscillator frequency. Similarly, the second harmonic of the local oscillator tunes 4 to 8 GHz and the third harmonic tunes 6 to 12 GHz. This means that, when the local oscillator is tuned to 3 GHz, for example, the analyzer can respond to six different signal frequencies (see Fig. 5.3-31). These responses, called *harmonic responses*, are a very useful way of extending

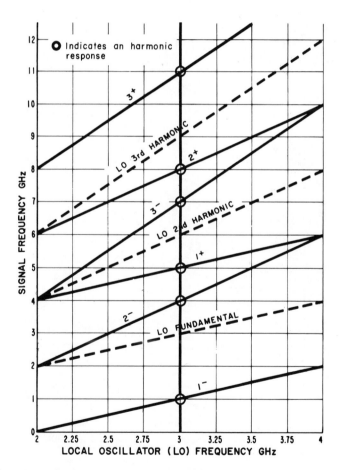

Fig. 5.3-31. Analyzer responses when LO is tuned to 3 GHz.

the frequency range of the analyzer display. They can all be readily identified, but they overlap on the display and can cause confusion in picking out any one special frequency. These responses are a result of the analyzer's ability simultaneously to display the entire spectrum from 10 MHz to 12.4 GHz. However, this may not be the most desirable mode of operation if you are interested only in the responses in one specific frequency range. The preselector allows you to observe just the one response you are looking for.

The analyzer can also mix with more than one local oscillator harmonic at a single input frequency. These display responses are called *multiple responses* and appear at different positions on the screen. For example, Fig. 5.3-32 shows the analyzer responding at LO frequencies of 2.33, 3.00, and

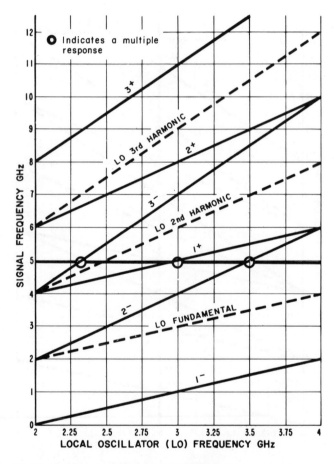

Fig. 5.3-32. Analyzer responses to a 5-GHz input
signal.

3.50 GHz to a 5-GHz input signal. Substitution into $f_{RF} = nf_{LO} \pm 2$ GHz produces:

$$5 = 3(2.33) - 2$$
$$= 1(3.00) + 2$$
$$= 2(3.50) - 2$$

Spurious responses are distortion products generated by the nonlinear behavior of the analyzer's input mixer when driven by signals greater than -30 dBm. In the case of intermodulation distortion (IM), these distortion products are due to the interaction of two or more strong input signals in the mixer.

These three types of responses can have a complicating effect on the spectrum analyzer display. If, however, a preselector is connected to the spectrum analyzer, the effects of the undesired responses can be virtually eliminated. Figure 5.3-33 shows how to connect the HP 8441A Preselector

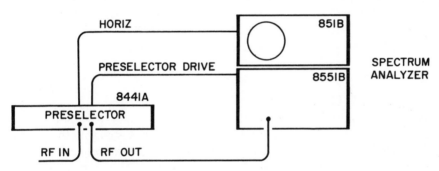

Fig. 5.3-33. 8441A/851B/8551B setup diagram.

to the HP 8551B/851B Spectrum Analyzer to achieve entirely multiple-response-free spectrum analyzer displays.

Consider that a 5-GHz and 8-GHz signal are simultaneously passed through a preselector and into the spectrum analyzer. If the preselector is set to track the 1+ mixing mode, it will pass only the 5-GHz signal when the LO is tuned to 3 GHz. The harmonic response to the 8-GHz signal that would occur at the same LO frequency has been blocked by the preselector.

The *multiple* responses of the analyzer to the 5-GHz signal would also be eliminated. Figure 5.3-34 indicates that multiples would occur at LO frequencies of 2.33, 3.00, and 3.50 GHz. Figure 5.3-34(a) is an oscillogram of these responses. Figure 5.3-34(b) shows the reduction of the 2− and 3− mixing responses when using the preselector. When the LO frequency is 2.33 GHz, the preselector is tuned to 4.33 GHz (tracking the 1+ response) and so rejects the 5-GHz signal. Similarly, the 5-GHz signal is rejected at

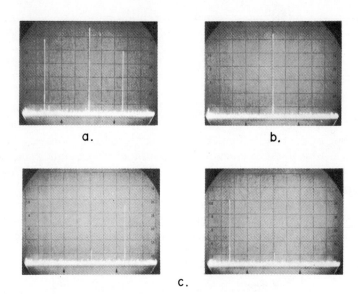

Fig. 5.3-34. (a) Multiple responses to 5-GHz signal,
(b) selection of 1⁺ response, (c) selection
of 2⁻ and 3⁻ responses.

the 2− mixing mode, since the preselector is then tuned to 5.5 GHz. The 2− and 3− mixing responses were selected in the same way, with the results shown in Fig. 5.3-34(c). The reduction of other than the desired mixing mode is typically greater than 35 dB.

A high-power signal (above −30 dBm) will generate *spurious responses* in the mixer, but the preselector will eliminate their effects from the display. With the 1+ mixing mode selected, the response to a 0-dBm, 5-GHz signal will be displayed when the LO is tuned to 3 GHz. When the LO is tuned so that the second harmonic (10 GHz) can produce a response, the preselector

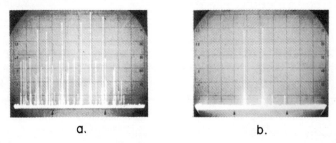

Fig. 5.3-35. (a) Intermodulation distortion products,
(b) reduction of intermodulation distortion.

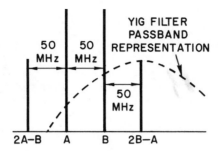

Fig. 5.3-36. Intermodulation distortion reduction of
closely spaced signals.

is tuned off of the fundamental (5 GHz) and therefore attenuates the fundamental by more than 50 dB. No harmonics are produced in the mixer, and therefore no spurious responses appear on the display.

Preselection can also aid in the reduction of *intermodulation* between several strong signals applied to the analyzer's mixer. Figure 5.3-35(a) shows signals at 4.9 and 5.1 GHz. Due to the high input level to the mixer (about 0 dBm), the display is complicated by strong intermodulation products as well as the multiple responses from the 3− and 2− mixing modes. Figure 5.3-37(b) shows the substantial reduction of these responses when using the preselector. Since the two input frequencies are far enough apart, the preselector passes only one signal to the mixer at any time. Since both signals are never present in the mixer simultaneously, intermodulation products cannot be produced. Use of the preselector produces noticeable improvement in IM distortion between signals as closely spaced as 35 MHz. For example, consider two signals separated by 50 MHz (A and B). These produce second-order distortion products at frequencies of 2A-B and 2B-A, both 50 MHz from the nearest fundamental (see Fig. 5.3-36). When the spectrum analyzer and the preselector are tuned to receive the 2B-A distortion signal, the passband of the preselector reduces the level of signal B by 10 dB and signal A by 25 dB. The distortion products are thus at least 25 dB below what they were without the preselector.

The preselector will improve the analytical capability of broadband signals, such as frequency combs, or very narrow pulses viewed on the spectrum analyzer. High-power signals must be reduced in the input attenuator so the spectrum analyzer mixer is not overdriven. (Mixer burnout will occur just over 1 mw.) With substantial attenuation, however, smaller signals are lost to the display.

This is especially significant when viewing broadband signals such as frequency combs from harmonic generators. Each spectral line contributes only a fraction of the total power to the mixer. In a broadband comb, each har-

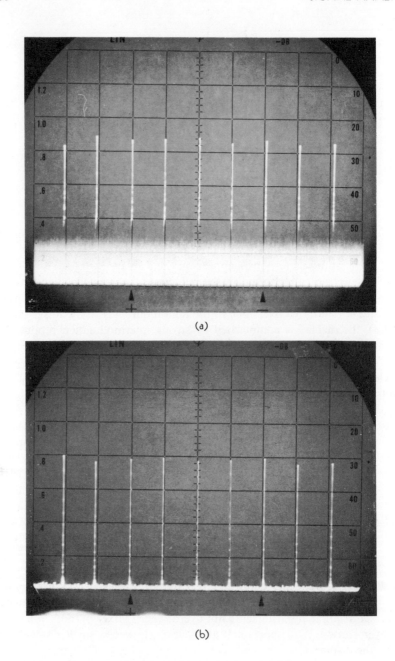

Fig. 5.3-37. (a) Degraded signal-to-noise ratio of frequency comb, (b) improved signal-to-noise ratio with a preselector.

This asymmetry in amplitude car
5.3-38(e) that one FM sideband e)
line when phase information is reta

The spectrum analyzer adds
displays the absolute magnitude of
the other sideband is reduced. The
two sidebands is twice the amplitud
the AM index of modulation and t
measured. This measurement depe
phase with the peak AM. That is
usually caused by the peak excursio
as a power amplifier reacting on a
the AM sidebands would be exactl
This level can then be used to calc
The ratio of the FM vector to the A
modulation will be the frequency
note that these measurements must
shows a carrier modulated with a
metrical sidebands spaced at the m
sidebands spaced at twice the modu
additional small sidebands are the d
of nonlinearities in the modulating
to the larger sideband is the percent
down from the fundamental modu
distortion of the modulation signal.

The wide sweep and flat fre
analyzers make it possible to do
itself. Figure 5.3-38(h) shows a carr
at 1600. If the 1600 MHz signal wer
signal, the carrier distortion would
must be taken to ensure that the
input mixer of the spectrum analyze
is called harmonic distortion. No s
the specifications of the analyzer b
band and from one frequency to an
tion is produced by nonlinearities in
oscillator, the level of the signal, and
factors in the level of this distortion

Harmonic distortion can be
into the spectrum analyzer at a low
distortion products are first observe
to −20 dBm in the fundamental b

monic's contribution may be small, but the total power may be quite large, necessitating a large value of input attenuation. As a result, the displayed spectral lines have a poor signal-to-noise ratio, and any IF gain introduced will not affect this. Figure 5.3-37(a) shows the degraded signal-to-noise ratio. The preselector reduces the total power level to the analyzer mixer to just those spectral lines falling within its nominally 35-MHz passband. This is much lower than the total comb power and allows the input attenuation and the IF gain to be reduced. Figure 5.3-37(b) illustrates the use of the preselector on the comb of Fig. 5.3-37(a) with the resultant increase in the effective signal-to-noise ratio.

The use of a preselector extends the effective distortion measurement capability of the spectrum analyzer by at least 30 dB. Spurious signals in the spectrum analyzer are produced almost entirely by nonlinear behavior of the input mixer. The level of these signals is typically more than 50 dB below the fundamental signal, as long as the total input power to the mixer is less than −30 dBm. Thus the smallest distortion component (harmonic or IM) that can be measured is limited by the residual distortion of the analyzer.

Spectrum Analyzer Displays, Modulation

In the everyday operation of a spectrum analyzer, it becomes the major part of the operator's job to interpret representative spectra displayed on the CRT face. It would be useful to become acquainted with some simplified concepts of common spectra. Figure 5.3-38 shows typical modulation and distortion spectra. Figure 5.3-38(a) shows the simple CW signal representation with the horizontal axis indicating frequency (as in all these types of measurements); the vertical scale displays amplitude information. Figure 5.3-38(b) shows that, if a CW signal is amplitude-modulated at frequency f_m, two sidebands will appear spaced f_m frequency from the carrier on both sides of the carrier. Figure 5.3-38(c) shows how amplitude-modulation sidebands would look if the carrier were modulated 100% (voltage or linear display). You will notice that the sidebands have an amplitude which is exactly half that of the carrier. In other words, as viewed in the spectrum analyzer, a 100% modulated carrier will have sidebands which are 6 dB down from the carrier. Thus the spectrum analyzer becomes an excellent tool for measuring percent modulation of an amplitude-modulated carrier.

For an arbitrary AM spectrum, as shown in Fig. 5.3-38(d), now in a logarithmic presentation, measure the number of dB the sidebands are down from the carrier. For 26 dB, subtract 6 dB from this to obtain the number of dB down the sidebands are from 100% modulation, in this case, 20 dB. Since 20-dB attenuation is equivalent to a voltage ratio of 0.1, the index of modulation m in this case is equal to 1. The log display permits very small indices to be resolved.

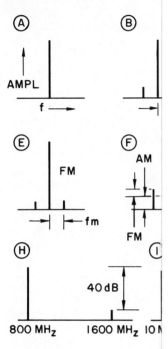

Fig. 5.3-38. Typical mo

Very small values of
analyzer display in exactly
is because the spectrum ana
information were retained,
Fig. 5.3-38(e), with one sid
dotted extending below the
from the carrier by the mo
can be determined in exac
number of dB the sideban
this figure, and convert the
modulation. Since phase m
lation for low indices of m
FM.

Figure 5.3-38(f) show
carrier. You will note that
The only way to have one si
FM or phase modulation to
frequency. Either one alon
duce a symmetrical spectru

distortion product level is a measure of the harmonic distortion. To keep this ratio as large as possible, the input signal level should be kept as low as possible and maximum sensitivity in the analyzer should be utilized.

The second important type of distortion which can occur in the spectrum analyzer is *second*-order intermodulation, shown in Fig. 5.3-38(i). This occurs when a high-frequency signal such as a carrier at 800 MHz mixes with an equal-amplitude low-frequency signal such as a carrier at 10 MHz to produce sidebands around the higher-frequency signal. These sidebands will be spaced, in this case, 10 MHz from the carrier. This type of distortion is measured as before by introducing a clean 10-MHz and 800-MHz signal of the same amplitude into the analyzer and increasing their amplitude together until the sidebands around the 800-MHz signal are observed. The ratio of the 800-MHz carrier to its sideband is a measure of the second-order intermodulation.

The third type of distortion important in the analyzer is called *third*-order intermodulation. This type of distortion occurs when two signals closely spaced together are of large enough amplitude to produce harmonics at twice their frequency. These harmonics then intermodulate with the fundamental signals to produce sidebands around the fundamental signals, as shown in Fig. 5.3-38(j). This third-order intermodulation is really a combination of the harmonic distortion and second-order intermodulation. It is measured the same way by introducing two clean signals at frequencies A and B and then increasing their amplitude until sidebands are first observed. The ratio of the carrier to these sidebands is the third-order intermodulation.

On important measurements, of course, it is recommended practice to use the RF attenuator to ensure that the input mixer is not being overdriven. By switching in 10 dB on the attenuator, any real distortion product from an external signal should move 10 dB on the screen.

A word of caution is due after this explanation, since, on multiband spectrum analyzers, at higher frequencies, many products of harmonic mixing can exist; at first it might be quite confusing for the user until he has a good understanding of the various mixing processes. If one starts using a spectrum analyzer, it is wise to become acquainted with the various mixing processes before passing judgment on a display. It may take, at first, quite a conditioning to expect and calculate the various products of mixing, mostly at higher frequencies. Bandpass filters will not help at all, since these signals do not enter from the source being tested but are generated in the harmonic mixing. Considering these facts, it is obvious that best results can be obtained by using the appropriate mixing harmonic and by restricting the general sweep of the local oscillator to areas where interfering modes are not present.

FREQUENCY MODULATION MEASUREMENTS. If a CW signal is *frequency-modulated*, modulation frequency, bandwidth occupied by significant sidebands, modulation index, and deviation are the most important questions to monic's contribution may be small, but the total power may be quite large, necessitating a large value of input attenuation. As a result, the displayed spectral lines have a poor signal-to-noise ratio, and any IF gain introduced will not affect this. Figure 5.3-37(a) shows the degraded signal-to-noise ratio. The preselector reduces the total power level to the analyzer mixer to just those spectral lines falling within its nominally 35-MHz passband. This is much lower than the total comb power and allows the input attenuation and the IF gain to be reduced. Figure 5.3-37(b) illustrates the use of the preselector on the comb of Fig. 5.3-37(a) with the resultant increase in the effective signal-to-noise ratio.

The use of a preselector extends the effective distortion measurement capability of the spectrum analyzer by at least 30 dB. Spurious signals in the spectrum analyzer are produced almost entirely by nonlinear behavior of the input mixer. The level of these signals is typically more than 50 dB below the fundamental signal, as long as the total input power to the mixer is less than -30 dBm. Thus the smallest distortion component (harmonic or IM) that can be measured is limited by the residual distortion of the analyzer.

Spectrum Analyzer Displays, Modulation

In the everyday operation of a spectrum analyzer, it becomes the major part of the operator's job to interpret representative spectra displayed on the CRT face. It would be useful to become acquainted with some simplified concepts of common spectra. Figure 5.3-38 shows typical modulation and distortion spectra. Figure 5.3-38(a) shows the simple CW signal representation with the horizontal axis indicating frequency (as in all these types of measurements); the vertical scale displays amplitude information. Figure 5.3-38(b) shows that, if a CW signal is amplitude-modulated at frequency f_m, two sidebands will appear spaced f_m frequency from the carrier on both sides of the carrier. Figure 5.3-38(c) shows how amplitude-modulation sidebands would look if the carrier were modulated 100% (voltage or linear display). You will notice that the sidebands have an amplitude which is exactly half that of the carrier. In other words, as viewed in the spectrum analyzer, a 100% modulated carrier will have sidebands which are 6 dB down from the carrier. Thus the spectrum analyzer becomes an excellent tool for measuring percent modulation of an amplitude-modulated carrier.

For an arbitrary AM spectrum, as shown in Fig. 5.3-38(d), now in a logarithmic presentation, measure the number of dB the sidebands are down from the carrier. For 26 dB, subtract 6 dB from this to obtain the number of dB down the sidebands are from 100% modulation, in this case, 20 dB. Since 20-dB attenuation is equivalent to a voltage ratio of 0.1, the index of modulation m in this case is equal to 1. The log display permits very small indices to be resolved.

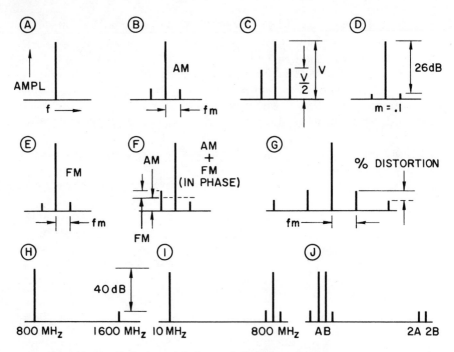

Fig. 5.3-38. Typical modulation and distortion spectra.

Very small values of frequency modulation appear in the spectrum analyzer display in exactly the same manner as amplitude modulation. This is because the spectrum analyzer does not retain phase information. If phase information were retained, the analyzer display would appear as shown in Fig. 5.3-38(e), with one sideband going up and the other sideband as shown dotted extending below the base line. Here again the sidebands are separated from the carrier by the modulation frequency, and the index of modulation can be determined in exactly the same manner as for AM. Measure the number of dB the sidebands are down from a carrier, subtract 6 dB from this figure, and convert the dB into a fraction. This fraction is the index of modulation. Since phase modulation is much the same as frequency modulation for low indices of modulation, its spectrum would appear identical to FM.

Figure 5.3-38(f) shows the spectrum of an AM- and FM-modulated carrier. You will note that the sidebands are not symmetrical in amplitude. The only way to have one sideband larger than the other is for both AM and FM or phase modulation to exist simultaneously and at the same modulation frequency. Either one alone, regardless of the wave shape, will always produce a symmetrical spectrum in amplitude and frequency about the carrier.

This asymmetry in amplitude can be easily explained by noting in Fig. 5.3-38(e) that one FM sideband extends above the line and one below the line when phase information is retained.

The spectrum analyzer adds together the AM and FM spectra and displays the absolute magnitude of the result. One sideband is added to and the other sideband is reduced. The difference in the amplitude between these two sidebands is twice the amplitude of the FM vector. Realizing this, both the AM index of modulation and the frequency index of modulation can be measured. This measurement depends on the peak FM deviation being in phase with the peak AM. That is the general case, since incidental FM is usually caused by the peak excursion of amplitude-modulating elements such as a power amplifier reacting on an oscillator. If the FM were not present, the AM sidebands would be exactly the average of the two sidebands shown. This level can then be used to calculate the percentage of AM modulation. The ratio of the FM vector to the AM vector multiplied by the AM index of modulation will be the frequency index of modulation. It is important to note that these measurements must be made on a linear scale. Figure 5.3-38(g) shows a carrier modulated with a pure sine-wave tone producing two symmetrical sidebands spaced at the modulating frequency and two additional sidebands spaced at twice the modulating frequency from the carrier. These additional small sidebands are the distortion products which resulted because of nonlinearities in the modulating process. The ratio of the smaller sideband to the larger sideband is the percent of distortion. Distortion sidebands 20 dB down from the fundamental modulating sidebands would indicate a 10% distortion of the modulation signal.

The wide sweep and flat frequency response of the new spectrum analyzers make it possible to do distortion measurement on the carrier itself. Figure 5.3-38(h) shows a carrier at 800 MHz and its second harmonic at 1600. If the 1600 MHz signal were, for example, 40 dB below the 800 MHz signal, the carrier distortion would be 1%. In making this measurement, care must be taken to ensure that the distortion shown is not produced in the input mixer of the spectrum analyzer. This type of distortion in the analyzer is called harmonic distortion. No specification for this distortion is given in the specifications of the analyzer because it varies so widely from band to band and from one frequency to another within the band. Since this distortion is produced by nonlinearities in the mixing process, the level of the local oscillator, the level of the signal, and the design of the mixer are all important factors in the level of this distortion.

Harmonic distortion can be measured by introducing a clean carrier into the spectrum analyzer at a low level and increasing the input level until distortion products are first observed. This typically will vary from -30 dBm to -20 dBm in the fundamental band. The difference in carrier level and distortion product level is a measure of the harmonic distortion. To keep this ratio as large as possible, the input signal level should be kept as low as possible and maximum sensitivity in the analyzer should be utilized.

The second important type of distortion which can occur in the spectrum analyzer is *second*-order intermodulation, shown in Fig. 5.3-38(i). This occurs when a high-frequency signal such as a carrier at 800 MHz mixes with an equal-amplitude low-frequency signal such as a carrier at 10 MHz to produce sidebands around the higher-frequency signal. These sidebands will be spaced, in this case, 10 MHz from the carrier. This type of distortion is measured as before by introducing a clean 10-MHz and 800-MHz signal of the same amplitude into the analyzer and increasing their amplitude together until the sidebands around the 800-MHz signal are observed. The ratio of the 800-MHz carrier to its sideband is a measure of the second-order intermodulation.

The third type of distortion important in the analyzer is called *third*-order intermodulation. This type of distortion occurs when two signals closely spaced together are of large enough amplitude to produce harmonics at twice their frequency. These harmonics then intermodulate with the fundamental signals to produce sidebands around the fundamental signals, as shown in Fig. 5.3-38(j). This third-order intermodulation is really a combination of the harmonic distortion and second-order intermodulation. It is measured the same way by introducing two clean signals at frequencies A and B and then increasing their amplitude until sidebands are first observed. The ratio of the carrier to these sidebands is the third-order intermodulation.

On important measurements, of course, it is recommended practice to use the RF attenuator to ensure that the input mixer is not being overdriven. By switching in 10 dB on the attenuator, any real distortion product from an external signal should move 10 dB on the screen.

A word of caution is due after this explanation, since, on multiband spectrum analyzers, at higher frequencies, many products of harmonic mixing can exist; at first it might be quite confusing for the user until he has a good understanding of the various mixing processes. If one starts using a spectrum analyzer, it is wise to become acquainted with the various mixing processes before passing judgment on a display. It may take, at first, quite a conditioning to expect and calculate the various products of mixing, mostly at higher frequencies. Bandpass filters will not help at all, since these signals do not enter from the source being tested but are generated in the harmonic mixing. Considering these facts, it is obvious that best results can be obtained by using the appropriate mixing harmonic and by restricting the general sweep of the local oscillator to areas where interfering modes are not present.

FREQUENCY MODULATION MEASUREMENTS. If a CW signal is *frequency-modulated*, modulation frequency, bandwidth occupied by significant sidebands, modulation index, and deviation are the most important questions to be answered. As shown in Fig. 5.3-38(g), the modulating frequency can easily be determined from the spectrum of a single-tone frequency-modulated signal, since the spectral lines (sidebands) will be spaced apart exactly by the modulating frequency. It was also mentioned before that, theoretically, an FM signal has an infinite number of sidebands; but the bandwidth occupied by the significant sidebands, that is, the bandwidth where sidebands have high enough amplitude, is of much interest. For a quick approximation of this bandwidth, multiply the sum of the carrier deviation and the modulating frequency by two. Mathematically,

$$BW = 2(\Delta F_c + f_a)$$

To determine modulation index and deviation, first let us acquaint ourselves with their relationship. By knowing the modulation frequency and the modulation indices where the carrier amplitude goes to zero (all energy in the sidebands), one can check the FM deviation. For single-tone FM, the modulation index is given by the formula

$$m = \frac{\Delta f_c}{f_a}$$

where m = modulation index,
Δf_c = carrier deviation,
f_a = modulation frequency.

If carrier deviation is the unknown,

$$\Delta f_c = m f_a$$

The values of m corresponding to zero carrier amplitude are listed in Table 5.3-1 for convenience. The table is valid for any combination of carrier deviation and modulation frequency producing the modulation indices listed.

Table 5.3-1. Values of Modulation Index for Which Carrier
Amplitude Is Zero

Order of Carrier Zero	Modulation Index
1	2.40
2	5.52
3	8.65
4	11.79
5	14.93
6	18.07
n ($n > 6$)	$18.07 + \pi (n - 6)$

Figure 5.3-39 shows a setup for accurately measuring the FM deviation of an FM signal generator with the spectrum analyzer. The generator is first placed in CW operation by disconnecting the audio oscillator from the

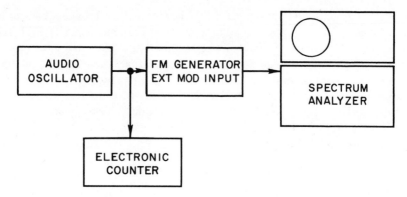

Fig. 5.3-39. Block diagram of FM deviation.

external FM jack of the generator. The spectrum analyzer and generator are tuned to the desired carrier frequency, producing a single CW response in the center of the analyzer display. The frequency of the analyzer is then stabilized and the attenuator and IF gain controls adjusted for a good display. The SPECTRUM WIDTH control is switched to the 10 kHz/CM position and its vernier used further to reduce spectrum width to about 2 kHz/CM. The IF bandwidth is switched to 1 kHz for good resolution of the carrier and sidebands.

The audio oscillator is connected to the FM generator, as shown in Fig. 5.3-39, and its frequency is accurately set to 6250 Hz, as indicated by the electronic counter. When multiplied by a modulation index of 2.40, this frequency results in a carrier deviation of 15 kHz. Then, with the audio oscillator amplitude control, the audio-modulating signal voltage is slowly increased from zero until the amplitude of the carrier response on the analyzer display first goes to zero. As the carrier amplitude is decreasing, sidebands begin appearing at intervals of 6250 Hz above and below the carrier; when the carrier amplitude is zero, all the energy is contained in the sidebands. At this point the generator deviation is 15 kHz and the deviation meter is checked for full-scale accuracy.

If the audio oscillator amplitude is slowly increased further, the carrier amplitude will increase again to some maximum and then begin decreasing until it goes to zero amplitude for the second time. At this point the modulation index is 5.52, as indicated in Table 5.3-1 for a second-order carrier zero. Note that any deviation, within limits of the generator, of course, may be set up by choosing the correct combinations of modulation index and modulation frequency. This technique is known as the Crosby Zero Method for measuring frequency deviation.

Using a modern spectrum analyzer with wide dynamic range makes it possible to determine accurately when the carrier is zero, because the IF gain may be increased, as carrier zero is approached, for maximum sensitivity without the large adjacent sidebands saturating the IF. Accurate carrier zero is essential for correct deviation measurements. It should also be stated that the analyzer must have a resolution at least three times better than the modulation frequency to be used to distinguish between the carrier and first sideband responses. Resolving capability of the analyzer, you recall, is largely determined by the IF bandwidth.

Residual FM of signal generators may also be checked, provided the peak-to-peak deviation is 10 kHz or greater. This is done by switching the SYNC control on the analyzer to INT and setting the SWEEP TIME slightly different than the power line frequency or submultiple (3 ms/cm for 60-Hz line). With the generator set for CW operation and the analyzer tuned to the operating frequency, residual FM will be indicated by a slow periodic movement of the CW response back and forth on the analyzer display. This movement will be at the differential rate of the line frequency and the analyzer's sweep time. The peak-to-peak deviation of the generator's residual FM is then measured by noting the maximum horizontal excursion of the CW response and reading the frequency from the calibrated spectrum-width control.

To measure *frequency modulation linearity* of voltage-tuned oscillators, like klystrons or similar microwave tubes, the Crosby Zero Method can be extended. A modern spectrum analyzer with wide dynamic range and calibrated controls is an excellent tool for such measurements, since the technique depends on discerning the point where the carrier amplitude has gone to zero in a spectrum display. A 60-dB logarithmic display provides extreme accuracy in this measurement. Changes in linearity as small as 0.1% can be measured by measuring small changes in the index of modulation. The technique depends on the fact that, at an index of modulation (carrier peak deviation/modulation frequency) of 2.405, the center frequency component of the spectrum goes to zero. By setting up the precise 2.405 index of modulation at various points on a tube-tuning curve, the modulation sensitivity can be measured. Thus a comparison of these modulation sensitivities gives the tuning linearity across the range.

The technique used is to modulate the microwave source with a low-frequency test signal with small deviation and then to position the modulating voltage along the tuning curve by dc voltage changes on the tube. The measurement procedure is as follows:

1. Set up the tube to be tested on CW, and position its frequency in the center of its band. Tune in the signal on the spectrum analyzer.
2. Connect a modulating oscillator whose frequency is monitored with an electronic counter to the FM input jack of the tube under test. The modulating oscillator must maintain a constant amplitude within

the accuracy of the linearity desired. Thus a 1 % change in amplitude of the oscillator results in a 1 % inaccuracy in the linearity measurement.

3. Pick a convenient modulating frequency, such as 10 kHz, which provides FM sidebands that are easily discerned with the analyzer at every 10-kHz spacing.

4. Increase the audio-modulating signal voltage until the amplitude of the carrier on the analyzer log display first goes down to zero. As the carrier amplitude is decreasing, the other sidebands are appearing on both sides of the carrier at the 10-kHz spacing. At the first carrier-equal-zero point, the modulating index is 2.405. Read and log the exact electronic counter readout of the audio frequency.

5. Now tune the repeller to another point on the tuning curve, using the appropriate dc repeller-voltage adjustment. The presentation will move off the analyzer display. After returning it back on the center of the analyzer, the carrier spectral line has moved up from zero.

6. Without readjusting the oscillator amplitude, reset the modulation frequency until the carrier zero is again realized, and again read the modulation frequency with the electronic counter.

7. Since at both conditions of measurement the modulation index has been set to $m = 2.405$, it can be shown that the modulation linearity is equal to the percentage change in modulation frequency. If the modulation signal amplitude was constant, the fractional change in slope (sensitivity) is calculated as follows:

$$\text{Linearity} = \frac{fm_1 - fm_2}{fm_1}$$

where fm_1 and fm_2 are the first and second modulating frequencies.

Thus, if the first modulation frequency is 10.00 kHz and the second modulation frequency is 10.100 kHz, the tuning linearity is 1 %.

The carrier zero method of measuring FM deviation is also a very powerful technique for use in calibrating FM signal generators and other FM deviation meters. For the analyzer, modulation frequencies in the 10-kHz range are most useful, since this is generally within the passband of most FM modulation circuits and, at the same time, provides a display that can have the spectral lines spaced by at least 1 cm for good resolution on the spectrum analyzer. Table 5.3-2 is a useful chart that provides the modulation frequency to be set on the counter for commonly used values of deviation for the various orders of carrier zeros.

The procedure for setting up a known deviation is as follows:

1. Select the column with the appropriate deviation required, such as, for example, 250 kHz.

Table 5.3-2. List of Modulation Frequencies to Be Used to Set Up
Certain Convenient FM Deviations

Order of Carrier Zero	Modu- lation Index	Commonly Used Values of FM Peak Deviation										
		7.5 kHz	10 kHz	15 kHz	25 kHz	30 kHz	50 kHz	75 kHz	100 kHz	150 kHz	250 kHz	300 kHz
1	2.40	3.12	4.16	6.25	10.42	12.50	20.83	31.25	41.67	62.50	104.17	125.00
2	5.52	1.36	1.81	2.72	4.53	5.43	9.06	13.59	18.12	27.17	45.29	54.35
3	8.65	.87	1.16	1.73	2.89	3.47	5.78	8.67	11.56	17.34	28.90	34.68
4	11.79	.66	.85	1.27	2.12	2.54	4.24	6.36	8.48	12.72	21.20	25.45
5	14.93	.50	.67	1.00	1.67	2.01	3.35	5.02	6.70	10.05	16.74	20.09
6	18.07	.42	.55	.83	1.88	1.66	2.77	4.15	5.53	8.30	13.84	16.60

2. Select an order of carrier zero number which gives a frequency in the table that is commensurate with the normal modulation bandwidth of the generator to be tested. For example, if an audio modulation circuit is provided in the 250 kHz example above, it will be necessary to go to the fifth carrier zero to get a modulating frequency within the audio passband of the generator (16.74 kHz).
3. Set the modulating frequency to 16.74 kHz. Monitor the generator output spectrum on the analyzer, and adjust the amplitude of the audio-modulating signal until the carrier amplitude has gone through four zeros and stops when the carrier is at its fifth minimum. With the modulating frequency of 16.7 kHz and the spectrum at its fifth zero, then a unique 250-kHz deviation is being provided by the setup. The modulation meter may then be calibrated. A quick check can be made by moving to the adjacent carrier zero and resetting the modulating frequency and amplitude (i.e., 13.84 at the sixth carrier zero in the above example).

Other intermediate deviations and modulation indices are settable using various orders of sideband zeros, but these are influenced by incidental amplitude modulation. Since it is known that amplitude modulation does not cause the carrier to change, but instead puts all the modulation power into the sidebands, incidental AM will not affect the Crosby Zero Method.

5.3.5 PULSE MEASUREMENTS

A CW signal modulated with a rectangular *pulse* train has been shown to have a spectrum containing information of the carrier frequency, pulse repetition frequency (PRF), and modulating pulse width. The carrier frequency is represented on the spectrum of a pulse display by the center, longest spectral line in the main lobe. PRF is represented by the spacing of the spectral lines, since each harmonic of the repetition frequency is included in the spectrum as a separate line spaced equally apart from each other line by exactly the repetition frequency. Spectral line frequencies may be expressed as:

$$F_1 = F_c \pm nf_r$$

where F_c = carrier frequency,

f_r = PRF (pulse repetition frequency),

n = 0, 1, 2, 3, ...

In a pulse spectrum display, if the pulses are perfectly rectangular (or for other functions whose derivatives are discontinuous at some points), the number of side lobes is infinite.

For a perfectly rectangular pulse, amplitude of the spectral lines forming the lobes varies as a function of frequency according to the expression

$$\frac{\mathrm{Sin}\ \omega \frac{\tau}{2}}{\omega \frac{\tau}{2}}$$

where τ represents pulse width.

Thus the points where these lines go through zero amplitudes are determined by the modulating pulse width only. As pulse width becomes shorter, minima of the envelope become further removed in frequency from the carrier, and the lobes become wider. The side-lobe widths Δf in frequency are related to the modulating pulse width by the expression $\Delta f = 1/\tau$. Since the main lobe contains the origin of the spectrum (the carrier frequency), the upper and lower sidebands extending from this point form a *main* lobe $2/\tau$ wide. Remember, however, that the total *number* of side lobes remains constant as long as the pulse quality, or shape, is unchanged and only its width is varied. Figure 5.3-40 compares the spectral plots for two pulse lengths, each at two repetition rates with carrier frequency held constant.

Notice in the drawings how the spectral lines extend below as well as above the base line. This corresponds to harmonics in the modulating pulse having a phase relationship of 180° with respect to the fundamental of the modulating waveform. Because present spectrum analyzers can only detect amplitude and not phase, the negative-going spectral lines will be inverted and all lines will only extend above the base line. Thus the phase information is lost in the display. This does not seriously limit the usefulness of the spectrum analyzer.

Spectrum analyzers were invented to measure the characteristics of pulsed magnetrons in radar, an application that continues to be of major importance. This involves measuring a pulse radar's magnetron output to ensure stable oscillation, free of moding and spurious signals. Present-day

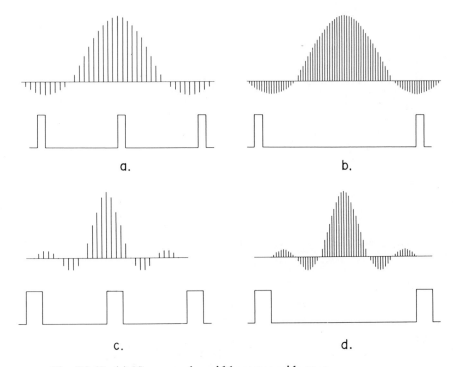

Fig. 5.3-40. (a) Narrow pulse width causes wide spectrum lobes, high PRF results in low spectral
line density, (b) wider pulse than (a) causes
narrower lobes but line density remains
constant since PRF is unchanged, (c) PRF
lower than (a) results in higher spectral
density. Lobe width is same as (a) since
pulse widths are identical. (d) Spectral
density and PRF unchanged from (c) but
lobe widths are reduced by wider pulse.

spectrum analyzers are ideal for these tasks for several reasons. The wide
range allows practically any radar frequency to be received. Broad spectrum
width enables complete displays of spectrum signatures. Wide-image separation (typically 4 GHz) and comparatively few spurious responses keep the
display uncluttered for accurate presentation of a radar spectrum.

To analyze or tune up a pulse radar's magnetron transmitter, connect
the analyzer RF input to a sample of the radar transmitter power through a
directional coupler of at least 30 dB coupling. Most radars have built-in
couplers providing a test output of about a milliwatt. Figure 5.3-41 shows this

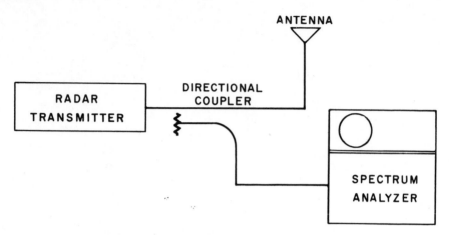

Fig. 5.3-41. Spectrum analyzer checking radar
magnetron operation.

setup. With the radar high voltage on, tune the analyzer to the transmitter
frequency which will center the main lobe of the radar's pulse spectrum on the
analyzer screen. Check for overloading of the analyzer's mixer by increasing
the RF input attenuator of the analyzer by 10 dB. The amplitude should uniformly decrease by 10 dB on the analyzer's display. If they do not, continue
increasing RF attenuation until this condition is reached. The IF gain may
be used to increase the display amplitude in accurate 10-dB steps as RF
attenuation is added.

The display should be symmetrical, approximating the envelope shown
in Fig. 5.3-42(a) (Figs. 5.3-42(a) through (f) are shown on a linear scale). If
incidental FM is present, it will show up as a loss of power in the main lobe
and increased power in the side lobes, as shown in Figs. 5.3-42(b) and (c). The
addition of linear AM causes the spectrum to become unsymmetrical, as
shown in Figs. 5.3-42(d) and (e). Incidental AM alone causes the side-lobe
amplitudes to decrease while the main lobe remains symmetrical, as illustrated
in Fig. 5.3-42(f). For closer observation of the side lobes, switch the analyzer
to a logarithmic display so the main lobe will be compressed and the side
lobes enlarged by the response of the analyzer. The photos in Fig. 5.3-43(a),
(b), and (c) are examples of good and bad spectra commonly encountered in
the field. Figure 5.3-43(d) points up the advantage of the accurate log display
used in Fig. 5.3-43(c) for good side-lobe detail.

With the equipment still connected as in Fig. 5.3-41, measure the frequency spread of the main lobe using the calibrated spectrum width control
and CRT graticule. Remember to measure this width at the minima of the

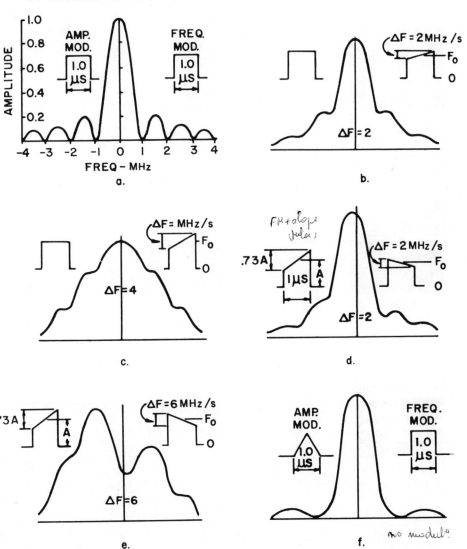

Fig. 5.3-42.

main lobe. Calculate the *modulating pulse midwidth* τ in microseconds from this by the equation

$$\tau(\mu s) = \frac{2}{f_{(MHz)}}$$

where $f_{(MHz)}$ = measured main lobe width, in megaHertz.

Pulse width may also be calculated from measurements of any of the

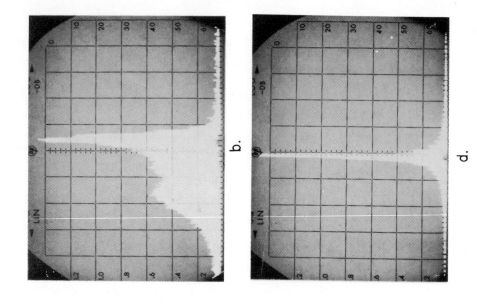

a.

b.

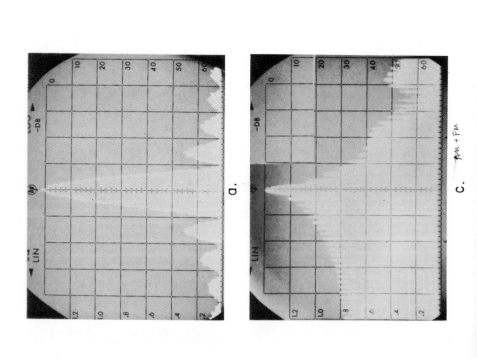

c. AM + FM

d.

side lobes, depending on the preference of the operator. Simply divide the right-hand member of the equation by 2, since the side lobes are exactly half the main lobe width.

There are occasions when it is desirable to check radar performance at a location where direct connection to the radar is impossible. In such instances, *pulse repetition frequency* (PRF) cannot be measured in the conventional manner with a crystal detector and oscilloscope, since there is insufficient power available from a pickup horn.

The Hewlett-Packard spectrum analyzer has the required sensitivity for good spectral displays using waveguide horns in proximity with the radar antenna. To measure PRF in remote locations, connect the analyzer's RF input to a suitable horn aimed at the radar antenna. Now connect the 20-MHz IF output from the rear of the spectrum analyzer to the input of a high-frequency oscilloscope and tune the analyzer to the radar center frequency. Now the calibrated sweep time of the oscilloscope can be used to measure the time between bursts of the 20-MHz IF output which correspond to the radar's pulse repetition period. The radar PRF is the reciprocal of the period measured. If the PRF were high enough, it could be measured directly by the spacing of spectral lines on the analyzer display. However, most pulse radars have a PRF of 1 kHz or less, resulting in spectral line spacing too close for even high-resolution analyzers to distinguish.

Another important behavior of a magnetron transmitter of high importance to be measured is the *frequency stability*. Two tests are to be made to determine its stability. *Frequency drift* can be measured with the same setup shown in Fig. 5.3-41 or through a horn, as described above. (When using the latter pickup technique, it is best to stop rotation of the antenna so its radiated power is toward the horn for constant spectrum amplitude on the display.)

By placing the main lobe of the display to a convenient reference mark along the X-axis of the CRT, the drift in frequency is easily determined by noting the total shift of the spectrum in centimeters on the X-axis and multiplying by the spectrum width setting of the analyzer.

Magnetron pulling is the other criterion of stability. The following check for magnetron pulling requires connecting to the radar, as in Fig. 5.3-41, since transmitter power must be sampled before it reaches the rotary joints of the antenna system. With the antenna scan system operating, observe the behavior of the spectrum. As the antenna rotates, watch for a periodic shift left and right or a breathing effect of the entire spectrum. This is magnetron "pulling," which is a shift in operating frequency as the phase of the load impedance reflection coefficient is varied through 360 degrees. The pulling effect can be measured by noting the maximum shift, in centimeters, of the main lobe on the screen and multiplying by the spectrum width setting, as was done for measuring drift. A moderate shift is normal for radars not employing ferrite isolators at the magnetron output; however, excess pulling may be due to an improperly tuned magnetron or high-load VSWR.

Spectrum Analyzer Response to Pulsed Carrier Signals—Radar

The responses that a spectrum analyzer, or any sweeping receiver, can have to a periodically pulsed RF signal can be of two kinds, which result in two different but similar displays. One response is called a "line" spectrum, and the other is called a "pulse" spectrum.

Keep in mind that these are both responses to a periodically pulsed RF input signal, and the "line" and "pulse" spectra refer to the response or display on the spectrum analyzer.

A LINE SPECTRUM occurs when the spectrum analyzer 3-dB bandwidth (BW) is less than the most closely spaced spectral components of the input signal. Since the individual spectral components are spaced by the pulse repetition frequency (PRF) of the periodically pulsed RF, this means that the spectrum analyzer BW must be substantially less than the PRF to get this type of display. All individual frequency components can be resolved, for only one is within the BW at a time. The display is truly a frequency domain display of the actual Fourier components of the input signal. Each component behaves as a CW signal would. The display has the normal true spectrum frequency domain characteristics:

1. The spacing between lines on the display will *not* change when the analyzer sweep time, display cm/s, is changed.
2. The amplitude of each line on the display, as measured by CW substitution, will not change as the BW is changed. (Of course, the BW must stay below the PRF to stay in this "line" spectrum mode.) The displayed height may change if the analyzer gain changes with BW, but the measured signal amplitude will not.

A PULSE SPECTRUM occurs when the spectrum analyzer BW is greater than the PRF. The spectrum analyzer in this case cannot resolve actual individual Fourier frequency domain components, since several lines occur within its bandwidth. However, if the spectrum analyzer BW is narrow compared to the spectrum envelope, then the envelope can be resolved. The display is not a true frequency domain display, but a combination time-and-frequency display. It is a time-domain display for the display pulse lines, since each pulse line occurs as each RF pulse occurs. The display lines occur at the actual PRF. It is a frequency-domain display of the spectrum envelope. The display has three distinguishing characteristics.

1. The spacing between the pulse lines on the display *will* increase linearly with the sweep speed, display cm/s. The pulse lines on the display occur at the PRF and are spaced in real time by 1/PRF. The shape of the spectrum envelope will not change with sweep speed.
2. The spacing between lines on the display will not change when the spectrum width, MHz/cm, is changed. The spectrum envelope will change horizontally, as one would expect.
3. The amplitude of the display envelope, as measured by CW substitution, will increase linearly as BW is increased. The linear increase means a 6-dB increase for a doubling of BW. This increase in amplitude with BW keeps up until the BW equals about 1/2 the width of the main lobe of the spectrum envelope; then no further increase occurs. At this point the BW has collected almost all the spectral components. At this point also, we have lost the ability to resolve the spectrum envelope. We can tell crudely in frequency where the whole spectrum is but get little information about its shape. The display is now time domain with respect to the spectrum. If the frequency scan were stopped by setting the spectrum width control to zero and one could sweep the display fast enough, one could begin to distinguish the time-domain shape of the pulse envelope. If the (BW) were increased further, becoming much wider than the main lobe width, the detailed shape in time domain of the pulse envelope could be observed.

In the pulse spectrum just described, the response of the spectrum analyzer to each RF input pulse is the impulse response of the analyzer's IF amplifier. The height of these impulse responses traces out the shape of the input spectrum as the front-end tuning is swept across the input spectrum. This impulse-type response is the explanation for the characteristics of this display. There are two requirements that must be met to have an impulse response, and both these are met when viewing a periodically pulsed RF signal on a spectrum analyzer in the pulse spectrum mode.

1. The system must respond to each pulse independently. The effects of one pulse must decay out in the system before the next pulse occurs. This recovery is assured by the requirement that the BW be greater than the PRF. The decay-time constant, time to decay to $1/e$ of its initial signal level, for an IF bandpass amplifier is approximately $0.3/(BW)$, where (BW) is the 3-dB bandwidth. So, for instance, if the (BW) = 1.7 (PRF), then the IF amplifier will have five time constants in which to recover between pulses and will decay to less than 1% of its response to one pulse when the next pulse occurs.
2. The input pulse must be short compared to the 10–90% rise time of

the system. The rise time of the system is that of the IF amplifier, and this is approximately 0.7/(BW). This condition is met by the requirement that the BW must be narrow compared to the spectrum envelope and small compared to the base width, $2\Delta F$, of the main lobe. The width, T_{eff}, of the RF pulse is $1/\Delta F$, so, if the BW is small compared to ΔF, then the rise time of the IF amplifier will be long compared to the duration of the RF input pulse. In all the relations, frequency is in Hz and time in seconds of consistent multiples, i.e., MHz and μs.

WHY USE A PULSE SPECTRUM RESPONSE? The spectrum envelope—not individual spectral component lines—is usually what is of interest. The use of the wider BW pulse spectrum display gives a greater response than with the line spectrum display. The display amplitude with a line spectrum is:

$$T_{\text{eff}}(\text{PRF})E_P'$$

where T_{eff} = pulse width,
   E_P' = peak amplitude of the pulse.

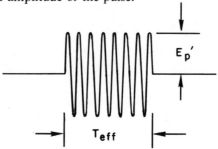

Fig. 5.3-44. "Line" spectrum.

With a pulse spectrum it is:

$$KT_{\text{eff}}(\text{BW})E_P'$$

where $K \approx \frac{3}{2}$,

so

$$\frac{\text{Pulse Spectrum Response}}{\text{Line Spectrum Response}} = \frac{K(\text{BW})}{(\text{PRF})} \approx \frac{3}{2}\frac{(\text{BW})}{(\text{PRF})}$$

Thus the amplitude of the response increases linearly with (BW), as is always the case with an impulse response.

The noise voltage level will increase as the $\sqrt{(\text{BW})}$, so the increase in signal-to-noise ratio increases as the $\sqrt{(\text{BW})}$. Thus, operating with a BW greater than the PRF increases sensitivity to pulsed RF signals. This, of course, also increases the dynamic range, since one gets the highest signal-to-noise ratio with the input level held to a value such that the peak-pulse signal does not overload the input mixer.

Pulse Desensitization[15]

There are some interesting and useful relationships that pertain to the formula for the amplitude of the pulse spectrum display. The equation is:

$$\frac{\text{Peak Amplitude of the Spectrum Display}}{\text{Peak Amplitude of the RF Pulse}^{16}} = K(\text{BW})T_{\text{eff}},$$

where $(\text{BW}) = \Delta f = $ 3-dB bandwidth,

$K = \frac{3}{2}$ (for a Gaussian or multistage synchronously tuned IF amplifier, which most spectrum analyzers approximate),

$K(\text{BW}) = $ the impulse bandwidth, IBW, of the spectrum analyzer. This is the same IBW that is commonly given for RFI receivers, etc.

$$T_{\text{eff}} = \frac{1}{\Delta F}$$

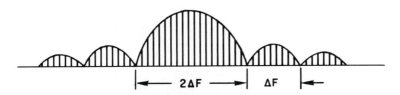

Fig. 5.3-45. Lobes on pulse spectrum.

where ΔF is the separation between minor lobe nulls of a trapezoidal-shaped RF pulse spectrum, so

$$\frac{\text{Peak Amplitude of Spectrum Display}}{\text{Peak Amplitude of Pulsed RF}} = \frac{\text{IBW}}{\Delta F}$$

if $\text{IBW} < \Delta F$.

This $KT_{\text{eff}}(\text{BW}) = \dfrac{\text{IBW}}{\Delta F}$ factor is sometimes called the pulse desensitization factor, since it tells how much the response of the analyzer or receiver is reduced because the input RF signal is a pulse compared to what the response would be if the signal were a CW signal of the same amplitude as the pulse. So this is indeed desensitization when one is determining the peak amplitude

---

[15] Private conversation with Rod Carlson, Hewlett-Packard Co., June 1967.

[16] By this is meant the display amplitude that the spectrum analyzer would have to an unmodulated signal of amplitude equal to the pulse peak.

of the pulse. This pulse spectrum mode of operation gives a sensitization with respect to the true spectrum, the line spectrum, that exists. This sensitization factor is the

$$\frac{K(\text{BW})}{(\text{PRF})} = \frac{(\text{IBW})}{(\text{PRF})} \approx \frac{3}{2}\frac{(\text{BW})}{(\text{PRF})}$$

given previously. So it is desensitization in one respect and sensitization in another.

Some Rules of Thumb for Choosing Bandwidth and Other Control Settings When Viewing Pulsed RF Spectra

SENSITIVITY VERSUS ENVELOPE RESOLUTION. The BW should be as wide as possible for the highest sensitivity, but not wider than $0.1/T_{\text{eff}}$, where T_{eff} is the pulse midwidth, or the resolution of the spectrum envelope will be seriously impaired. Another way of saying this is that the BW should be less than 5% of the width of the spectrum envelope main lobe. If higher resolution is required to look at faster-falling spectrum envelope skirts or to look more deeply into nulls, narrower BW is required. To resolve 20–30 dB into nulls, the rule would be

$$\text{BW} < \frac{0.03}{T_{\text{eff}}}$$

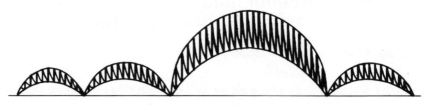

Fig. 5.3-46. Pulse spectrum display with PRF too fast and inadequate BW.

AVOIDANCE OF BASE LINE LIFTING. This phenomenon, shown in Fig. 5.3-46, occurs with a pulse spectrum when the response from one pulse has not fully decayed in the spectrum analyzer when the next pulse occurs. As mentioned before, the IF amplifier decay-time constant is $\approx 0.3/(\text{BW})$, and for the pulse to decay down to 1% requires five time constants. So to meet this requirement, the rule is

$$(\text{BW}) > 1.7(\text{PRF})$$

This will keep the lifted base line at least 40 dB below the spectrum envelope.

Actually, less than 1 dB error in the observed spectrum envelope seems to occur right up to the point where BW = PRF, at which point the base line is about 25 dB below the envelope. Base line lift should not be a concern up to this point, as it does not affect the validity of the usual spectrum envelope.

AVOIDANCE OF ENVELOPE PEAK FLUCTUATION. Fluctuation of the spectrum envelope peak on the display occurs when the pulse does not occur during the time the analyzer is sweeping through the peak of the spectrum envelope on each sweep. This occurs with fast sweeps, low PRF, and wide pulses, and the fluctuation looks almost random at times and like something is cutting on and off. To avoid this, the time to scan through 10% of the main lobe width must be greater than 1/PRF. So the rule is

$$\left(\frac{df}{dt}\right) < 0.2(\text{PRF})\Delta F$$

or

$$\left(\frac{df}{dt}\right) < \frac{0.2(\text{PRF})}{T_{\text{eff}}}$$

where $\left(\frac{df}{dt}\right)$ = analyzer frequency sweep velocity in Hz/sec,

ΔF = minor lobe null separation in Hz (the main lobe base width = $2\Delta F$),

T_{eff} = RF pulse width in sec.

This rule ensures that the peak will be displayed on each sweep and avoids having to watch several sweeps to catch the peak value of the displayed spectrum.

LOSS IN SENSITIVITY DUE TO SWEEPING TOO FAST. If one sweeps a spectrum analyzer past a CW signal too fast, a loss in response occurs. This is a well-known effect. The rule to avoid this was given before and is

$$\left(\frac{df}{dt}\right) < (\text{BW})^2$$

where $\frac{df}{dt}$ = frequency sweep rate.

This rule is important in pulse work when calibrating with CW signals. It applies to any line spectrum situation.

Loss in sensitivity can also occur from sweeping too fast through the pulse spectrum of a periodically pulsed RF signal. The rule for less than 1 dB loss in sensitivity is

$$\left(\frac{df}{dt}\right) < \frac{2.5}{T_{\text{eff}}^2}$$

This relationship comes from an article in the June 1954 issue of the then *Proceedings of the IRE*. This means that with a pulse spectrum, one can sweep much faster than he might expect from the CW rule. Where the

(BW) $< 0.1/T$, one could sweep more than 2,500 times as fast as the CW rule would say! This phenomenon was verified on the HP spectrum analyzer. It is an interesting effect but practically not a problem, since one will always run into the preceding effects explained above before this effect sets in. This is because

$$\text{PRF always} \ll \frac{1}{T_{\text{eff}}}$$

MEASURING PEAK POWER. When measuring peak power, it is convenient to have sufficient BW so that the analyzer will respond fully during the pulse, and peak power can be measured by direct CW substitution with no BW corrections. This requires that the analyzer rise time be short compared to the pulse duration.

If $(BW) > 1/T_{\text{eff}}$, at least $1\frac{1}{2}$ rise times are allowed for the analyzer to rise to the peak value, and the error will be less than 3%. This BW is useful, of course, only for peak-pulse amplitude measurement, since little spectrum resolution can be made with it. If sufficient BW is not available to meet this requirement, peak power is, of course, measured by applying the pulse de-sensitization factor, $K(BW)T_{\text{eff}}$.

FREQUENCY DOMAIN. Most of the preceding has dealt with the pulse spectrum mode of response. The area between it and the line spectrum mode of response is a gray area which does not have quite the characteristics of either type of display. In this area, interpretation of the display is difficult, and it is best to avoid it.

The criterion for being in the line spectrum response mode when viewing a periodically pulsed RF signal is that

$$(BW) < \frac{(PRF)}{10}$$

It is required in this case that there be no decay in response between pulses, and the above allows less than 3 dB decay. The above also results in an intersection of the IF BW filter skirts, that is, more than 25 dB down.

The criterion for being in the pulse spectrum mode was that $(BW) > (PRF)$, so there is a decade of gray area to avoid.

5.3.6 RADIO FREQUENCY INTERFERENCE (RFI/EMI)

In the 1920's there was only a handful of radio broadcast stations in the whole United States. Very few families had radio receiver sets in their homes. However, electronics went through a very rapid growth, and the electromagnetic spectrum consequently grew more intense with addition of more communications: TV, facsimile, radar, navigation aids. The prevention of radio frequency interference (RFI) became extremely important. In the United States the Federal Communications Commission and in Europe The Hague (the Netherlands) Convention had to assign channel allocations very carefully, limiting bandwidth and radiated power of each channel to prevent interference of stations with one another.

Radio frequency interference is not generated only by radio transmitters; there are many sources which cause serious RFI problems. Spurious signals, harmonics, RF leakage, and transients are only some of the forms of interference caused by an electronic or electrical device. The U.S. government establishes standard methods for determining RFI emanating from devices and documents them in the form of military standards. Any instrument or device needs careful testing and control of radiation to be supplied for government installations.

Interference may occur anywhere in the spectrum, and the problem of checking for its presence and intensity is manifold. The common approach has been to tune a number of receivers, with various antennas or probes, through their frequency ranges, searching for undesired signals indicated on a calibrated signal-strength meter. The process is long and tedious, with band-switching, slideback detector adjustments, mechanical cranking, and the constant concern that one might have "blinked" when going through a signal, missing its meter deflection completely. Some improvement has been made, however, by using a motor drive and mechanical linkage to alleviate the drudgery of manual tuning, and X-Y recording of the detector output has solved the "blinking" problem. But what about the case of transients or closely spaced signals swept through too fast for the X-Y recorder to respond to their true amplitudes? Many times the closing of relay contacts or a switch will produce serious transients, and continuous monitoring of the entire spectrum of interest is the only way it can be detected. A modern spectrum analyzer with its rapid electronic spectrum sweep over wide ranges at high sensitivities has an important role in RFI/EMI testing and spectrum surveillance work. This single instrument presents continuous visual coverage of the spectrum from 10 MHz to 10 GHz for rapid location of interference. External waveguide mixers extend the frequency range into higher frequencies.

Although there is a wide variety of specifications, limits, and procedures, much common information underlies many of the tests and may be useful. MIL-STD-826A was written to standardize the various test methods which had previously been in existence. A measurement procedure has been written[17] which divides RFI/EMI measurement into five categories.

[17] Hewlett-Packard, *EMI Measurement Procedure*, 02464-1, February 1968.

A. Method 3002, Conducted Interference (14 kHz to 100 MHz)
B. Method 3003, Antenna, Terminal Conducted, Low Power
   (14 kHz to 10 GHz)
C. Method 3004, Antenna, Terminal Conducted, High RF Power
   (14 kHz to 10 GHz)
D. Method 4001, Radiated Interference (14 kHz to 10 GHz)
E. Method 6001, Radiated Susceptibility (14 kHz to 12.4 GHz)

These methods make use of modern wide-sweeping spectrum analyzers.

All the measurements require that the displayed signals be read directly in terms of decibels above one microvolt (dBμv) if they are narrow-band signals and in terms of decibels above one microvolt per megahertz bandwidth (dBμv/MHz) if they are broadband signals. The spectrum analyzer used is not calibrated in these terms. Consequently a precalibrating procedure has to be performed to achieve fast, low-in-error measurements. Usually a calibration figure and a bandwidth figure are determined for both the control settings and the graticule display scale.

A ● Conducted Interference (Method 3002) 14 kHz to 100 MHz

The objective of this method is to measure the conducted interference on wires and cables running between instruments. In this test the entire frequency range of 14 kHz to 100 MHz is covered with one step.

Since this measurement does not cover microwave frequencies, it is not dealt with in this text. Further information can be found in the reference cited above.

B ● Antenna, Terminal Conducted, Low Power (Method 3003) 14 kHz to 10 GHz

The objective of this test is to measure the interference appearing at the antenna terminals on such instruments as receivers, transmitters, and RF amplifiers. The frequency range of 14 kHz to 10 GHz is covered in four steps. The test setup, instrument connections, control settings, and calibration are different in each band.

The 14-kHz to 100-MHz region can use a low-frequency spectrum analyzer and is not discussed here. The 100-MHz to 1.8-GHz band uses the setup shown in Fig. 5.3-47.

A calibration and CRT marking procedure are shown in the example below: Calibration of narrow-band signal scale in dBμv due to setting controls.

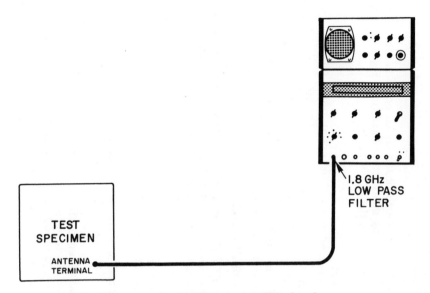

Fig. 5.3-47. Test setup for 10 MHz to 1.8 GHz band
(Method 3003).

EXAMPLE: $C_2 = 158$ dB

$$C_2 - 90 = 158 - 90 = 68 \text{ dB} = \text{IF GAIN setting}$$

Calibration of broadband signal scale in $dB\mu v/MHz$ is obtained by subtracting Bandwidth Figure B (in dB MHz) from the $dB\mu v$ scale.

EXAMPLE:

Measured Impulse Bandwidth $BW_i = 80$ kHz

$$\text{Bandwidth Figure B} = 20 \log \frac{80 \text{ kHz}}{1 \text{ MHz}} = -22 \text{ dB MHz}$$

For the 30 $dB\mu v$ point on the narrow-band scale, the broadband scale will read

$$30 \text{ dB}\mu v - (-22 \text{ dB MHz}) = 52 \text{ dB}\mu v/MHz, \text{ etc.}$$

(*Note:* The exact labeling of the broadband scale depends on the actual bandwidth of your analyzer and has to be measured beforehand.) Narrow-band and broadband signal limits are copied from MIL-STD 826A. Labeling of frequency scale:

Left end is 0, right end is 2.0 GHz

Mark frequencies, specification limits, and method number directly on CRT face or use overlay. A photo of the spectrum will contain all the information.

Calibration of narrow band signal scale in dBμV due to setting controls

Example: $C_2 = 158$ dB

$C_2 - 90 = 158 - 90 = 68$ dB = IF GAIN setting

Calibration of broadband signal scale in dBμV/MHz is obtained by subtracting Bandwidth Figure B (in dBMHz) from the dBμV scale.

Example:
Measured Impulse Bandwidth $BW_i = 80$ kHz
Bandwidth Figure B = $20 \log \dfrac{80 \text{ kHz}}{1 \text{ MHz}} = -22$ dBMHz

For the 30 dBμV point on the narrow band scale, the broadband scale will read

30 dBμV - (-22 dB MHz) = 52 dBμV/MHz, etc.

NOTE: The exact labeling of the broadband scale depends on the actual bandwidth of your analyzer and has to be measured beforehand (see Section II D).

Narrow band and broadband signal limits are copied from MIL-STD 826A.

Labeling of frequency scale:
Left end is 0, right end is 2.0 GHz.

Mark frequencies, specification limits, and method number directly on CRT face or use overlay. A photo of the spectrum will contain all the information

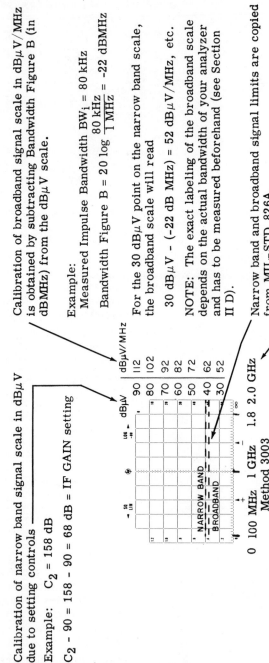

dBμV	dBμV/MHz
90	112
80	102
70	92
60	82
50	72
40	62
30	52

NARROW BAND
BROADBAND

0   100 MHz   1 GHz   1.8   2.0 GHz
Method 3003

Fig. 5.3-48.

The frequency dial reading is usually accurate to within ±1% of its own local oscillator and can be read directly. Greater accuracy, if needed, can be realized through the use of a comb generator or other accurately known reference signal.

It is a good practice to set the input attenuator of the spectrum analyzer to maximum, to prevent burning out the mixer diodes in case of excessively high signals. Then reduce the attenuator setting in steps until the signal is visible or the setting is reached where the spectrum analyzer is calibrated. Spurious responses are always at a lower level than the signals that generate them. If all levels of displayed signals are below the specification limits, there is no need to identify them as spurious responses. Spurious responses with displayed levels equal to the specification limits are generated by signal levels in the order of 50 dB above specification level for narrow-band signals.

In the 1.8-GHz to 5.6-GHz frequency band, in addition to the spectrum analyzer system, a preselector is connected instead of the low-pass filter. The

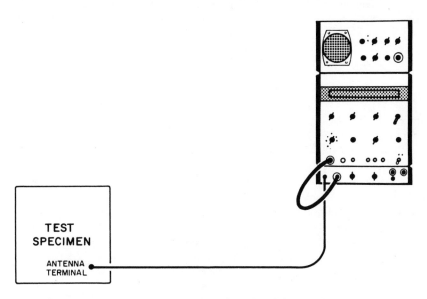

Fig. 5.3-49. Test setup for 1.8 to 5.6 GHz band (Method 3003).

setup is shown in Fig. 5.3-49. Calibration and marking of the display are done similarly to the technique described above.

In the 5.2- to 10-GHz band, an additional low-noise traveling-wave tube amplifier is connected into the setup as shown in Fig. 5.3-50. As another option (Option 02) the TWT amplifier can be left out.

The minimum signals that can be measured with the Option 02 setup (signal 16 dB above noise) are 50 dBμv and 70 dBμv/MHz if the spectrum

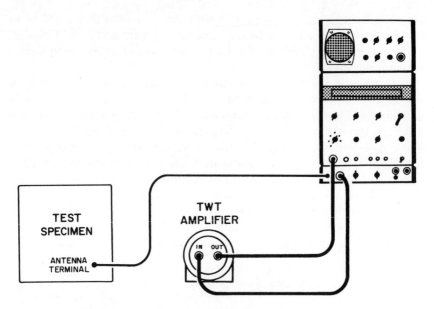

Fig. 5.3-50. Test setup for 5.2- to 10-GHz band.

analyzer RF unit has a noise figure equal to the sales specification. However, the noise figure of an average RF unit is better than the specification and allows measurements to be made at levels 4 to 8 dB lower.

Narrow-band signals of levels down to 37 dBμv can be measured with the Option 02 method by using either long exposure photographs of the display (20 seconds) or a display unit with variable-persistence CRT display instead of the standard display unit. With this method, the displayed noise appears as a solid bar (instead of the grasslike, ragged normal display), out of which the signals rise.

The Option 02 method has the advantage of better front ends pulse overload range than Option 01. Hence it is the better method when signals of amplitudes above the levels mentioned above have to be measured.

Option 01 must be used for measurements of broadband pulse signals and levels below 70 dBμv/MHz.

C ● Antenna, Terminal Conducted, High RF Power (Method 3004) 14 kHz to 10 GHz

The spurious output of a transmitter or an RF amplifier is measured in this method. The frequency range is covered in three steps. The 14-kHz to 100-MHz band is not discussed in this text, since it covers only frequencies below the interest of microwaves. In the range of 100 MHz to 1.8 GHz, the

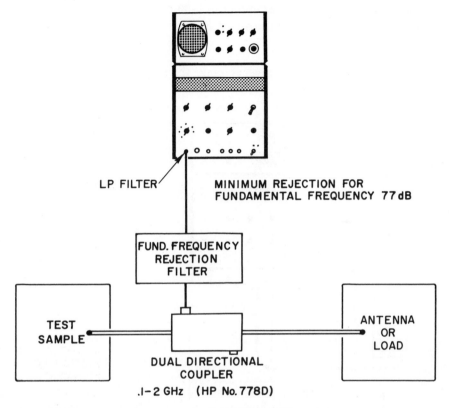

LP FILTER   **MINIMUM REJECTION FOR**
                       FUNDAMENTAL FREQUENCY 77 dB

FUND. FREQUENCY
REJECTION
FILTER

TEST
SAMPLE

ANTENNA
OR
LOAD

DUAL DIRECTIONAL
COUPLER
.1−2 GHz   (HP No. 778D)

Fig. 5.3-51. Test setup (Method 3004, 100 MHz to 1.8
GHz).

test setup shown in Fig. 5.3-51 is used. The procedure to calibrate and mark
the display is as follows:

To set IF gain:

1. Replace test sample with signal generator.
2. Adjust signal generator output power to −50 dBmw.
3. Set IF gain control so that displayed signal peak is at the −50 dBmw
   line of the display.
4. Sweep frequency through measurement range. If passband of fundamental frequency reject filter is not flat, mark the −50 dBmw
   signal line directly on the CRT face.

To calibrate fundamental frequency level:

1. Remove fundamental frequency rejection filter.
2. Replace test sample with signal generator.

3. Adjust signal generator output to −50 dBmw.
4. Mark signal height directly on CRT as −50 dBmw for fundamental frequency.

In the 1.8 to 10 GHz frequency range the test setup is shown in Fig. 5.3-52.

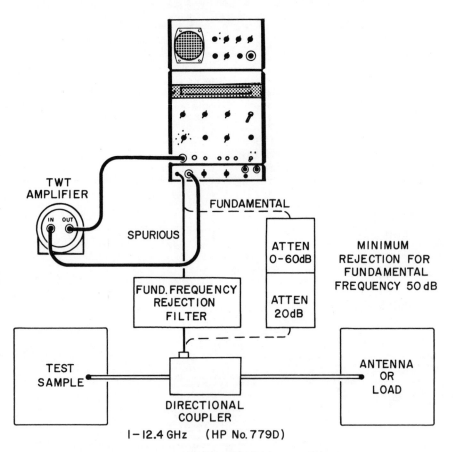

Fig. 5.3-52. The test setup (Method 3004, 1.8 to 10 GHz).

The procedure to calibrate and mark up the CRT face is the same as in the lower-frequency procedure. The measurement procedure for making such tests may be performed in the following way:

1. Replace the fundamental frequency rejection filter with a variable, step attenuator, where the attenuator is set to maximum attenuation

value. Usually the range of step attenuations is not adequate (60 dB), so an attenuating pad may be cascaded with it.

2. Reduce the attenuator setting until fundamental frequency is readable.

3. The calibration of the CRT face is now as marked before, plus the variable attenuator setting.

   Example: With an attenuator setting of 60 dB, the line marked −30 dBmw represents a level of −30 + 60 + 20 = +50 dBm.

4. Read the level of the fundamental off display.

5. Replace the attenuator combination with the fundamental frequency rejection filter.

6. The CRT calibration is now the same as before.

7. Read the spurious output frequencies and levels of test samples off display.

D ● Radiated Interference (Method 4001) 14 kHz to 10 GHz

Electromagnetic interference radiated from a test specimen is measured in this method using appropriate antennas and the spectrum analyzer. The frequency range is covered in three bands. The test setup, instrument connection, control settings, and calibration are different with each band.

For the 14-kHz to 25-MHz and 25- to 200-MHz bands the technique is described in detail in the reference cited above and will not be dealt with, since this frequency region is not in the direct interest of microwaves.

In the frequency region of 200 MHz to 10 GHz the basic test setup is shown in Fig. 5.3-53.

E ● Radiated Susceptibility (Method 6001) 14 kHz to 12.4 GHz

In this method the instrument under test is subjected to a specified RF field, and the effects of this field on performance are noted. The objective is to determine whether the instrument will perform as desired in the expected environment. The frequency range of 14 kHz to 12.4 GHz is covered in two bands. The RF fields are set up by an antenna/signal generator combination. The field level is determined using the spectrum analyzer, as in Method 4001. The instrument under test is then placed in the field, and the signal generator output is monitored with the spectrum analyzer. The test is conducted as specified in MIL-STD 826A.

The 14-kHz to 100-MHz band is not covered in this text, since it is below the range of microwave frequencies.

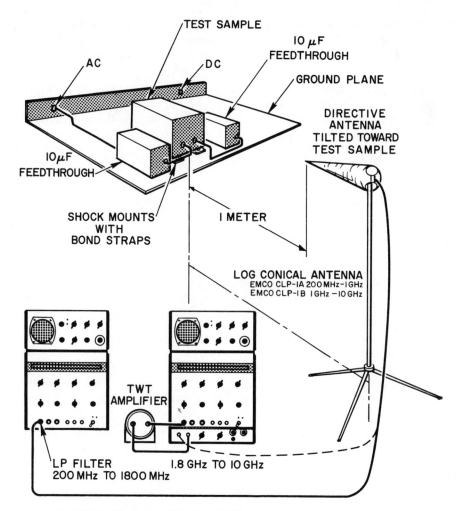

Fig. 5.3-53. Test setup (Method 4001).

The basic setup to make measurements in the 100-MHz to 12.4-GHz frequency range is shown in Fig. 5.3-54. The calibration and CRT face-marking procedure and the measurement are described below.

1. Copy specification limits from MIL-STD 826A.
2. Copy antenna factor curves supplied with your antennas.
3. Subtract antenna factors (in dB) from specification limits at as many frequencies as required to get a detailed curve. This is the input voltage of the spectrum for the specified field strength.
4. Record the values obtained in (3).

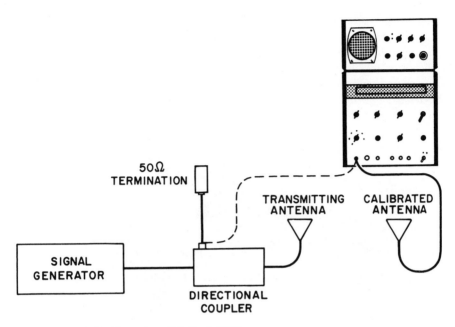

Fig. 5.3-54. Test setup (Method 6001).

5. Mark the curves obtained on your spectrum analyzer display (as shown in examples for Method 4001). Your analyzer is now calibrated for the required field strength.
6. Connect the calibrated antenna of proper frequency range to spectrum analyzer.
7. Place calibrated antenna where the test sample will be placed.
8. Adjust signal generator output power until displayed signal reaches the marked curve on your spectrum analyzer display.
9. Disconnect spectrum analyzer from calibrated antenna, disconnect 50Ω termination from directional coupler.
10. Set spectrum analyzer attenuator (frequency range 100 MHz to 12.4 GHz) to 60 dB.
11. Connect spectrum analyzer to directional coupler.
12. Read and record voltage of displayed signal.
13. Repeat procedure until the whole frequency range is covered.
14. Replace calibrated antenna by test sample.
15. Follow procedure given in MIL-STD 826A, set output level of signal generator to values recorded in steps (12) and (13).

SPECTRUM SIGNATURE. An extension of RFI/EMI measurement is the vast *spectrum signature* work performed for the government by various companies and institutions. Spectrum signature refers to the energy distribution.

both desired and undesired, emanating from a device. For example, it would include the main and side lobes of a pulsed transmitter or signal generator *plus* any undesired outputs. This information is useful in statistical prediction of interference caused by the outputs of various devices. Previous spectrum analyzers could not look at complete spectrum signatures because of their restricted sweep width and narrow image separation. Figure 5.3-55 (shown below) is a photograph of the spectrum signature of a 0.5-microsecond pulse

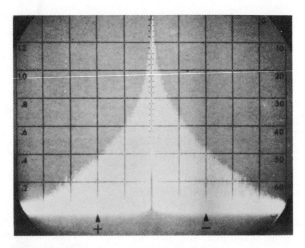

Fig. 5.3-55. Spectrum signature of a 0.5 μs pulse.

with a rise time of less than 20 nanoseconds as seen on the analyzer display. Here the spectrum width of the analyzer is set to 30 MHz/cm, giving a 300-MHz-wide display of the main and side lobes. The logarithmic vertical display has been selected so that the side lobes down to 60 dB below the main lobe are visible. Notice that the spectrum of the pulse is more than 200 MHz wide 60 dB down.

A wide-sweeping spectrum analyzer with wide-image separation and preselection is invaluable for *spectrum surveillance* work also, where the relative amplitudes and frequencies of all electromagnetic radiation at a location are desired. An application of this would be in a missile or space-craft launch site or tracking station where multiple command, communications, and radar-tracking signals are transmitted at various times, often simultaneously.

The far-ranging sidebands of radar transmitters, intermodulation products of telemetry and communication transmitters, or spurious signals can blot out reception of valuable data. It can also be responsible for accidental triggering of control links for detonation devices in missiles and retrorocket firing in spacecraft. To spot such signals and eliminate them before they cause trouble, a wide-scan spectrum analyzer with preselector can be set up with appropriate antennas in the launching vicinity and used as a spectrum monitor.

5.3.7 ANTENNA PATTERN MEASUREMENTS

An interesting application of the spectrum analyzer is its use not as a wide-sweeping device but rather as a CW single-frequency receiver for antenna field pattern measurement. The general qualifications of a pattern receiver are high sensitivity with low noise, good frequency stability, wide dynamic range, and good shielding. As we saw earlier, the analyzer meets these requirements well and has the added features of selectable vertical displays for linear, log (dB), or square (power) patterns plus X-Y recorder output signals. The logarithmic display is by far the most useful in this application, since it allows simultaneous on-scale viewing of the main lobe of an antenna's field pattern along with side lobes that are down 60 dB.

To make rectilinear pattern plots with the spectrum analyzer, connect the output of the antenna in test to the analyzer input, as shown in Fig. 5.3-56. Connect the X-Y recorder output of the analyzer into an X-Y recorder.

The illustration shows the antenna in test mounted on a rotary platform

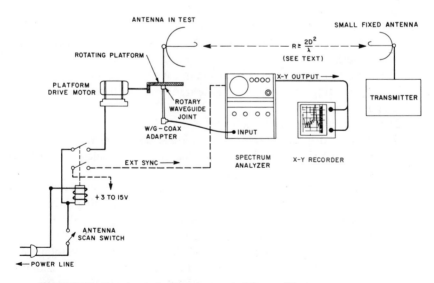

Fig. 5.3-56. Spectrum analyzer connected for rectilinear antenna pattern plots.

which is common at antenna range installations. This platform allows the antenna in test to be rotated through 360 degrees in azimuth while coupling the antenna coaxial transmission line or waveguide through a rotating joint to the analyzer input.

A smaller directional-transmitting antenna capable of being aimed for maximum signal at the receiving antenna is located at a range or distance R away. For minimum error in received power due to mutual coupling effects of the two antennas, this range should satisfy the equation

$$R \geq \frac{2D^2}{\lambda}$$

where R = range (distance between antennas),
$\quad\quad D$ = largest aperture dimension of antenna,
$\quad\quad \lambda$ = free-space wavelength of test frequency.
$\quad\quad$ (All dimensions are in identical units.)

The two antennas should be set up on towers above flat terrain to minimize multipath reception due to ground waves and reflections from buildings, trees, etc. At the transmitter location:

1. Set CW frequency and power of transmitter as desired.
2. Stabilize frequency, preferably phase-lock or AFC system.
3. Aim antenna toward receiving antenna.

At the receiver location:

1. Tune spectrum analyzer to approximate transmitter frequency with spectrum width at 1 MHz/cm, SYNC at INT, and IF bandwidth at *auto-select*.
2. Rotate test antenna for a maximum CW response on CRT. Vertically position test antenna for desired angle with respect to transmitting antenna.
3. Stabilize analyzer local oscillator.
4. Select log display on analyzer so vertical scale is now calibrated in dB. Adjust IF gain and input attenuator for full-scale indication of test signal.
5. Reduce spectrum width to zero, keeping the response centered on the display with fine tuning. The analyzer is now operating at only the test signal frequency. Retune analyzer frequency slightly to peak vertical response and touch up IF gain so response is just 7 cm high.
6. Adjust sweep time to some value slightly longer than the total time required to rotate the test antenna through 360 degrees of azimuth. For example, to X-Y record the pattern, a relatively slow sweep time is required; therefore, set sweep time to 3 sec/cm and rotate antenna

at $2\frac{1}{3}$ rpm. This results in a horizontal display of 42 degrees rotation per centimeter of display.

7. Set X-Y recorder gain and position controls to coincide with analyzer horizontal sweep and vertical deflection limits.

8. Now rotate antenna 180° away from the maximum reception point just set up. This will position the antenna in the correct relationship with the start of the sweep in the analyzer to cause the main lobe of the pattern to appear in the center of the display when the plot is made.

9. Set analyzer SYNC to single sweep and turn on X-Y recorder servo and pen-down switches.

10. Simultaneously push the *single sweep* button on analyzer and start antenna scan drive motor. The trace on the CRT and the X-Y recorder will now move horizontally at the 3 – s/cm rate set up with analyzer sweep time and be deflected vertically by strength of the received signal as the antenna rotates through 360 degrees. Thus a rectilinear plot of the antenna field pattern is obtained. The analyzer sweep may also be triggered from an external +3 to +15 volts which can be synchronized with the antenna scan motor switch, eliminating the need to operate manually the single-sweep button. Exact synchronism is not particularly important, since the sweep time of the analyzer is longer than one rotational cycle of the antenna in test. This allows the operator to locate the identical contours on the left and right edges of the plot denoting the 360° boundaries.

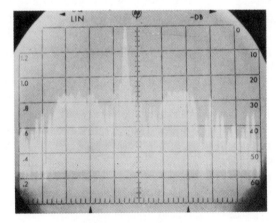

Fig. 5.3-57. Antenna pattern obtained from spectrum analyzer. Vertical and horizontal outputs on rear panel also allow X-Y recording of display.

Figure 5.3-57 is a time-exposed photo of an antenna pattern plot on the analyzer (the alternate method of trace recording). Notice the deep minima in the plot, some extending nearly 60 dB below the main lobe. This illustrates the importance of the logarithmic display and 60-dB dynamic range.

The above test did not permit direct correlation of horizontal display with the test antenna azimuth. The newer-type spectrum analyzer display units have an additional sweep drive capability which permits such a function. By applying 0 - 15 volts to the sweep drive input, the trace is made to move across the screen of the CRT.

The spectrum width should be set to zero. Thus a voltage ramp taken from the test antenna pedestal azimuth indicator may be used to drive the horizontal display plates of the CRT. This provides for direct calibration of the horizontal scale of the rectilinear plots of patterns as the analyzer is used as a fixed-tuned receiver.

QUESTIONS

1. In what domain does the spectrum analyzer display a signal?
2. What kind of mathematical manipulation is done electronically in a spectrum analyzer to display signals in the frequency domain?
3. A square wave is built of what harmonics?
4. How many spectral lines describe a single-tone 50%-amplitude-modulated carrier? How far will the spectral lines be located from each other, and what will be the relative magnitude of each to the other?
5. What information can be determined from the spectra of a modulated signal?
6. How, approximately, is the bandwidth of frequency modulation determined?
7. What information comes from knowing the spacing of nodes between lobes on the spectrum of a pulse?
8. What is defined as the maximum spectrum width in a spectrum analyzer?
9. What technique is used in modern spectrum analyzers to overcome frequency instability due to inherent instability and FM of the local oscillator?
10. What is the major advantage of using a 2-GHz IF frequency?
11. What is the usable spectrum width?
12. What are spurious responses?
13. What is the most serious limitation to resolution in a spectrum analyzer?
14. What are the contributing factors to the overall noise figure of a spectrum analyzer?

15. What happens to the spectrum analyzer display if a signal identical in frequency with the IF is connected to the input?

16. When and how does second-order intermodulation occur? How does the spectrum analyzer display look?

17. What is the carrier deviation of a frequency-modulated carrier when the modulation index is 5.52 and the modulation frequency is 50 kHz?

18. What is the electronically tuned element in a preselector?

19. If a preselector is connected before a spectrum analyzer and swept simultaneously with the local oscillator, what kind of distortions or responses can be eliminated?

20. How many responses are there for the first three harmonics, and at what frequencies will they appear if a spectrum analyzer has a 2-GHz IF and the local oscillator is tuned to 2.5 GHz?

21. What provides the information about PRF of a pulse on a spectrum?

22. What is the modulating pulse midwidth if the side lobes on the spectrum are spaced 25 MHz apart?

23. What are the different kinds of interference tests (RFI/EMI) included in present-day standards?

24. What is the critical distance for making antenna pattern measurements if the largest aperture dimension of the antenna is 18 in. and the frequency of measurement is 8.5 GHz?

6
NETWORK ANALYSIS

6.1 MEASUREMENTS OF SCATTERING PARAMETERS

Scattering (s) parameters were discussed in Chap. 3. A multiport network can be characterized by a set of parameters describing both the complex reflection coefficients of each port versus frequency and the complex forward-and-reverse transmission coefficients of each pair of ports versus frequency. Although several techniques can be used to obtain a set of parameters that fully describe the behavior of a multiport network, s-parameters are used because of their simplicity. Short- and open-circuit terminations must be applied to the ports when measuring other parameter sets. Errors or even parasitic oscillations can result if these circuits are imperfect. In making s-parameter measurements, the ports are terminated with accurately achievable resistances only; and, by using the set of transform equations given in Chap. 3, these measured s-parameters can be changed into terms of any other parameter.

There are basically two kinds of measurements to make: transmission coefficient and reflection coefficient. For example, the transmission coefficient (forward) on a general two-port network (Fig. 6.1-1) is defined by s_{21};

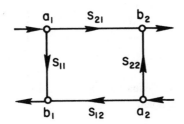

Fig. 6.1-1. Flow graph of a two-port network.

this means that if the second port is terminated with its characteristic impedance and a signal is applied to the input port (No. 1), the voltage ratio measured between nodes b_2 and a_1 gives the value of the transmission coefficient. The reflection coefficient is defined under the same conditions as described above, but the voltage ratio is measured between nodes b_1 and a_1, describing s_{11}. These values are complex numbers having magnitude and phase information.

6.1.1 TRANSMISSION MEASUREMENTS

Attenuation

When making *transmission measurements*, one is often interested only in the magnitude of gain or attenuation of a device. Phase information is immaterial in many applications. The reader is acquainted with attenuation from study of transmission line theory. Since attenuation is really negative gain, it does not require much imagination to understand gain also. *Magnitude of attenuation* measurements are discussed first.

According to Beatty,[1] attenuation is defined as the decrease in power level (at the load) caused by inserting a device between a Z_0 source and load. Figure 6.1-2 shows the basic idea of such an attenuation measurement.

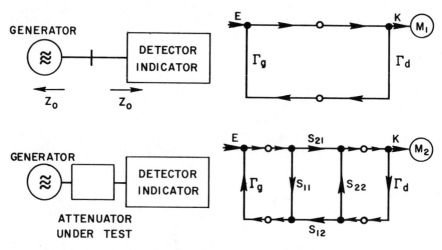

Fig. 6.1-2. Measuring attenuation: (a) calibration,
(b) measurement.

Basically, making a measurement means comparing some unknown value to a known calibrated value. This can easily be interpreted as making a substitution measurement. Three kinds of substitution techniques are generally used for attenuation measurements:

1. RF substitution,
2. AF or dc substitution (square-law detection),
3. IF substitution or linear detection.

As a general rule, if one compares an unknown value to a known value, the accuracy consideration in making the comparison is such that the "standard" must be at least three times, preferably ten times, more accurate than the desired measurement accuracy. Attenuators are used as standards for

---

[1] Beatty, R. W., "Insertion Loss Concepts," *Proc. IEEE*, Vol. 52, No. 6 (June 1964), 663.

making substitution attenuation measurements, dropping high power to lower power in transmission systems, padding out undesirable reflections, etc.

Attenuators

Attenuators can be classified in many different ways. It would be advisable to classify them first by whether they are built into coaxial or waveguide transmission lines. Further classification can be done by calibration accuracy. Attenuators in the "standard" category are precision built and calibrated; then, of course, there are the attenuators generally called pads. Another type of classification depends on whether they are fixed or variable in value.

The important properties of attenuators (in other words, their specifications) will dictate other ways of classifying them. Important properties of attenuators are frequency range, attenuation accuracy, attenuation variation versus frequency, input and output match (reflection coefficient), power-handling capability, and phase linearity. Some of these properties are more important in the microwave region than at lower frequencies, and they are emphasized more when selecting an attenuator. This fact will be more apparent to the reader later in this chapter.

Coaxial attenuators generally have multi-octave bandwidths to accommodate wider usage. In coaxial transmission lines where the propagation is done in the principal (TEM) mode, an obvious solution is to build an attenuator of lumped-circuit elements, keeping the size of the elements small enough to prevent an excessive reactance that would vary the value of attenuation and cause noticeable mismatch at the ports. Such a structure can be achieved by a "T" (tee) attenuator network (Fig. 6.1-3), in which attenuation can be set to different values by varying the values of R_1 and R_2.[2]

$$R_1 = Z_0 \frac{K - 1}{K + 1}$$

$$R_2 = \frac{2Z_0 K}{K^2 - 1}$$

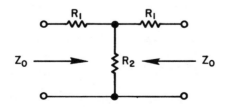

Fig. 6.1-3. "Tee" attenuator circuit.

[2] *Reference Data for Radio Engineers*, 4th ed. (New York: ITT), p. 252.

where K is the transmission coefficient or, in other words, the voltage ratio between output and input. Mathematically,

$$\text{Attenuation (dB)} = 20 \log_{10} K.$$

A tee network is easily placed in a coaxial structure. Figure 6.1-4 shows

RESISTIVE RODS (R_1)

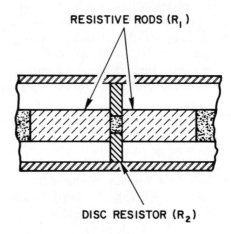

DISC RESISTOR (R_2)

Fig. 6.1-4. Coaxial tee attenuator.

such a layout. The series resistances (R_1) are composed of resistive rods, preferably thin film deposited on tubular or rod substrate, the parallel element (R_2) forms a disc resistor that makes contact all the way around its outer diameter with the outer conductor; its center is in contact with the ends of the resistive rods.

A π attenuator scheme can also be realized by using two disc resistors and a rod resistor in between them. Figure 6.1-5 shows the schematics of

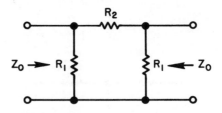

Fig. 6.1-5. "π" attenuator circuit.

such attenuators. The calculations of attenuation values are similar to those for T attenuators.[3]

---

[3] *Reference Data for Radio Engineers*, 4th ed. (New York: ITT), p. 252.

$$R_1 = Z_0 \frac{K + 1}{K - 1}$$

$$R_2 = Z_0 \frac{K^2 - 1}{2K}$$

where K is the transmission coefficient.

At higher frequencies where the geometry of T and π attenuators becomes comparable to the wavelength of the applied signal, appreciable amounts of reactance degrade their performance. Distributed *lossy-line attenuators* proved to have very favorable performances at higher frequencies. E. Weber[4] thoroughly explains how these attenuators operate. In essence, if the center conductor of a transmission line is made of lossy material (preferably thin film deposited on circular cross-section substrate), and if the attenuation per unit length is not exceedingly high, flat attenuation response with low reflections can be achieved for multi-octave bandwidths. Such attenuators are being produced in standard 7-millimeter coaxial line with very respectable SWR specifications and flat attenuation-versus-frequency response from 2–18 GHz frequency band. The lower-frequency limit for such attenuators is given as follows:[5]

$$\frac{R\lambda_0}{2\pi Z_0} \ll 1$$

and the attenuation is

$$\text{attenuation} = 4.34 \frac{R}{Z_0} \text{ (dB per meter)}$$

where λ_0 stands for wavelength in the equivalent lossless line (free-space wavelength), R denotes the resistance per meter, and Z_0 is the characteristic impedance of the equivalent lossless line.

Such high-frequency attenuator elements must have a very thin layer of resistance to maintain their resistivity constant with frequency changes due to skin depth. In other words, the thickness of the resistive elements in these microwave attenuators must be less than the skin depth to keep the resistivity of each element in the attenuator constant with varying frequency. It was shown in the text above that T- or π-type attenuators have serious limitations at higher frequencies; it was also shown that the distributed lossy-line attenuators have good high-frequency behavior, but they give inferior lower-frequency performance. However, the two kinds of attenuators complement each other to cover all the practical coaxial frequency range from dc to 18 GHz.

[4] Weber, E., *Precision Metalized Glass Attenuators*, MIT Radiation Lab Series, Vol. II, pp. 751–74.
[5] Weinschel, B. O., *Microwave Attenuation Standards* (Gaithersburg, Md.: Weinschel Engineering *Internal Report 90-110*, June 1965).

In the last few years a *distributed thin-film attenuator*[6] has been developed. These attenuators have a two-dimensional attenuating "card" that provides both series and shunt losses of a distributed nature. The card consists of a thin rectangular sheet of resistive material deposited on a dielectric substrate. Figure 6.1-6 illustrates such a card attenuator.

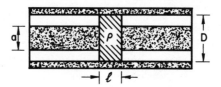

Fig. 6.1-6. Distributed-thin-film attenuator card.

The resistive film is vacuum deposited. Substrates are made of aluminum oxide (high-alumina) ceramics. Strip transmission lines are added to the input and output terminals, as well as narrow ground electrodes from both longitudinal sides. Figure 6.1-7 shows how the attenuator cards are loaded

Fig. 6.1-7. Distributed-thin-film attenuator cartridge.

into cylindrical cartridges for use in coaxial transmission lines when coax-to-strip-line transitions are added.

From transmission line theory, characteristic impedance is

$$Z_0 = \sqrt{\frac{R + j\omega L}{G + j\omega C}}$$

It was found that $\sqrt{L/C} = Z_0$ for lossless lines, which means that the characteristic impedance is frequency independent if the line is lossless.

If one assumes that there is no reactive component in a short piece of transmission line being used as an attenuator, characteristic impedance may be rewritten as follows

[6] Adam, S. F., "Precision Thin-Film Coaxial Attenuators," *Hewlett-Packard Journal*, June 1967, pp. 12–19.

$$Z_0 = \sqrt{\frac{R}{G}} \text{ ohms}$$

This attenuator would have a characteristic impedance independent of frequency. Furthermore, the propagation constant is

$$\gamma = \alpha + j\beta = \sqrt{(R + j\omega L)(G + j\omega C)}$$

If no reactive components are considered, the attenuation constant becomes

$$\alpha = \sqrt{RG} \text{ nepers per unit length}$$

If the resistivity of the film is ρ (ohms per square), the series resistance is

$$R = \frac{\rho}{a} \text{ ohms per unit length}$$

and shunt conductance is

$$G = \frac{4}{\rho(D - a)} \text{ mhos per unit length}$$

Then,

$$Z_0 = \rho \sqrt{\frac{D - a}{4a}} \text{ ohms}$$

and

$$\alpha = \sqrt{\frac{4}{a(D - a)}} \text{ nepers per unit length}$$

The last equation proves that the attenuation is independent of the resistivity of the film, provided, of course, that the film is homogenous. The attenuation in dB depends on the geometry of the resistive film. Since the attenuation is independent of resistivity, the attenuator is insensitive to temperature changes.

It was assumed above that no reactive components exist or are appreciable in size; however, at very high frequencies there are some ill effects due to the actual current lines flowing in the card and causing small frequency dependence. Compensation can be achieved by using lumped-circuit techniques that enable one to correct for these frequency responses.

These types of attenuators (T or π and distributed) replace both of the types mentioned above, since they are reasonably inexpensive to manufacture and since their performances are superior to any of the others mentioned.

VARIABLE COAXIAL ATTENUATORS can be made in two different ways. One approach is the step attenuator, whereby fixed attenuators of the kinds described above are switched by some mechanical means into the transmission line. The other way is the continuously variable attenuator approach. One may wonder why step attenuators are made at all, since continuously variable ones are available. With the present state of art, continuously variable coaxial attenuators cannot cover so wide a frequency range, holding accuracies, and other vital properties of attenuators as the fixed-value attenuators do. For limited bandwidth, all kinds of techniques are used to make continuously variable coaxial attenuators.

The *cutoff attenuator* is one of the oldest continuously variable coaxial attenuators, and it is still very widely used.[7] This type of attenuator utilizes a waveguide operating below cutoff. It is well known from transmission line theory that higher-order modes can also be launched beyond their cutoff frequencies, but their fields decay in an exponential manner along the line of propagation. It is quite possible to excite the desired higher-order modes and to couple them selectively. The attenuation beyond cutoff can then be calculated. In this manner an attenuator can be built that follows a mathematically predictable law. A simple but very precise attenuator, the *piston attenuator*, can be built by choosing a geometry of circular cross-section coupling into the lowest cutoff higher-order mode (TE_{11}). Figure 6.1-8 shows

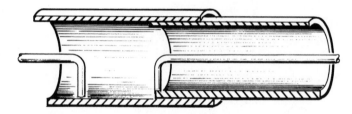

Fig. 6.1-8. Cross-section of a piston attenuator.

such an arrangement, in which two magnetic coupling loops are used in two sliding concentric cylinders that provide a physical displacement adjustment to change the distance of the loops, thus varying the attenuation in an exponential manner. The cutoff wavelength in the circular waveguide is

$$\lambda_c = \frac{2\pi r \sqrt{\epsilon_r}}{S_{mn}}$$

where r is the inner radius of the circular waveguide used, ϵ_r is the relative dielectric constant of the media filling the waveguide (it is 1 for free space), and S_{mn} is a dimensionless quantity (1.841 for TE_{11} mode).

The attenuation of a beyond-cutoff attenuator can be given by

$$\alpha = \frac{2\pi}{\lambda_c} \sqrt{1 - \left(\frac{\lambda_c}{\lambda}\right)^2} \text{ nepers per unit length}$$

It can be seen from the expression of attenuation that, if the operating fre-

[7] Montgomery, C. G., *Technique of Microwave Measurements*, M.I.T. Radiation Lab Series, Vol. XI, p. 685.

quency is reasonably close to cutoff, the attenuation is quite frequency dependent. But, if the operating frequency is much lower than the cutoff frequency, the frequency dependence of a cutoff attenuator will be quite small. This fact led designers of such attenuators to use small-diameter tubes to place the cutoff frequency many octaves above the operating frequency. This makes the expression $\left(\dfrac{\lambda_c}{\lambda}\right)^2$ negligibly small for the operating bandwidth and consequently achieves a flat attenuation-versus-frequency performance.

A great disadvantage of these attenuators is their behavior at low attenuations, which is caused by the presence of other higher-order modes that are excited and attenuate faster than the desired one. This fact generally impairs the use of these attenuators much under the attenuation value of 20 dB and means that they can be used only as attenuators having at least 20 dB residual loss. Some models of these attenuators provide some lower usable residual losses by using some very troublesome mode-suppressing techniques to achieve a few dB more usable range. But the piston attenuators are used as standards for high values of attenuation, since attenuation in decibels is linearly proportional to the physical displacement of the probes.

Other types of continuously variable attenuators can be made by using variable resistances of either the T or π pad type or the distributed long-line attenuator type. These attenuators have frequency limitations due to their physical nature.

A resistive "bridged-tee" network can be realized in transmission lines that carry TEM waves, and they have quite good behavior up to a few GHz; their performance deteriorates very sharply beyond this, since many disadvantageous discontinuities are in existence. Double potentiometer techniques with tapered resistances coupled to coaxial structures (variable pads) are also often used for lower-frequency microwave networks.

The long lossy-line distributed attenuator can be made variable by moving a shielding contact over the long resistive element, effectively changing its length and varying the attenuation. Figure 6.1-9 shows such an ar-

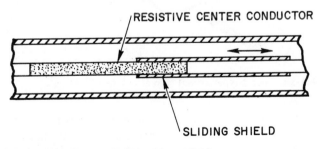

Fig. 6.1-9. Lossy line distributed-film variable
attenuator.

rangement. There are numerous other approaches to make continuously variable coaxial attenuators. The few mentioned above are the most common.

STEP ATTENUATORS can be categorized into two families. One of them is the turret type, in which the desired attenuators are usually placed in a cylindrical arrangement and are switched with a rotary motion in and out between junctions of the transmission line. The attenuator elements can be any kind of the fixed coaxial type. Figure 6.1-10 shows the schematics and a simplified diagram of such an attenuator.

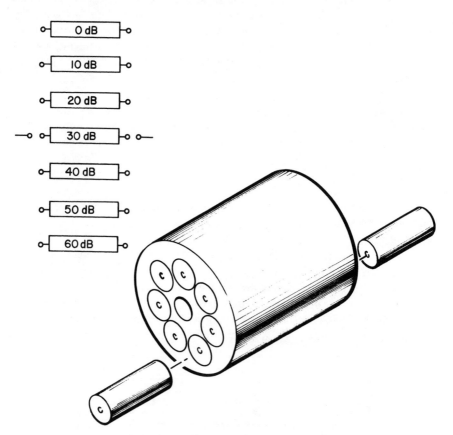

Fig. 6.1-10. Turret-type coaxial step attenuator.

The other type of step attenuator can also use any kind of fixed coaxial elements. This type uses cascaded elements arranged so that an attenuator can be bypassed if it is not desired. The schematic of such an attenuator is shown in Fig. 6.1-11. In 10-dB steps, 10-, 20-, 40-, 80-dB attenuators allow

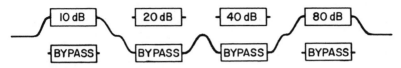

Fig. 6.1-11. Cascaded coaxial step attenuator.

setting values of attenuations up to 150 dB. The example shows 90-dB attenuation. Each of the contacts can be moved or the attenuators slid.

The usable frequency band in *waveguide transmission lines* is limited to not quite half an octave. When wideband attenuation characteristics are mentioned, the full waveguide bandwidth is meant. The attenuators in question will be wideband type, covering the entire usable waveguide frequency band. Since the voltages and the current flow in waveguides are not generally that explicit, and since they vary their position with varying frequency, it is advisable to consider waveguide attenuators as interfering with the electric or magnetic fields. Attenuators can be made to interfere with either the electric or the magnetic field or, if it is preferred, with both. A simple, continuously variable attenuator can be made by inserting a resistive card in line with the electric field. Figure 6.1-12 shows an attenuator of that kind—the *flap attenuator*.

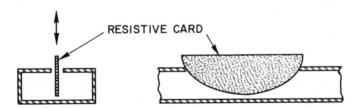

Fig. 6.1-12. Waveguide flap attenuator.

By inserting the card further into the guide, more electric field lines will be weakened; consequently, attenuation increases. The continuous curve of the card, shown on the side view of the attenuator, is designed to decrease the magnitude of reflection due to that discontinuity as it is distributed over a longer path. These attenuators can be made fairly flat in frequency response, but they are not phase invariant.

Another attenuator of the same class is the *side-vane attenuator*, in which the attenuating card is always entirely inside the guide, but it is moved from the side toward the center for increasing attenuation. This happens because the electric field distribution of the rectangular waveguide is such that highest

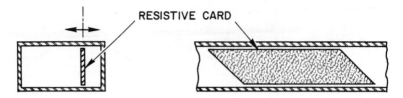

Fig. 6.1-13. Side-vane attenuator.

field intensity is in the center of the waveguide. Figure 6.1-13 shows the cross section of such an attenuator. This is also not phase-invariant.

The card is tapered to decrease the input and output reflection of the attenuator. The mathematical analysis of such attenuators is quite involved. Knudsen[8] handles the problem and has developed design curves for thin-film attenuators of this nature used in waveguide transmission lines.

A precision waveguide attenuator,[9] the *rotary-vane attenuator*, was developed in the early 1950's and was described by Southworth.[10,11] This device is distinguished by the fact that its attenuation follows a predictable, mathematical law and is not dependent on frequency.

Basically, this attenuator consists of three sections of waveguide in tandem. In each section a resistive film is placed across the guide, as shown in Fig. 6.1-14. The middle section is a short length of round guide that is free to rotate axially with respect to the two fixed-end sections. The end sections are rectangular-to-round waveguide transitions in which the resistive films are normal to the E-field of the applied wave. The construction is symmetrical, and the device is bidirectional.

When all films are aligned, the E-field of the applied wave is normal to all films. No current flows in the films then, and no attenuation occurs. If the center film is now rotated to some angle θ, the E-field can be considered to be split into two components: $E \sin \theta$ in the plane of the film and $E \cos \theta$ at right angles to it. The $E \sin \theta$ component will be absorbed by the film, whereas the $E \cos \theta$ component, oriented at an angle θ with respect to the original wave, will be passed unattenuated to the third section. When it encounters the third film, the $E \cos \theta$ component will be split into two components. The $E \cos \theta \sin \theta$ component will be absorbed, and the $E \cos^2 \theta$ component will emerge at the same orientation as the original wave. The attenuation is thus

[8] Knudsen, H. L., "Champs Dans un Guide Rectangulaire à Membrane Conductrice" (Field in a Rectangular Waveguide with a Conductive Slab), *L'Onde Electrique*, April 1953.

[9] Hand, B. P., "A Precision Waveguide Attenuator Which Obeys a Mathematical Law," *Hewlett-Packard Journal*, Vol. 6, No. 5, January 1955.

[10] Southworth, G. C., *Principles and Applications of Waveguide Transmission* (Princeton, N.J.: D. Van Nostrand, 1950), p. 374.

[11] The inventor of this type of attenuator is understood to be the late A. E. Bowen.

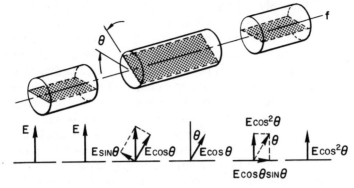

ATTENUATION = 20 log cos²θ = 40 log cos θ

Fig. 6.1-14. Functional drawing indicating operating principle of a rotary-vane variable attenuator.

ideally proportional only to the angle to which the center film is rotated and is completely independent of frequency. In dB terms, the attenuation is equal to 40 log cos θ. Figure 6.1-14 is a functional drawing that indicates the operation of the rotary-vane attenuator.

Phase-shift variations in the attenuator are very small. For settings between 0 and 40 dB, variations in phase shift are less than one degree. This small value makes the attenuators valuable in applications where it is important that applied power be varied independently of phase. Such requirements occur, for example, in measurements on multi-element antennas where the drive to the various elements must be varied to obtain the desired antenna pattern. By inserting rotary attenuators in cascade with the appropriate elements, the excitation can be varied over wide ranges.

Since attenuation is virtually unaffected by frequency, these attenuators, besides being valuable in general-purpose applications, offer a solution to the problem of providing signal generators with precision attenuators at frequencies where cutoff attenuators have excessive slope. Accuracy of the attenuator does not depend on the stability of the resistive films; as long as their attenuation is high and remains high, performance is not affected.

FIXED-WAVEGUIDE ATTENUATORS are usually built of the above kinds of attenuators, leaving the moving mechanism out of the structure and providing only fixed positioning of the attenuating element.

A DIRECTIONAL COUPLER is a device which, when inserted in a transmission line, will respond to a wave traveling in a particular direction on the transmission system and yet remain unaffected by a wave traveling in the opposite direction. Since the wave induced in the auxiliary arm of a properly operating coupler is proportional only to a wave traveling to the right on the transmission system, the power represented by this induced wave is a definite fraction of the power associated with the wave traveling to the right. This ratio of induced power to total power is known as the coupling factor and is normally expressed in decibels. As an example, the power in the secondary arm of a 20-dB coupler is 20 dB below (or one-hundredth the power) entering the primary arm of the coupler. Since couplers cannot be made to respond only to a wave traveling in the forward direction, some measure of the response to a wave traveling in the reverse direction must be indicated. This factor, known as directivity, is the ratio (expressed in dB) of the power in the secondary arm (when power is flowing in the forward direction of the main arm) to the power in the secondary arm (when an equal power is flowing in the reverse direction in the main arm). In other words, a directional coupler can be inserted into a transmission line in two different ways, forward and reverse, as far as the main line is concerned. Figure 6.1-15 shows the schematics of these two ways.

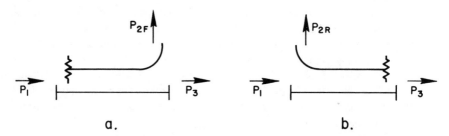

Fig. 6.1-15. Schematics of (a) forward, (b) reverse
directional coupler.

The coupling coefficient can be determined from the forward-connected coupler case, as in Fig. 6.1-15(a).

$$\text{coupling factor} = 10 \log \frac{P_1}{P_{2_F}} \text{ dB}$$

Directivity is determined as the ratio, in dB, of the power appearing at the auxiliary port when the coupler is in the forward direction to the power appearing at the auxiliary port when the coupler is in the reverse direction and the main arm terminated in its characteristic impedance.

$$\text{directivity factor} = 10 \log \frac{P_{2_F}}{P_{2_R}} \text{ dB}$$

A simple type of directional coupler is illustrated in Fig. 6.1-16. The coupling loop (L) is so designed that both the electric and the magnetic fields of the

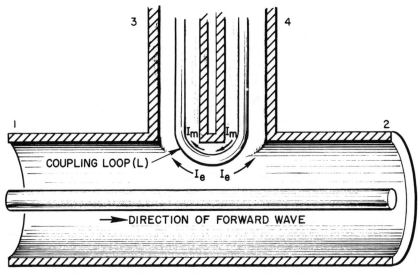

I_m = INDUCED CURRENT DUE TO MAGNETIC FIELD
I_e = INDUCED CURRENT DUE TO ELECTRIC FIELD

Fig. 6.1-16. A simple coaxial direction coupler.

wave on the transmission line will induce waves in both terminals of the loop, as shown. The magnetic field of the wave on the transmission line produces a current flowing from terminal 4 to terminal 3 of the loop. By designing the coupling loop properly, the current in terminal 4, induced by the magnetic flux, will cancel the current induced by the electric field; the two currents will add in terminal 3. This means that a wave traveling to the right on the transmission line will produce a response in terminal 3, while a wave traveling left will be canceled. For waves traveling to the left, a reverse action will take place; there will be no response in terminal 3 and a response in terminal 4. If terminal 2 is terminated in the characteristic impedance of the line, the reverse-induced wave (due to the wave traveling left) will be absorbed, making the system respond only to a wave traveling to the right.

The most widely used directional coupler *in waveguide*, the multihole coupler, is also an interference-type device. That is, directivity is achieved by producing, in the auxiliary arm of the coupler, two or more signals of such phase and magnitude that the signals traveling in one direction add, whereas the signals traveling in the opposite direction cancel each other. The simplest form of interference coupler consists of two holes spaced $\frac{1}{4}$ wavelength apart and connects two parallel sections of waveguide (Fig. 6.1-17). As shown in the figure, the signal in the main arm propagates through both holes. At each hole the signal splits into two components, one traveling in the forward

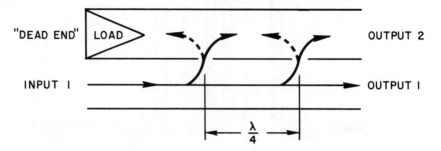

Fig. 6.1-17. Two-hole simple interference waveguide
directional coupler.

direction (to Output 2) and one traveling in the reverse direction (to Dead
End). The two wave components traveling from Input 1 to Output 2 move
the same electrical distance and are in phase; hence they add together. At
the same time, the two wave components traveling from Input 1 toward Dead
End move over paths that are electrically half a wavelength (180 deg) different; consequently they are exactly out of phase and they cancel each other.
This reasoning applies only when the coupled power is small compared with
the Input 1 power (weak coupling), for only then are the two coupled signals
of equal magnitude.

The two-hole coupler provides good directivity over a narrow bandwidth, but at wavelengths other than the design wavelength the two reverse-
coupled signals do not completely cancel because they are not 180 degrees
out of phase. To increase the useful bandwidth, the theory of the two-hole
device is extended to the design of a multihole array. An analysis of a large
array is beyond the scope of this text; however, the design of the multihole
array is similar in theory to that used for end-fire antennas. In both cases the
radiation from a number of elements is combined to provide an equal-ripple
type of response.

VARIABLE DIRECTIONAL COUPLERS are constructed in a manner similar
to coaxial directional couplers, which were discussed earlier. These attenuators are operated as waveguide-beyond-cutoff-type attenuators[12] in which the
power decreases exponentially with distance from the point of excitation. In
this type of attenuator, the point of excitation (the RF input line) is a strip
line at one end of a cavity. The pickup line is on a movable plunger assembly

[12] A discussion of the operation of waveguide beyond cutoff can be found in Terman
and Pettit's *Electronic Measurements* (New York: McGraw-Hill Book Company, 1952),
pp. 656–58.

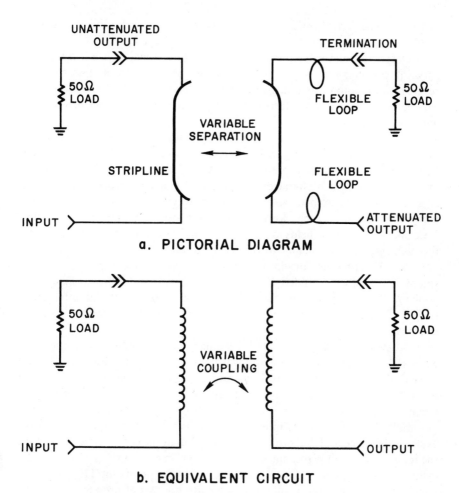

Fig. 6.1-18. Schematic representation of a variable attenuator of the waveguide-beyond-cutoff type.

that is moved by a rack and gear arrangement calibrated with a front panel control. Figure 6.1-18 is a schematic representation of this type of attenuator.

Nonreciprocal Attenuators

There also exist two-port or multiport networks in which the transmission coefficients in the forward direction are different from those in the reverse direction. The *ferrite isolator* is such a two-port network. In other words, using the scattering parameter notation, s_{21} is not equal to s_{12}.

A ferrite isolator is intended to be a unidirectional device in which the forward transmission is practically lossless, or very low in loss, but the reverse transmission acts as a good load that has very high attenuation. This can be achieved by using ferrite materials with gyromagnetic behavior at microwave frequencies.

Magnetic materials that are formed by replacing some of the iron atoms in magnetite (an iron oxide) with other metallic atoms, such as zinc, cobalt, nickel, manganese, or aluminum, are known as ferrites. This type of magnetic material exhibits such characteristics as good magnetic properties, high resistivity, and transparence to electromagnetic waves. Ferrites are non-current-carrying material, in other words, insulators, and have magnetic properties. If dc and RF magnetic fields are applied to ferrites at the same time, magnetic flux will result perpendicular to the RF field intensity.

Since ferrites are insulators with very low conductances, microwave energy can easily be guided through without suffering extreme losses. By applying RF energy and propagating it through ferrite material, interaction typical of magnetic materials will take place between the RF and the spinning electrons.

Michael Faraday discovered a similar effect in optical frequencies as early as 1845, and the phenomenon was named for him: the Faraday effect.[13] A typical Faraday rotation isolator and the transmission of energy in the foward and reverse direction are shown in Fig. 6.1-19. This device has a rectangular waveguide input and output. Both input and output parts are essentially rectangular-to-cylindrical waveguide transitions providing the necessary impedance matching to the ferrite section in the center area. Note that the input and output waveguides are rotated 45° in respect to each other. The cylindrical waveguide is designed to carry only the dominant TE_{11} mode. A pencil-shaped ferrite rod is placed in line with the axis of propagation. Both ends of the ferrite are tapered to minimize reflections. A strong permanent magnet is placed around the ferrite, magnetizing it in the axial direction. The ferrite in this state will rotate the electric field by some amount that depends on the geometry, magnetic properties of the ferrite, and the magnetic field strength of the permanent magnet. In this particular case, the rotation was set for 45°. The direction of rotation is determined only by the polarity of the applied dc magnetic field. In this case it was clockwise in the forward direction. The output transition is rotated exactly by this amount to enable one to take the energy passing through out of the device. If one applied energy in the reverse direction, the ferrite would again rotate the wave in the

---

[13] Ginzton, E. L., *Microwave Measurements* (New York: McGraw-Hill Book Company, 1957), p. 73.

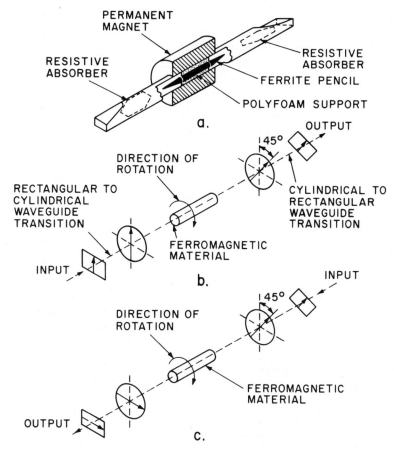

PERMANENT
MAGNET

RESISTIVE
ABSORBER

RESISTIVE
ABSORBER

FERRITE PENCIL

POLYFOAM SUPPORT

a.

OUTPUT

45°

DIRECTION OF
ROTATION

RECTANGULAR TO
CYLINDRICAL
WAVEGUIDE
TRANSITION

CYLINDRICAL TO
RECTANGULAR
WAVEGUIDE
TRANSITION

FERROMAGNETIC
MATERIAL

INPUT

b.

INPUT

45°

DIRECTION OF
ROTATION

FERROMAGNETIC
MATERIAL

OUTPUT

c.

Fig. 6.1-19. Functional diagrams of a Faraday rotation isolator: (a) schematic representation of the main elements, (b) transmission of energy through the isolator in the forward direction, and (c) transmission in the reverse direction. Reprinted from Edward Ginzton, *Microwave Measurements* (New York: McGraw-Hill Book Company, © 1957) p. 74 by permission of the publisher.

same rotational direction, turning the wave 90° away from the waveguide transition and not allowing it to pass. Of course, this polarization of the electric field does not satisfy the boundary conditions of propagation; it would reflect. In the transition sections resistive absorber cards are placed

parallel with the wide walls of the rectangular waveguide. These cards, perpendicular to the electric field lines when normal polarization exists, will not
absorb any energy, since they are coinciding with equipotential surfaces. But,
when signal is applied in the reverse direction and the electric field is rotated
90°, the absorber will be in line with the electric field lines, absorbing the
energy and not allowing it to be reflected. As will be seen later in the text,
this is quite important where separation of signal source from transmission
devices is necessary to eliminate "signal source pulling" and to decrease errors
due to multiple mismatch losses, etc. There are quite a few other structures
which use the Faraday rotation effect to produce isolation.[14] Construction of
a four-port Faraday rotator circulator is illustrated in Fig. 6.1-20.

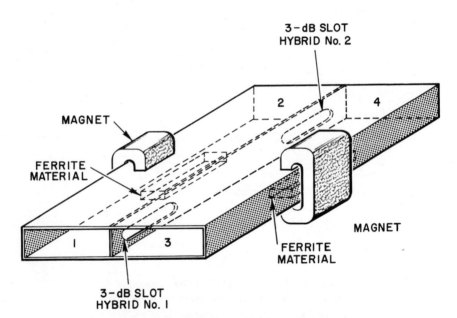

Fig. 6.1-20. Four-port Faraday circulator. Reprinted
from A. L. Lance, *Introduction to Micro-
wave Theory and Measurements* (New York:
McGraw-Hill Book Company, 1964) p. 174
by permission of the author and publisher.

When a vertically polarized wave enters port 1 of the circulator, it will
be divided equally in each arm of the waveguide by the first 3-dB hybrid.
Energy in arm 1 will undergo a net phase shift of 180° with respect to the

[14] Lance, A. L., *Introduction to Microwave Theory and Measurements* (New York:
McGraw-Hill Book Company, 1964), Chap. 10.

energy in the other waveguide. When it reaches the second 3-dB hybrid, it will recombine with the energy through arm 2. Since energy entering arm 2 is traveling in the opposite direction, it will not undergo any differential phase shift after being split at the second 3-dB hybrid and will recombine at the first 3-dB hybrid in such a manner that all the energy will leave through port 3. Similarly, energy entering port 3 will leave at port 4, and energy entering port 4 will leave at port 3. Such a device is extremely useful for coupling an antenna to a receiver-transmitter system.

As with coaxial ferrite devices, isolators, attenuators, phase shifters, and switches can also be constructed using ferrite materials.

Besides the use of Faraday rotation, another nonreciprocal effect is also used to make circulators. A device that uses *nonreciprocal phase shift* in rectangular waveguide depends upon different phase shift of the device in an opposite direction of propagation. If ferrites are placed in two adjacent guides, but oppositely polarized, nonreciprocal phase characteristics can be observed.

Attenuation Measurements

As was mentioned at the beginning of this section, measurement of attenuation magnitude can be divided into three categories: (1) RF substitution, (2) audio frequency (AF) or dc substitution, and (3) IF substitution using linear detection.

Figure 6.1-21 shows the block diagram of an *RF substitution* mea-

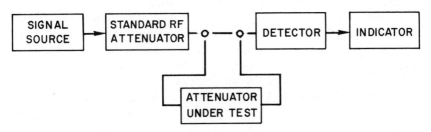

Fig. 6.1-21. Attenuation measurement using RF substitution.

surement arrangement. Since measurements generally compare an unknown quantity of something to a known quantity of the same thing, this technique merely substitutes attenuation. As a good practice, the standard attenuator is a variable calibrated attenuator set to a high attenuation value and connected directly to the dectector. The indicator is adjusted to a well-defined position. Upon insertion of the attenuation under test, the standard attenuator is adjusted by decreasing its attenuation until the indicator registers the same position as before. The difference in attenuation settings of the standard attenuator is substituted for the attenuator under test in order to maintain the same indicator level; consequently the attenuation value of the attenuator under test equals the difference of attenuation settings of the standard attenuator. For example, at the first operation, which is called calibration, the standard attenuator is set to 50 dB dial reading. When the attenuator under test is inserted (this operation is called the test), the dial reading is 27.8 dB, so the attenuator under test has a value of 22.2 dB.

Figure 6.1-22 is a block diagram of the *audio frequency or dc substitution*

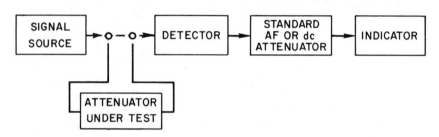

Fig. 6.1-22. Audio frequency or dc substitution.

techniques. This technique uses a low-frequency or dc attenuator as a standard. The attenuation measurement is done exactly as the one above, but it uses the audio frequency or dc attenuator for calibration. Although this technique takes advantage of the fact that low-frequency attenuators can be calibrated to a much higher degree of accuracy, detector law has to be considered. In the RF substitution technique, the RF level into the detector did not change, so the detector was not questioned. This basic difference determines the value of each for different uses, which will be dealt with in more detail later in this section under error analysis.

Figure 6.1-23 is a block diagram of a simple IF substitution scheme. Superheterodyning achieves high sensitivity; consequently it is needed in large dynamic range attenuation measurements. The two kinds of techniques discussed above employ square-law detection in systems that are capable of measuring medium-range attenuation up to about 50 dB. With linear detection, heterodyne-type measurements, dynamic ranges in excess of 150 dB can be achieved. The substitution is done with a precisely calibrated IF attenuator, and IF frequency is the difference of local oscillator and signal source frequency. To understand how linear detection is achieved by the use of a mixer, which basically is a nearly square-law device, let us write the two signals entering the mixer that is being squared:

$$(E_1 + E_2)^2 = E_1^2 + 2E_1E_2 + E_2^2$$

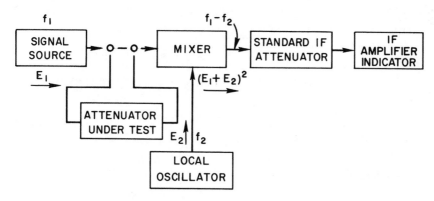

Fig. 6.1-23. Attenuation measurement with IF substitution.

If

$$E_1 \ll E_2 \quad \text{then} \quad E_1^2 \lll E_2^2$$

Hence it can be said that, if E_1 is at least 30 or 40 dB smaller than E_2, the E_1^2 term can be neglected. The remaining portion of the expression,

$$2E_1E_2 + E_2^2$$

shows that E_1 is a linear function, since E_2 (local oscillator signal level) is held constant.

If one desires to sweep-frequency test the attenuation, RF or AF (dc) substitution techniques are quite simple to apply with the use of a swept-frequency signal source using either an oscilloscope or an X-Y recorder as indicator. Calibration lines can be drawn and marked in increments, as the resolution of the system allows. Then the measurement will provide the actual frequency response of attenuation. Such an attenuation-measuring system is shown in Fig. 6.1-24. Figure 6.1-25 shows the calibration lines taken when the standard attenuator is connected directly to the directional coupler. The heavier line represents the frequency response of the attenuator under test. Directional coupler 1 serves as part of the ALC feedback loop. Directional coupler 2 separates the reflections from the crystal detector to the attenuator under test, lowering the ambiguity of the measurement caused by multiple mismatch errors. In other words, the directional coupler serves only as a very well-matched pad. Waveguide directional couplers generally have extremely low mismatches.

Several IF attenuation measurement systems are available to use for making a swept-frequency, large dynamic range attenuation system. One of these techniques is shown in the block diagram of Fig. 6.1-26.

The two sweepers are offset by the IF frequency, and the sweep ref-

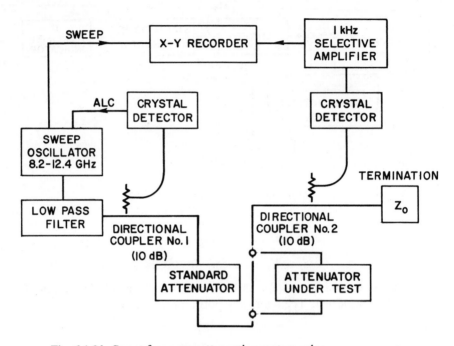

Fig. 6.1-24. Swept-frequency attenuation system using
RF substitution.

erence of one sweeper drives the other. The output of the synchronous
detector is the reference of the phase-lock box developing the error signal
connected to one of the sweepers VTO voltage. This achieves a constant
IF frequency that allows the use of quite narrow bandwidths in the IF amplifier, resulting in very wide dynamic range with increased sensitivity of the
amplifier.

   Many other high-attenuation schemes can be developed, all using the
basic linear detection technique that allows large dynamic range attenuation
measurements.

Errors in Attenuation Measurements

   Since measurement generally is a comparison of an unknown value to
an already calibrated known value, the sources of errors in such measurements can be divided into three categories: (1) scalar errors, (2) uncertainty
errors, and (3) vectorial errors.

   SCALAR ERRORS are the ones that usually can be easily calibrated out, or
at least alleviated and cut down in size. One such error is the calibration error

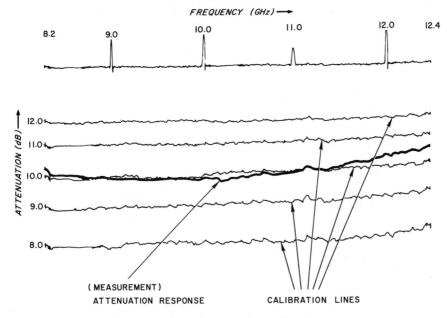

Fig. 6.1-25. X-Y recording of attenuation calibration
and measurement.

of the standard being used. (More accurate calibration of the standard will achieve a higher degree of accuracy.) Drift of the signal source and/or indicator can be understood as a scalar error, since drift is not vectorial in its nature. Drift can easily be checked and compensated for by constant recalibration.

UNCERTAINTY ERRORS are not reduced by special calibration because they are random in nature. Such errors are caused by such factors as connector repeatability, readability, and resetability of the standard's dial and by readability of the indicator.

VECTORIAL ERRORS are such that they can be represented as vectors rotating with frequency. If they are individually determined at each frequency and vectorially summed, corrections can be made to cancel their effect. Such errors are the multiple mismatch errors caused by two or more reflections offset from each other along a transmission line and beating together. The example in Chap. 3 dealt with this case. If the magnitude and phase of the reflection coefficient of source, detector, and attenuator are known at all the frequencies of interest, the error term derived in Chap. 3 can be used to eliminate these types of errors, which are usually the largest source of ambiguity. The following examples of attenuation measurement cases will ex-

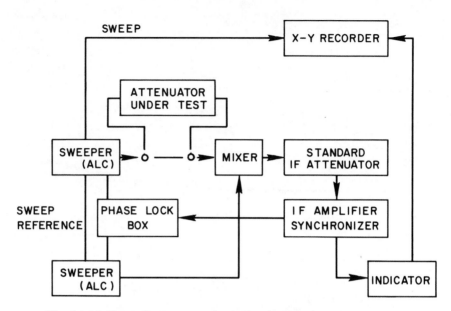

Fig. 6.1-26. Swept-frequency attenuation measuring
system using heterodyne IF substitution
technique with synchronous detection and
phase-locked sweepable sources.

plain in detail how this kind of error analysis can be done. Figure 6.1-27
shows a block diagram of an RF substitution attenuation-measuring system.

Sources of scalar and uncertainty errors are listed below for a typical
coaxial system.

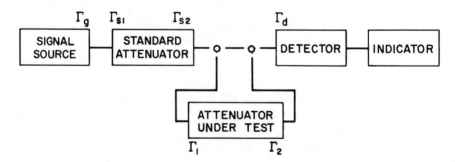

Fig. 6.1-27. Block diagram of an RF substitution attenuation measuring system showing the re-
flection coefficients.

Standard attenuator calibration accuracy	$\pm 2\%$ of dB
Indicator readability	± 0.05 dB
Connector repeatability	± 0.05 dB
System drift	± 0.05 dB
Maximum possible error due to uncertainty and scalar errors if 20-dB attenuation is measured	± 0.55 dB

Multiple mismatch errors are vectorial errors. One must consider all those junctions where the reflection coefficient might vary either in magnitude or phase (or both). Furthermore, any port, even though unchanged, looks into a different port than it did before. Γ_{s2} looks into Γ_d during calibration, but at test, Γ_{s2} with Γ_1 and Γ_2 with Γ_d form different junctions. The flow graphs of the measurement are shown in Fig. 6.1-28.

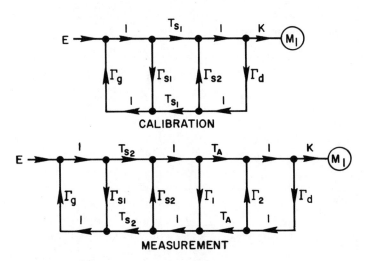

Fig. 6.1-28. Flow graphs of an attenuation measuring system using RF substitution.

Assuming no reflections, the two flow graphs will read

$$M_1 = E T_{s1} K \qquad M_1 = E T_{s2} T_A K$$

Hence

$$T_A = \frac{T_{s1}}{T_{s2}}$$

The attenuation

$$20 \log T_A = 20 \log T_{s1} - 20 \log T_{s2}$$

In other words, the attenuation of the attenuator under test is equal to the attenuation difference read on the standard attenuator.

Solving the flow graphs with the reflections included, employing the nontouching loop rule, provides calibration,

$$\frac{M_1'}{E} = \frac{T_{s1}'K}{1 - \Gamma_g\Gamma_{s1} - \Gamma_{s2}\Gamma_d - T_{s1}^2\Gamma_g\Gamma_d + \Gamma_g\Gamma_d\Gamma_{s1}\Gamma_{s2}}$$

and measurement,

$$\frac{M_1'}{E} = \frac{T_{s2}'T_AK}{1 - \Gamma_g\Gamma_{s1}' - \Gamma_{s2}'\Gamma_1 - \Gamma_2\Gamma_d - T_{s2}^2\Gamma_g\Gamma_1 - T_A^2\Gamma_{s2}'\Gamma_d - T_{s2}^2T_A^2\Gamma_g\Gamma_d + \Gamma_g\Gamma_1\Gamma_2\Gamma_dT_{s2}^2}{\quad + \Gamma_1\Gamma_{s1}'\Gamma_{s2}'\Gamma_dT_A^2 + \Gamma_g\Gamma_{s1}'\Gamma_{s2}'\Gamma_1 + \Gamma_{s2}'\Gamma_1\Gamma_2\Gamma_d + \Gamma_g\Gamma_{s1}'\Gamma_2\Gamma_d - \Gamma_g\Gamma_{s1}'\Gamma_{s2}'\Gamma_1\Gamma_2\Gamma_d}$$

At this point it is worthwhile to take a closer look at the higher-order terms in the denominator. Most of the time, when attenuation measurements higher than 6 dB are involved and the reflections are reasonably small, the fourth- and higher-order terms will be several orders of magnitude smaller than the second-order terms and can be neglected. In the case above, when 20-dB attenuation is measured, the value of

$$T_A = 0.1 \quad \text{and} \quad T_A^2 = 0.01$$

Neglecting the fourth- and higher-order terms, calibration

$$\frac{M_1'}{E} = \frac{T_{s1}'K}{1 - \Gamma_g\Gamma_{s1} - \Gamma_{s2}\Gamma_d}$$

and measurement

$$\frac{M_1'}{E} = \frac{T_{s2}'T_AK}{1 - \Gamma_g\Gamma_{s1}' - \Gamma_{s2}'\Gamma_1 - \Gamma_2\Gamma_d}$$

Hence attenuation

$$T_A = \frac{T_{s1}'}{T_{s2}'}\frac{1 - \Gamma_g\Gamma_{s1}' - \Gamma_{s2}'\Gamma_1 - \Gamma_2\Gamma_d}{1 - \Gamma_g\Gamma_{s1} - \Gamma_{c2}\Gamma_d}$$

The error is

$$\Delta = \frac{\dfrac{T_{s1}}{T_{s2}}}{\dfrac{T_{s1}'}{T_{s2}'}\dfrac{1 - \Gamma_g\Gamma_{s1}' - \Gamma_{s2}'\Gamma_1 - \Gamma_2\Gamma_d}{1 - \Gamma_g\Gamma_{s1} - \Gamma_{s2}\Gamma_d}}$$

Assume $T_{s1} = T_{s1}'$ and $T_{s2} = T_{s2}'$

Error in terms of dB is

$$\Delta_{(dB)} = 20\log\left|\frac{1 - \Gamma_g\Gamma_{s1} - \Gamma_{s2}\Gamma_d}{1 - \Gamma_g\Gamma_{s1}' - \Gamma_{s2}'\Gamma_1 - \Gamma_2\Gamma_d}\right| \, dB$$

For maximum possible error determination, it is a good policy to take into account the maximum possible reflection coefficient of the devices. Assuming a leveled waveguide signal source with a precision attenuator as a standard and a detector of the standard kind, the reflections will be as follows:

$$|\Gamma_g| = 0.02$$
$$|\Gamma_{s1}| = |\Gamma_{s2}| = 0.07$$
$$|\Gamma_1| = |\Gamma_2| = 0.07$$
$$|\Gamma_d| = 0.11$$

To obtain the maximum error, the numerator has to have maximum value and the denominator, minimum value; for minimum error, the opposite is true.

To obtain maximum value of the numerator,

$$\Gamma_{s1} = \Gamma_{s2} = 0$$

To obtain all other values to maximum,

$$\Delta_{\max} = 20 \log \left| \frac{1}{1 - (0.02 \times 0.07) - (0.07 \times 0.07) - (0.07 \times 0.11)} \right|$$
$$= +0.12 \text{ dB}$$

To get the maximum value of the denominator,

$$\Gamma'_{s1} = \Gamma'_{s2} = \Gamma_2 = 0$$

To obtain all other values to minimum,

$$\Delta_{\min} = 20 \log \left| \frac{1 - (0.02 \times 0.07) - (0.07 \times 0.11)}{1} \right| = -0.09 \text{ dB}$$

The positive and negative signs do not take into account the fact that the error is of attenuation, which may change the signs of the error terms.

For maximum possible error, all the errors taken into account can be expressed as follows:

$$\Delta_{\text{total}} = \frac{+0.55 + 0.09}{-0.55 - 0.12} = \frac{+0.64}{-0.67} \text{ dB}$$

if 20-dB attenuation is measured.

To make this measurement more accurate, several errors can be reduced. The standard attenuator calibration accuracy of 0.4 dB can be cut down if special calibration of the standard is performed by a standards laboratory. For example, the National Bureau of Standards can provide 0.02-dB per 10-dB calibration accuracies at single frequencies. Many times the decision to take better standards or cut down ambiguities in such measurement systems is made by taking economic factors into account also. The example analyzed above used the RF substitution technique for making attenuation measurements.

Audio or dc substitution technique uses basically the same setup, except the standard used for substitution is in low frequencies or in dc where the standard can be calibrated for accuracies that are better by several orders of magnitude.

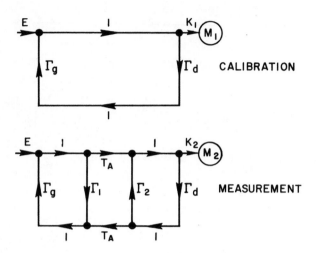

Fig. 6.1-29. Flow graphs of an attenuation measuring
setup using audio or dc substitution.

Figure 6.1-29 shows the flow graphs of an attenuation-measuring setup employing audio or dc substitution technique. M_1 and M_2 mean different meter readings, including the calibrated attenuator. K_1 and K_2 represent the possible different detection laws involved when the detector is subjected to different power levels from calibration to measurement.

We solve the problem in the same manner used above. Assuming no reflections,

$$M_1 = EK_1, \qquad M_2 = EK_2T_A$$

We divide the two expressions, expressing T_A as

$$T_A = \frac{M_2}{M_1} \frac{K_1}{K_2}$$

Further assuming that $K_1 = K_2$ for perfect nonchanging detector (square) law,

$$T_A = \frac{M_2}{M_1}$$

we solve the flow graphs with the sources of errors included as follows:

$$M_1' = \frac{EK_1}{1 - \Gamma_g\Gamma_d}$$

$$M_2' = \frac{EK_2T_A'}{1 - \Gamma_g\Gamma_1 - \Gamma_2\Gamma_d - T_A^2\Gamma_g\Gamma_d + \Gamma_g\Gamma_d\Gamma_1\Gamma_2}$$

Attenuation from the above two equations is

$$T_A' = \frac{M_2'}{M_1'} \frac{K_1}{K_2} \frac{1 - \Gamma_g\Gamma_1 - \Gamma_2\Gamma_d - T_A'^2\Gamma_g\Gamma_d + \Gamma_g\Gamma_d\Gamma_1\Gamma_2}{1 - \Gamma_g\Gamma_d}$$

The error is

$$\Delta = \frac{T_A}{T_A'} = \frac{\dfrac{M_2}{M_1}}{\dfrac{M_2'}{M_1'}\dfrac{K_1}{K_2}\dfrac{1 - \Gamma_g\Gamma_1 - \Gamma_2\Gamma_d - T_A'^2\Gamma_g\Gamma_d + \Gamma_g\Gamma_d\Gamma_1\Gamma_2}{1 - \Gamma_g\Gamma_d}}$$

$$= \frac{M_2}{M_1}\frac{M_1'}{M_2'}\frac{K_2}{K_1}\frac{1 - \Gamma_g\Gamma_d}{1 - \Gamma_g\Gamma_1 - \Gamma_2\Gamma_d - T_A'^2\Gamma_g\Gamma_d + \Gamma_g\Gamma_d\Gamma_1\Gamma_2}$$

It can be seen that, if the detector can be kept within its square-law region, K_2/K_1 does not differ too much from 1. This technique is quite a bit more accurate than the one previously described. However, a word of caution is due: One must not forget that square-law errors can be a great deal larger, if not analyzed carefully, than those that RF substitution would yield. Many times these two techniques are used together. A certain amount of attenuation is substituted in RF; then audio or dc substitution takes over when the signal level is low enough for square-law operation.

The above techniques provide attenuation measurements to about a 50-dB dynamic range. For higher values of measurements, linear detection gives wide dynamic range where the technique for error analysis is the same as for the other two techniques. To keep the mixer in the linear-law region, care should be exercised not to overdrive the input. With heterodyne and similar linear detection schemes, attenuation measurements well above 100-dB dynamic range can be performed.

6.1.2 IMPEDANCE MEASUREMENTS

At this point the reader already is familiar with the importance of impedances and the basic concepts of impedance (transmission line theory, Chap. 2). Furthermore, a working knowledge has been gathered of how to use the Smith impedance chart calculator. Impedance measurements have to be made to determine the characteristics of a termination connected at the end of a lossless, uniform transmission line. Impedance is a vectorial quantity,

$$Z = R + jX \tag{6.1-1}$$

where R, the real component, is the resistive part and X, the imaginary component, is the reactive component of impedance. If reactance is positive,

$$X = \omega L \tag{6.1-2}$$

is inductive; if it is negative,

$$X = -\frac{1}{\omega C} \tag{6.1-3}$$

is capacitive.

It is also already known to the reader that impedance has to be normalized to be plotted on a Smith Chart:

$$\frac{Z}{Z_0} = \frac{R}{Z_0} + j\frac{X}{Z_0} \tag{6.1-4}$$

Reflection Coefficient and Voltage Standing-Wave Ratio

If a load terminating a lossless, uniform transmission line having Z_0 characteristic impedance has an impedance differing from that characteristic impedance, reflections will be set up that produce standing waves on the line. The deviation of the impedance of that load from the characteristic impedance of that line can be measured by the ratio of the reflected voltage (E_r) and incident voltage (E_i), producing the reflection coefficient (Γ_L).

$$\Gamma_L = \frac{E_r}{E_i} \tag{6.1-5}$$

The reflection coefficient, being a vectorial quantity having magnitude and phase information, fully describes that load impedance terminating the transmission line.

The voltage standing-wave ratio (VSWR), as a figure of merit, provides scalar information about that load, measurable along the transmission line, probing the field strength variation, receiving information on the ratio of voltage maximums (E_{max}) and voltage minimums (E_{min}).

$$\text{VSWR} = \sigma = \frac{E_{max}}{E_{min}} \tag{6.1-6}$$

The relationship between VSWR and the magnitude of reflection coefficient ($|\Gamma| = \rho$), as derived in Chap. 2, is

$$\sigma = \frac{1+\rho}{1-\rho} \quad \rho = \frac{\sigma-1}{\sigma+1} \tag{6.1-7}$$

Return loss is defined as the ratio of reflected voltage over the incident voltage in terms of decibels:

$$\text{return loss} = -20\log_{10}\rho \, (\text{dB}) \tag{6.1-8}$$

The impedance of a load on a uniform transmission line is given by

$$Z_L = \frac{1 + j\sigma\tanh\gamma x}{\sigma + j\tanh\gamma x} \tag{6.1-9}$$

The impedance of a load on a lossless, uniform transmission line is

$$Z_L = \frac{1 + j\sigma\tan\beta x}{\sigma + j\tan\beta x} \tag{6.1-10}$$

where σ denotes the voltage standing-wave ratio established on the transmission line by the load, x is the distance from the plane of the load, and β is the propagation constant $(2\pi/\lambda)$, as shown in Fig. 6.1-30.

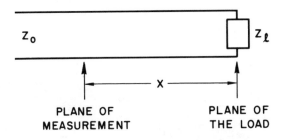

Fig. 6.1-30. Impedance of a load.

Slotted Lines

In Chap. 2 the reader became acquainted with Smith Charts and acquired a working knowledge of how to plot the impedance of a load with a slotted line. A slotted line is a section of uniform, lossless (very-low-loss) transmission line with a longitudinal aperture to provide access to a pickup probe (usually electric field pickup antenna) sliding along the line that detects voltage variations prevailing on the transmission line.

Slotted lines are made in coaxial and waveguide configurations. Typically, a waveguide slotted line is mounted on a universal carriage of the type shown in Fig. 6.1-31. Waveguide sections are provided with a tapered slot

Fig. 6.1-31. Universal probe carriage.

to allow the pickup probe to penetrate inside the transmission to sample the electric field distribution, as in Fig. 6.1-32. Coaxial slotted lines can be

Fig. 6.1-32. Waveguide slotted section.

made of a slab line design, where the coaxial line is transformed (through a conformal mapping) to a slab line configuration (Fig. 6.1-33). Such a slotted section is shown in Fig. 6.1-34.

Pickup probes can be broadband (Fig. 6.1-35), in which an external

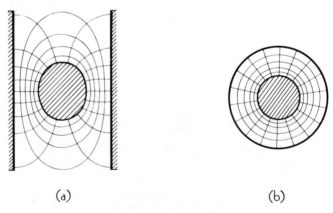

(a)   (b)

Fig. 6.1-33. Cross section and field configuration of coaxial slotted lines: (a) slab line, (b) simple coax line. Reprinted from W. Bruce Wholey and W. Noel Eldred, "A New Type of Slotted Line Section," *Proceedings of the I.R.E.*, Vol. 38, No. 3 (March 1950), p. 244 by permission of the publisher.

Fig. 6.1-34. Coaxial slab line slotted section.

detector has to be mounted, or untuned (Fig. 6.1-36), in which a detector has already been built into the probe.

There are several methods for measuring voltage standing-wave ratio with a slotted line. In the most straightforward method, the VSWR can be read directly on a standing-wave indicator, but for a VSWR higher than about 10, this method can result in erroneous readings. When the VSWR is high, probe coupling must be increased if a reading is to be obtained at a

Fig. 6.1-35. Broadband probe.

Fig. 6.1-36. Untuned probe.

voltage minimum. However, at the voltage maximum this high coupling may result in a deformation of the pattern with consequent error in reading. In addition to the error caused by probe loading, there is also danger of error resulting from change in detector characteristics at high RF levels.

Two other techniques, the double minimum method and the calibrated attenuator method (in other words, RF substitution method), can be used to obtain greater accuracy. In the double minimum method it is necessary to establish the electrical distance between the points at which the output is double the minimum (see Fig. 6.1-37). In the calibrated attenuator method,

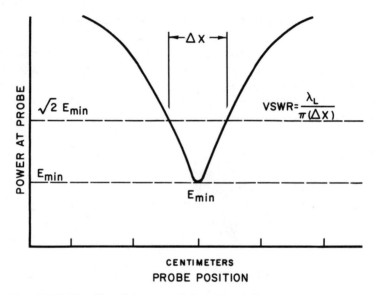

Fig. 6.1-37. Double minimum technique measuring
high VSWR.

a calibrated variable RF attenuator is used between the signal source and the slotted section and is adjusted to keep the rectified output of the crystal detector equal at the voltage minimum and voltage maximum points. In this case, the VSWR is merely the difference in the attenuator settings. This can be easily converted from decibels to VSWR by taking the antilog of that value and dividing by 20. Figure 6.1-38 shows the calibrated attenuator method measurement setup.

Signal sources can introduce at least three undesirable characteristics that will affect slotted-line measurements. These include presence of RF harmonics, FM, and spurious signals. Signal sources used for standing-wave measurements should have relatively low harmonic content in their output. The VSWR at the harmonic frequency may be considerably higher than at the

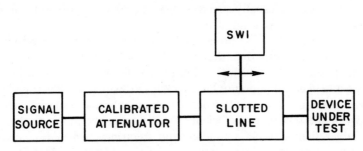

Fig. 6.1-38. Attenuator method of measuring VSWR.

fundamental. Spurious frequencies in the signal source are also undesirable, for, unless they are very slight, they will obscure the minimum points at high VSWR values.

Figure 6.1-39 shows the plot of a VSWR pattern made with a signal-source producing unwanted FM. Instances are common in which the presence of RF harmonics have led to very serious errors in VSWR measurements.

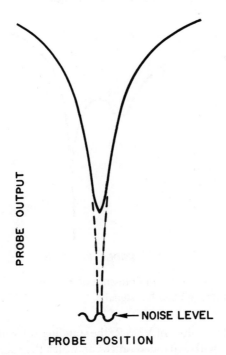

Fig. 6.1-39. Minimum area of standing wave pattern in presence of FM.

Such harmonics are usually present to an excessive degree only in signal sources that have coaxial outputs.

Coaxial pickups of a broadband type will often pass harmonic frequencies more efficiently than they pass the fundamental. In waveguide systems, signal sources such as internal-cavity klystrons have more or less fixed coupling and in addition do not have pickups extending into the tuned cavity to cause agitations of the cavity fields. Harmonics become especially troublesome when the reflection coefficient of a load at a harmonic frequency is much larger than it is at a fundamental frequency, which is a common condition. When the harmonic content of a signal source is high, the large reflection coefficient of the load at the harmonic frequency can cause the harmonic standing-wave fields to be of the same order of magnitude as the field of the fundamental frequency. Thus the device having a VSWR of 2.0 at the fundamental frequency will often have a VSWR of 20 or more at the second harmonic frequency. If such a device is driven from a signal source having about 15% second harmonic content, the peaks of standing waves of second harmonics will be about one-quarter the amplitude of the peaks at the fundamental frequency. Figure 6.1-40 shows a typical VSWR pattern obtained when the RF signal contains harmonics.

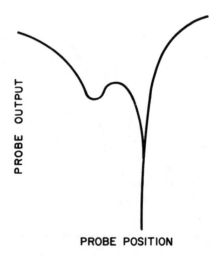

PROBE POSITION

Fig. 6.1-40. VSWR pattern in presence of harmonics
in the applied RF signal.

To measure low values of VSWR, the standing-wave indicators available today are equipped with expanded ranges on which the voltage standing-wave ratio can be read directly to quite high accuracies. However, a word of caution is due at this time concerning the possible sources of errors when making VSWR measurements. Especially at low VSWR values, the slotted line's own residual mismatch is a very probable cause of error. Another cause is slope, including mechanical and electrical slope, of the slotted line, which means that the peaks of the standing-wave pattern change as the slotted line is moved from one end toward the other.

Reflections from the probe itself can cause errors, as can the attenuation in the slotted line and reflection from the source. Then, of course, detector law comes into effect when high-voltage standing-wave ratios are being measured. Last but not least, the instrumentation calibration error of the standing wave indicator is also a source of error.

The VSWR of a low-loss, two-port device is not so simple to measure as in the examples cited. As was mentioned in the introduction to scattering parameters, it is well known that a low-loss, two-port device's input reflection coefficient, in other words, s_{11}, can be measured accurately only if the other port is terminated in its characteristic impedance. That means that, if one desires to measure the input reflection coefficient or voltage standing-wave ratio of a low-loss, two-port device, a termination of characteristic impedance has to be provided for the other port. Unfortunately, at the present state of art, perfect terminations at microwave frequencies cannot be made easily, and it is almost impossible to match them perfectly across wide frequency ranges. Another technique has to be used to maintain characteristic impedance at the far end of the device. A sliding load can be used for this purpose. A setup showing this technique is provided in Fig. 6.1-41. Moving a sliding

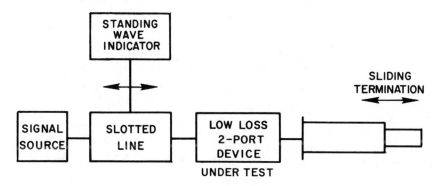

Fig. 6.1-41. Low-loss two-port network VSWR measurement setup using sliding termination.

termination along the line enables the user to place the plane of reflection due to that termination anywhere on the line and to measure all its phases.

Figure 6.1-42 is a simple vector diagram showing how the reflection

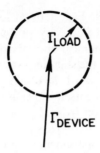

Fig. 6.1-42. Vector diagram of the reflection coefficients of the device and load, when the
load is being phased.

coefficient of the device and the reflection coefficient of the load are vectorially
placed in relationship to each other. Phasing the load means that this vector
is being rotated while the reflection coefficient of the device stays still. This
suggests that a technique can be developed to measure both vectors when they
are added and to make another measurement when they are subtracted;
taking the mean of these measurements would give the device's actual reflection coefficient. This technique can be quite cumbersome; the so-called
coupled sliding-load technique is an easier way of getting rid of the error due
to the sliding load.[15]

Figure 6.1-43 shows the block diagram for the coupled sliding-load
technique. This technique provides constant phase angle between the inci-

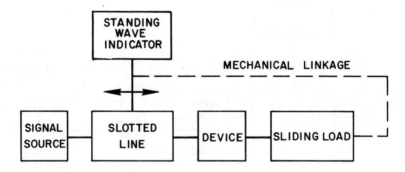

Fig. 6.1-43. Coupled sliding load VSWR technique
measuring VSWR of low-loss two-port
networks.

[15] Weinschel, B. O., G. U. Sorger, S. J. Raff, J. E. Ebert, "Precision Coaxial VSWR
Measurements by Coupled Sliding-load Technique," *IEEE Transactions on Instrumentation
and Measurement*, Vol. IM-13, No. 4, December 1964.

dent voltage (E_i) and the reflected voltage due to the sliding load (E_L). E_r is the voltage reflected by the device under test at the detector's location. Equation (6.1-11) shows the mathematical relation

$$\frac{|E_i + E_L| + |E_r|}{|E_i + E_L| - |E_r|} \qquad (6.1\text{-}11)$$

which clearly shows that, if mechanical linkage is placed between the slotted line and the sliding load, only the device's mismatch will come into the picture. Resolving Eq. (6.1-11) would give the voltage standing-wave ratio information.

The procedure to plot the impedance of a load involves taking a standing-wave ratio measurement and the location of the minimum of the standing wave pattern; then the shift of minimum is measured when the plane of load is short-circuited. This information gives the necessary polar coordinates to plot the impedance on the Smith Chart. An example follows.

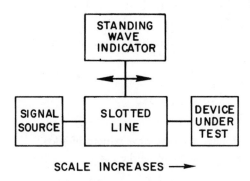

Fig. 6.1-44. Impedance measuring setup with a slotted line.

Equipment was set up as shown in Fig. 6.1-44, and the following data were gathered.

Null₁	Null₂	Min	VSWR
8.35 cm	12.35 cm	9.72 cm	2.2:1

Nulls were taken with the line terminated in a short circuit. The two null positions were adjacent nulls; it can be clearly seen from their spacing (which is a half-wavelength) that the wavelength is 8 cm. Substituting these data into Eq. (6.1-10), it is necessary to determine the sign of X. X is taken as a positive number if the minimum moves toward generator when it is shorted and negative when it moves toward load. $\beta = 2\pi/\lambda = 2\pi/8 = 0.785$. Then, the normalized impedance

$$\frac{Z_L}{Z_0} = \frac{1 + j2.2 \tan\,[0.785(-2.63)]}{2.2 + j \tan\,[0.785(-2.63)]} = 1.18 + 0.86j \qquad (6.1\text{-}12)$$

Plotting the voltage standing-wave pattern on the slotted line, and labeling each point of interest (Fig. 6.1-45), we can easily solve this problem with the help of the Smith Chart.

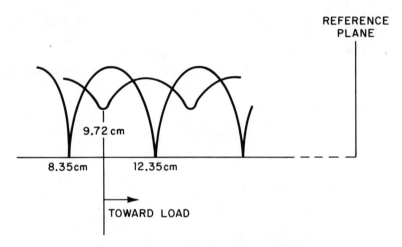

Fig. 6.1-45. Voltage standing-wave pattern of a load
and short circuit on the line.

Making up a table showing data and performing the necessary calculations give this information:

λ	Null₁	Null₂	Min	Δcm	Δλ	VSWR
8 cm	8.35 cm	12.35 cm	9.72 cm	2.63 cm	0.329	2.2:1

In Chap. 2 it was stated that, if the line is shorted, the impedance has to be plotted in the direction where the minimum shifted, becoming a null. Null₂ was drawn in Fig. 6.1-45, and the minimum was taken into account; consequently the plot is made toward the load. Figure 6.1-46 shows the impedance plot on the Smith Chart. The normalized impedance can be read

$$\frac{Z_L}{Z_0} = 1.18 + 0.86j$$

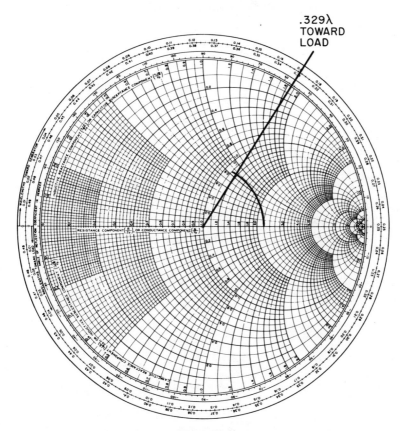

.329λ
TOWARD
LOAD

Fig. 6.1-46. Smith Chart plot of an impedance
measured with a slotted line.

which is exactly the same as calculated before. If the transmission system has
a characteristic impedance of 50 ohms, the impedance

$$Z_L = 50(1.18 + 0.86j) = 59 + 43j \qquad (6.1\text{-}13)$$

One of the useful properties of the Smith Chart is that a normalized impedance value may be graphically converted to admittance simply by plotting the
image point λ/4 around the constant SWR circle from the known impedance
point (Fig. 6.1-47). This characteristic of the chart is extremely useful in
design work with parallel impedances, because admittance values are very
convenient for parallel-circuit considerations.

Consider the circuit shown in Fig. 6.1-48. The load is an antenna which,
because of design considerations, can be matched to only a normalized

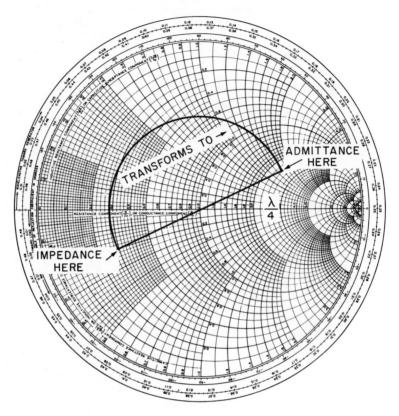

Fig. 6.1-47. Impedance—admittance relationship on the Smith Chart.

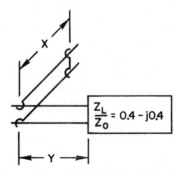

Fig. 6.1-48. Impedance matching of an antenna with a stub.

impedance of $0.4 - j0.4$. It is desired to place a matching stub on the transmission line at distance Y (physically removed from the antenna) and to use this stub to match or "flatten" the transmission line. The frequency is 3,000 MHz, and the line is a coaxial air line. The stub of length X is a purely reactive element having an adjustable, sliding short on its end. What is the design distance Y from the antenna to the stub, and what is the length X of the stub for proper matching?

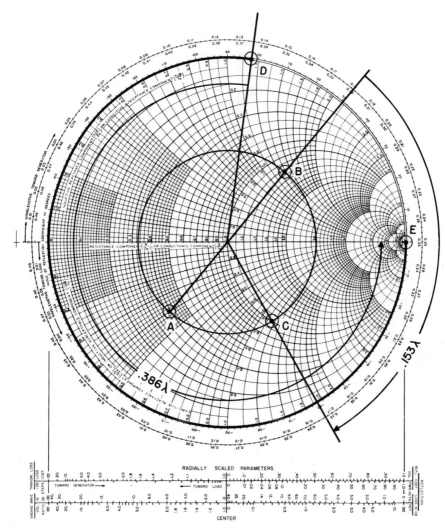

Fig. 6.1-49. Smith Chart plot of antenna matching
with a stub.

The solution, shown in Fig. 6.1-49, is detailed as follows:

1. Plot the antenna impedance $0.4 - j0.4$ (point A) and the SWR circle, and convert the antenna (load) impedance to admittance (point B). This admittance value is $1.23 + j1.23$ mhos.
2. Move around the SWR circle toward the generator to the constant-conductance circle which passes through the center of the chart (point C). At point C the admittance of the line looking back toward the antenna is $1.0 - 1.15$ mhos. If a susceptance of $+j1.15$ mhos is added in parallel at this point, the combination will equal a resistive component of 1.0, and the line will be perfectly flat from that point toward the generator. This means the input admittance of stub X should look like $+j1.15$ capacitive susceptance.
3. The distance Y, which can be easily computed from the phase angle between points B and C, is found to be 0.153, or, in this case, 1.53 cm from the antenna to the matching stub.
4. To determine the length of stub X, it is necessary only to enter on the Smith Chart the desired capacitive susceptance of $+j1.15$ (point D). Note that this point is on an infinite SWR circle precisely at the circumference of the chart, because the matching stub is completely reactive and has no resistive components in it. From point D, move around the Smith Chart toward the load until a shorted point is obtained at point E. This movement requires 0.386 cm, or, in the case of $3,000$ MHz, 3.86 cm.

By using a shorted matching stub 3.86 cm long placed 1.53 cm from the mismatched antenna, the entire system can be made to appear "flat," minimizing losses in the line caused by high SWR.

DOUBLE-STUB TUNERS. Because of the relative ease in varying stub length, compared with changing the distance from the stub to the load, tuning is often accomplished by using a double-stub tuner. Operation of

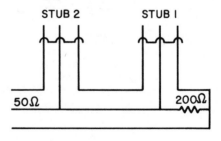

Fig. 6.1-50. Double stub tuner connected to a
mismatched barretter.

coaxial double-stub tuners may be easily explained and visualized by using the Smith Chart.

Consider the following example (Fig. 6.1-50) of a double-stub tuner used to match out a mismatched barretter mount. The assembly represents an instrument in which it is desired to match a 200-ohm barretter or instrument fuse into a 50-ohm transmission line for maximum power transfer in a power-measuring circuit. As shown in Fig. 6.1-50, stub 1 is located a short fixed distance from the 200-ohm load and has an adjustable shorting element. Stub 2, attached still farther down the line, also has an adjustable shorting element. Figure 6.1-51 illustrates a typical operation of the double-stub tuner in matching the 200-ohm load to the 50-ohm line at a particular frequency.

The 200-ohm resistive load represents a normalized impedance of 4.0, which is equivalent to a normalized admittance of 0.25 (plotted as point A on the Smith Chart). The admittance is transformed around the constant SWR circle from point A to point B. At point B, parallel susceptance is added from the first stub. The susceptance represented is purely reactive, since it is assumed that there are no losses in the stub. This parallel combination results in movement along a constant-conductance circle on the chart to point C, since no dissipative element is present in tuning stub 1. From point C the movement is toward the generator on a new constant SWR circle (existing in the center portion of the 50-ohm line toward point D). At point D another parallel susceptance is added by stub 2, which again moves the admittance along a constant-conductance circle into the center of the diagram, or a 1.0 normalized admittance.

The addition of the two susceptances from the double stubs provides a transformation from a badly mismatched line to a line which is matched and limited only by the precision of the tuning procedure. Mismatch loads of any impedance or phase could likewise be tuned out by using a single tuner having two types of adjustment (stub length and position). Another method could use two fixed-length stubs that are shorted on the ends and movable along the line. *Note:* As the frequency increases and the spacing between the two stubs comes near $\lambda/2$, a limitation of tuning range is encountered.

Other Smith Chart Scales

A number of radially scaled parameters are provided on Smith Charts and are scaled in such a manner that they may be set off from the center point of the Smith Chart by using, for instance, a pair of dividers. Application of the more useful parameters will be briefly discussed.

1. STANDING WAVE. The voltage standing-wave ratio (VSWR), shown as the lowest left-hand scale, is a translation of the scale on the horizontal

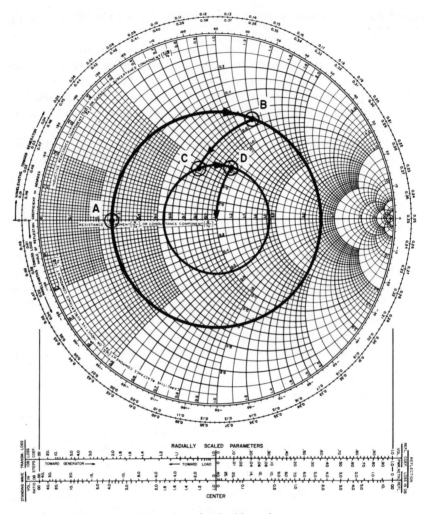

Fig. 6.1-51. Smith Chart plot of a double stub tuner
matching out the mismatched barretter.

center line of the Smith Chart (Fig. 6.1-52(a)). Note that the lower scale progresses from a reading of 1 at the center to infinity at the left-hand margin and agrees with all the axis crossings on the horizontal center line. Voltage standing-wave ratio is commonly expressed in dB; conversion to dB can be conveniently made on the adjacent "in dB" scale. The relation can be expressed as

$$SWR_{dB} = 20 \log_{10} SWR_{numerical}$$

For example, an SWR of 2.0 occurs opposite the 6.0 dB point on the adjacent scales. Consequently, 6 dB equals $20 \log_{10} (2.0)$, where 2.0 is the ratio of voltages.

2. REFLECTION COEFFICIENT. The voltage reflection coefficient (ρ), shown as the upper-right-hand scale in Fig. 6.1-52(b), is related to the SWR by the simple equation

$$\text{SWR} = \frac{1 + \rho}{1 - \rho} \qquad (6.1\text{-}14)$$

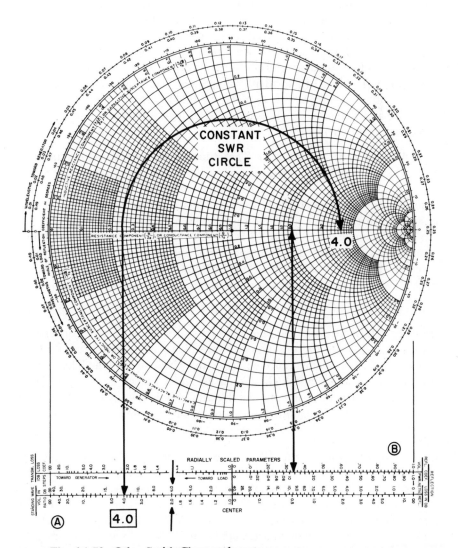

Fig. 6.1-52. Other Smith Chart scales.

Note that a reflection coefficient of 0.333 is very nearly equal to a SWR of 2.0. The adjacent power-reflection-coefficient scale is little used; mathematically, it is simply a squared function of adjacent voltage reflection coefficient.

3. LOSS IN dB (RETURN). This very important scale is located near the lower-right-hand corner. Return loss is the relation between the power returning down the line from a mismatched load to the power incident to that load. Thus a return loss of 10 dB means that the reverse power traveling in a line from a mismatched load is 10 dB below the reference power incident on that mismatched load. For example, the SWR with a certain mismatched load is 2.0, which is equal to a return loss of about 9.5 dB (i.e., the reflected power is 9.5 dB down from the incident power). To confirm that quantity for the same SWR, we reason as follows: the voltage vector of the backward-flowing power is 0.333 (if incident voltage was 1.0). Power returning down the line is then $(0.333)^2$, or 0.111. Note that 0.111 is approximately -0.5 dB. Mathematically, return loss is equal to $-20 \log_{10} \rho$ (where ρ is the voltage reflection coefficient).

4. LOSS in dB (REFLECTED). The reflection loss in dB, also commonly known as mismatch loss, is the relation of the power being transferred in the forward direction to the incident power on a mismatched plane. Consequently mismatch loss is the power which actually arrives at the load and is dissipated, or the power which is transmitted in an antenna, etc.

For example, when the SWR is 2.0, the reflected loss (scale adjacent to return loss scale) is 0.5 dB. In other words, the power actually being dissipated in a 2-to-1 normalized load is 0.5 dB lower than the incident reference power. Mathematically, reflection loss is equal to

$$-10 \log_{10} (1 - \rho^2) \text{dB}$$

Both the return loss and the reflected loss scales are extremely convenient, because they describe the amount of power which is being transferred forward and that which is being reflected from the given mismatched load. Surprisingly, even large mismatches, such as 3-to-1 (which cause quite large standing waves on a transmission line), result in only slightly more than 1 dB power loss in the transmitted forward direction power (as read on the reflected loss scale).

5. TRANSMISSION LOSS (1-dB STEPS). So far, all of our analysis has considered only lossless lines, in which a given line is described by a constant-radius SWR circle. In the case of a line with finite attenuation, however, a correction must be made, because the power (and hence the voltage) does not remain constant in amplitude as it travels along a transmission line; the forward and reverse interaction results in an SWR no longer constant with position along the line (Fig. 6.1-53). Graphically, a lossy line is represented

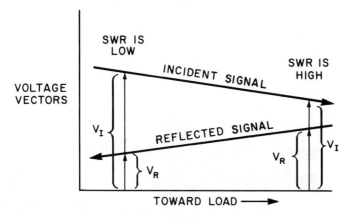

Fig. 6.1-53. Voltage relations on a lossy transmission
line.

by a spiral on the Smith Chart, because the power being transferred toward
the load in a lossy system and being reflected back toward the generator
experiences an attenuation in both directions. Therefore the nearer the
observer moves to the generator, the smaller the ratio between the reflected
voltage and the incident voltage. The 1-dB transmission loss scale is used for
SWR measurement, as shown in the following example.

Example: A slotted line is used to measure the SWR in a system which
consists of an unknown load at the end of a remote cable that has 2 dB of
insertion loss (Fig. 6.1-54). The SWR obtained is 2.0, and the null shift

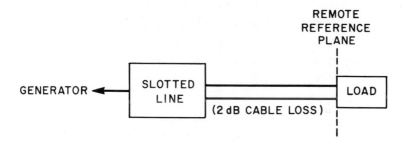

Fig. 6.1-54. VSWR of a load through a lossy line.

caused by going from the unknown load to a short at the remote reference
plane is 0.2 λ toward the load. We will calculate the true impedance of the
unknown load at the remote reference plane.

The solution is shown in the Smith Chart in Fig. 6.1-55. The procedure

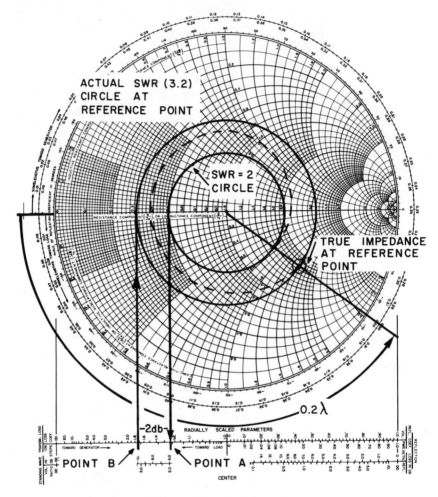

Fig. 6.1-55. Smith Chart plot of a load through a
lossy line.

is to plot the measured SWR and to determine the phase angle from the null shift. However, the intersection of the radial line and the SWR circle does not yet result in the remote impedance. To obtain the remote impedance, it is necessary to apply a 2-dB correction to compensate for the cable loss. The correction is made by marking point A on the 1-dB transmission loss scale to correspond with the 2.0 SWR circle. A 2-dB correction is made toward the load to point B. The radius related to point B is that of the SWR circle at the remote reference point. By swinging this radius around, it is seen that the new SWR circle has a radius equivalent to an SWR of 3.2. Note that the actual impedance relationships occurring along the line are now described by the approximate spiral shown. Incidentally, it can be noted that the "transmission loss in 1-dB steps" scale is related to the "return loss in dB" scale by a factor of 2. This relationship can be justified by considering that the return loss includes the two-way attenuation through a given piece of cable, whereas the transmission loss is defined as merely a one-way attenuation loss.

6. TRANSMISSION LOSS (COEFFICIENT). This scale is a correction factor relating to the additional line losses in a transmission line in which the SWR is greater than unity. The simplified reason for such a situation is that standing waves cause increases of current at certain points on the line and increases of voltage at other points. Since the resistive losses are related to the current squared and the dielectric losses are related to the voltage squared, the average line losses of a line with a large SWR are significantly larger at large SWR. Therefore the line losses should be used to correct the calculated attenuation factor when a very large SWR is encountered.

Swept Slotted-Line VSWR Measuring Technique

Slotted-line measurements described so far only accommodate single-frequency measurement. In 1966 a new swept-frequency slotted-line VSWR measuring technique was introduced.[16,17] The measurement setup is shown in Fig. 6.1-56. The sweep oscillator output is connected to the input of the slotted-line sweep adapter, which is essentially a short piece of slotted line with a stationary detector probe. The output of the adapter's probe is connected to the ALC input of the sweep oscillator, forming a power-leveling feedback loop. The slotted line is placed between the slotted-line sweep adapter and the device whose SWR is being measured, and the output of the detector probe of the slotted line goes to the vertical input of the oscilloscope. The horizontal input of the oscilloscope is taken from the sweep output of the sweep oscillator.

To permit the slotted-line probe output to be displayed on the oscilloscope with sensitivities as high as 0.5 dB/cm, the sweep oscillator output must be held reasonably constant as the frequency varies. The function of the slotted-line sweep adapter is to level the oscillator power output in such a way that the voltage output of the slotted-line probe remains constant with frequency, except for the variations caused by the SWR being measured. The adapter consists of a short length of slotted line, a well-matched 6-dB attenuator, and two *matched* detector probes, one for the adapter and one for

[16] Sorger, G. U., and B. O. Weinschel, "Swept Frequency High Resolution VSWR Measuring System," Weinschel Engineering Co. *Internal Report 90–117, 723–3/66*, March 1966.

[17] Adam, S. F., "Swept VSWR Measurement in Coax," *Hewlett-Packard Journal*, Vol. 18, No. 4, December 1966.

the slotted line. Matching the two probes makes the frequency response of the adapter probe, which samples the oscillator power, exactly equal to the frequency response of the slotted-line probe. Thus the oscillator power is adjusted to keep the output of the slotted-line probe constant with frequency. The 6-dB attenuator improves the frequency response, probe isolation, and impedance match of the adapter.

Although any oscilloscope can be used for displaying the slotted-line output, the variable persistence and storage feature of the one shown in Fig. 6.1-56 are particularly useful, because these factors permit the unknown

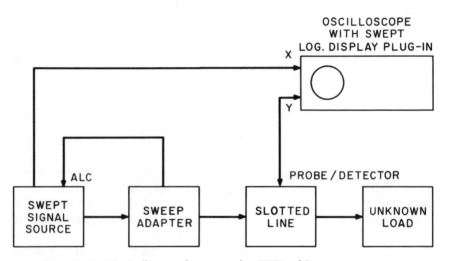

Fig. 6.1-56. Block diagram for measuring SWR with swept-frequency slotted line technique.

SWR to be read directly from the display. If a nonstorage oscilloscope is used, the SWR data have to be photographed, using a time exposure. The swept-frequency indicator plug-in is also a great convenience, because it has a logarithmic vertical amplifier which makes it possible to read SWR directly in dB.

Although in operation the sweep oscillator will be swept internally, the new SWR-measuring technique can be explained best by pointing out what happens at a fixed frequency. Figure 6.1-57 is a series of oscillograms taken with the equipment shown in Fig. 6.1-56. Oscillogram points on the horizontal axis correspond to frequencies between 8.2 and 12.4 GHz. The vertical scale factors are all 0.5 dB/cm. The following references are to Fig. 6.1-57(b). Part (a) shows what happens at a single frequency when the slotted-line carriage is moved over at least one-half wavelength; the oscillo-

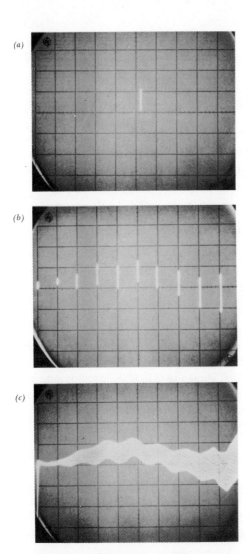

Fig. 6.1-57. Oscillograms made with setup of Fig. 6.1-56 showing measurement of SWR of a load. (a) At single frequency, trace moves up and down as slotted-line carriage moves over at least one-half wavelength. With logarithmic display unit, length of vertical line is SWR in dB. (b) Multiple exposure showing SWR measurements at several fixed frequencies across band. (c) Typical pattern produced by swept-frequency measurement. Vertical: 0.5 dB/cm; horizontal: 8.2 to 12.4 GHz.

scope traces out a vertical line whose length is equal to the SWR in dB of the device being tested. This can be proven true as follows.

Transmission line theory tells us that a uniform, lossless line terminated in an impedance which is not equal to its characteristic impedance will have two waves traveling on it in opposite directions. Besides the incident wave E_i traveling toward the load, there will be a reflected wave E_r going in the opposite direction. The incident and reflected waves will interfere and form a standing-wave pattern on the line. If the voltage on the line is measured, it will be found that there are points of maximum voltage,

$$E_{\max} = |E_i| + |E_r|$$

and points of minimum voltage,

$$E_{\min} = |E_i| - |E_r|$$

The maxima and minima will be one-half wavelength apart.

Standing-wave ratio is defined as

$$\text{SWR} = \frac{E_{\max}}{E_{\min}}$$

If the slotted-line carriage is moved over at least one-half wavelength, the oscilloscope spot will move up and down between $E_{\max}$ and $E_{\min}$ and will trace out a line like that shown in (a). If the oscilloscope has a linear vertical amplifier, $E_{\max}$ and $E_{\min}$ can be read from the display, and the SWR can be calculated. However, it is much better if the oscilloscope has a logarithmic vertical amplifier, because it will then display a vertical line with length

$$\log_{10} E_{\max} - \log_{10} E_{\min} = \log_{10} \frac{E_{\max}}{E_{\min}} = \log_{10} \text{SWR}$$

If the vertical amplifier is calibrated in dB/cm, the SWR in dB is simply the length of the vertical line traced out on the display as the slotted-line carriage is moved over one-half wavelength or more. The SWR can easily be read from the trace and then converted to a voltage ratio by the formula

$$\text{SWR} = \log^{-1} \frac{dB}{20}$$

For the single frequency of (a), the SWR is about 0.5 dB or 1.06. Notice that when the display is logarithmic, only the vertical length of the trace is significant, and the base line does not have to be displayed. Part (b) of the figure shows a series of traces corresponding to SWR measurements at several fixed frequencies.

If the sweep oscillator is swept internally several times per second across the frequency band, and if at the same time the carriage of the slotted line is moved manually over at least one-half wavelength of the lowest frequency in the band, either a time exposure of the oscilloscope display or a stored pattern will look like (c). This technique, i.e., manually moving the slotted-line carriage while the oscillator sweeps automatically, is the normal one for making swept-frequency SWR measurements. It yields results like (c), in which the width of the pattern as a function of frequency corresponds to the SWR in dB of the device being tested.

Reflectometers

From transmission line theory the reader already knows that, when an incident signal travels on a uniform lossless transmission line down to a termination where the impedance of the termination is not equal to characteristic impedance, a reflected wave will be launched from that termination in a direction opposite to the incident wave. Furthermore, the ratio of the reflected voltage over the incident voltage gives the reflection coefficient,

$$\Gamma = \frac{E_r}{E_i}$$

The reflection coefficient can be measured directly if the incident wave on a uniform lossless transmission line is separated from the reflected wave and sent through to a so-called ratiometer. Directional couplers separate signals going in a particular direction on the transmission line. If two directional couplers, connected so that the secondary arms go in opposite directions, are cascaded, a so-called reflectometer can be realized. Figure 6.1-58 shows the original reflectometer setup introduced in 1954.[18] The forward-connected

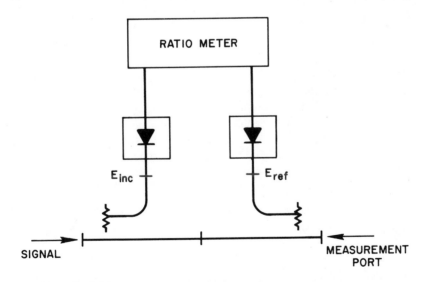

Fig. 6.1-58. Reflectometer.

[18] Hunton, J. K., and N. L. Pappas, "The H-P Microwave Reflectometers," *Hewlett-Packard Journal*, Vol. 6, No. 1–2, Sept.–Oct. 1954.

directional coupler samples the incident voltage, and the reverse-connected directional coupler samples the reflected voltage. Since the two directional couplers must track close to each other, the two matched crystal detectors have to be connected on the auxiliary ports of the directional couplers. The ratiometer will then measure the detected voltages of the incident and reflected wave. By definition, the ratiometer will then give the reflection coefficient directly. This technique allows swept-frequency measurements since, regardless of the signal level variation, the auxiliary ports of both directional couplers will provide a ratio that is the same at any signal level. Effectively, this technique is not sensitive to signal level variations.

Sources of errors from these types of measurements are directivity of the directional couplers being used, the main line mismatch of the directional couplers, the square law of the detectors used, the match of the crystal detectors, the tracking of the couplers and the detectors together, and the calibration of the ratiometer.

The reflectometer provides a rapid, accurate means of gathering information on microwave device performance over broad frequency ranges, in particular the information on reflection coefficient as shown in Fig. 6.1-58. There are different methods for displaying swept-frequency measurements while using the ratiometer. The first method is the X-Y recorder. An X-Y graphic recorder is a convenient means of obtaining a permanent record of swept-frequency measurements. Voltage to operate the X-Y system of the recorder is provided from the ratiometer. The X-drive is obtained from the sweep oscillator. For reference purposes, markers indicating frequency limits can be added to the test data record.

The oscilloscope can be substituted for the X-Y recorder. Oscilloscope display of swept-frequency measurements is particularly useful when the device under test is to be tuned or adjusted—for instance, filters, directional couplers, or amplifiers. Response limitation of the ratiometer requires a direct-coupled oscilloscope and a long-persistence cathode-ray tube. Calibration lines are drawn directly on the oscilloscope graticule prior to testing.

The third method uses the front panel of the ratiometer as a most convenient indicator for single-frequency measurements. In swept-frequency applications where optimum accuracy is not necessary or essential, limits can be marked on the meter and the swing of the meter pointer observed.

Swept-Frequency Measurement of Reflection Coefficient with the Waveguide Reflectometer

Setup for swept-frequency measurement of reflection coefficient at waveguide frequencies is shown in Fig. 6.1-59. This system is a standard reflectometer. The sweep oscillator output is 1-kHz-modulated and applied to a waveguide transmission system that contains two broadband directional

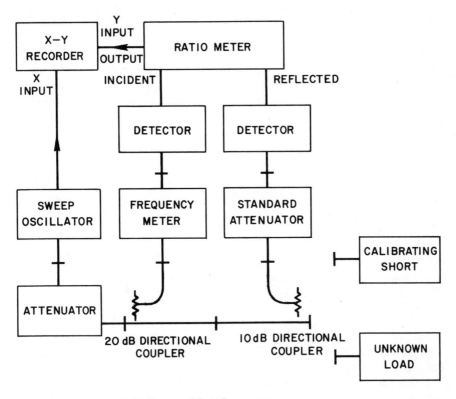

Fig. 6.1-59. Typical waveguide reflectometer.

couplers—one to sample incident signal and the other to sample reflected signal. Detectors at the secondary output of each coupler detect the modulation voltages from the signal samples. These voltages are applied to separate inputs of the ratiometer, which displays their ratio on a meter scale calibrated in units of reflection coefficient. As the frequency range of interest is swept, a permanent part of the reflection coefficient is recorded by the X-Y recorder. The output connection of the ratiometer provides a voltage proportional to meter deflection for operation of the Y-system of the recorder while the sweep oscillator supplies the X-drive. A frequency meter is used to place reference markers on the frequency axis of the X-Y recording. The attenuator between the sweep oscillator output and the 20-dB directional coupler establishes the proper working level for the detectors and optimizes impedance match between the sweep oscillator and the transmission system. Calibration of the reflectometer is accomplished by using a short circuit and a standard attenuator.

Calibration procedure is as follows: Place a calibrating short circuit at the measuring port of the reflectometer. Set the standard attenuator to the return loss equal to the maximum expected reflection coefficient. For example, if the maximum anticipated reflection coefficient is 0.13, set the standard attenuator for 17.7 dB of attenuation. Figure 6.1-60 provides a nomograph which facilitates conversion of reflection coefficient to return loss. At this time the ratiometer's 1-kHz modulation is to be set up for maximum output from the sweep oscillator. Since an X-Y recorder is being

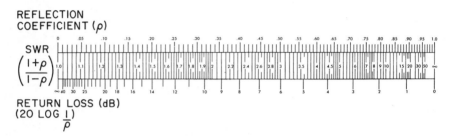

REFLECTION
COEFFICIENT (ρ)

SWR
$\left(\dfrac{1+\rho}{1-\rho}\right)$

RETURN LOSS (dB)
$\left(20 \text{ LOG } \dfrac{1}{\rho}\right)$

Fig. 6.1-60. Nomograph for conversion of reflection
coefficient and SWR to return loss.

used, a slow sweep rate has to be set by the sweeper to accommodate the slow response of the X-Y recorder. In the trigger mode of the sweep oscillator, trigger the sweep oscillator to sweep the frequency range of interest, and observe the entire sweep to be sure that the system is set up correctly. When all equipment is set up properly, frequency markers can be set by making several sweeps across the band to get a range of vertical lines on the X-Y recorder to set frequency limits of interest. A calibration grid of different reflection coefficients can be drawn with the different return loss values as shown in Fig. 6.1-61. The solid heavier line is the plot of reflection coefficient versus frequency. This was achieved by replacing the short circuit at the measurement port with the unknown load to be measured and then setting the standard attenuator's attenuation to 0 dB.

In coaxial transmission line systems, directional couplers or even dual-directional couplers can be set up in the very same system to perform measurements. Accuracy is limited because coaxial directional couplers do not have so high directivity and because the connectors are not so good as waveguide flanges.

With the development of sweep oscillators that could be leveled, the need for ratiometers was eliminated and a standing-wave indicator was used in the reflected arm instead; the incident signal formed the automatic level control's loop. This provided a more modern type of reflectometer system. Figure 6.1-62 shows a block diagram of a typical modern reflectometer system.

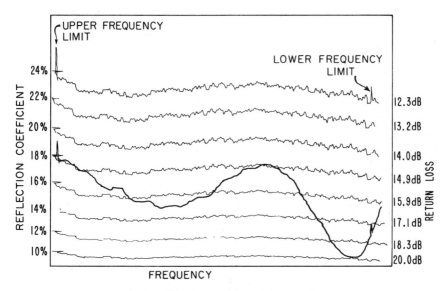

Fig. 6.1-61. Typical calibration grid and swept-frequency reflection coefficient measurement.

Figure 6.1-63 shows an RF pre-insertion X-Y recorder setup for reflection coefficient measurements. An accurate method for X-Y-recording swept-reflection tests is to use a precision variable attenuator for calibration, as shown in Fig. 6.1-63. The sweeper output, amplitude-modulated at 1 kHz, drives the swept-frequency indicator and is adjusted for a level of indication by sweeping the frequency band. With the reflectometer output shorted, the

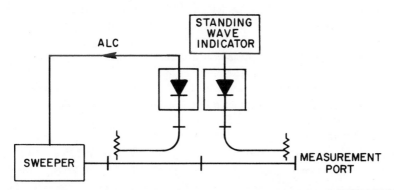

Fig. 6.1-62. Leveled reflectometer.

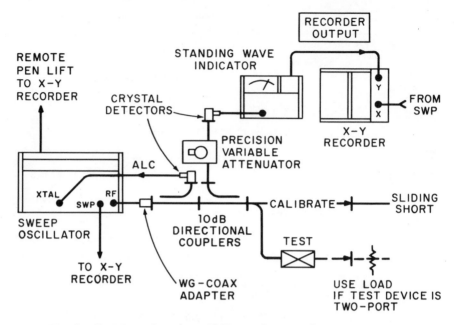

Fig. 6.1-63. RF pre-insertion—X-Y recorder setup for
reflection coefficient measurement.

precision variable attenuator is adjusted to pre-insert specific values of
return loss to the reverse arm detector. Grid lines are sequentially plotted on
the X-Y recorder for each attenuator setting by manually triggering a single
sweep on the sweeper. When using a sliding short, rapid phasing is possible
during the slow calibration sweeps. By continuously phasing the short
circuit during each calibrating sweep, all phases of source mismatch error
signals are encountered at the reverse arm detector. The result is a fine-grain
variation on the grid lines which defines the source mismatch limits within the
ambiguity limits of the directional coupler's directivity. Using the mean of
the fine-grain variations as the true return loss for each grid line source,
mismatch error is removed and better accuracy obtained in the measurement.
The attenuator is then returned to zero and the short replaced with a test
device. The final sweep is triggered, and return loss of the test device is
recorded over the calibration grids shown in Fig. 6.1-64. The calibration
grid, once plotted, can be used as an underlay for many swept-frequency
tests by using transparent paper for the actual recording of each device
tested. Grid line calibration should be checked periodically during long
hours of testing and a new grid always plotted when the system has been
turned off. This system, while slower than the oscilloscope readout, has the
advantage of greater dynamic range due to amplification of the reflected

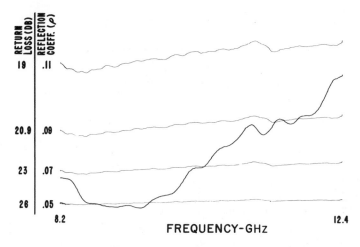

Fig. 6.1-64. X-Y plot of return loss using RF
pre-insertion technique.

signal by the standing-wave indicator. Return losses to 40 dB can be measured, limited primarily by the system rather than by resolution. Another advantage of this system is that it does not rely on the square-law characteristics of the detector, since the crystal operates near the same RF level for both calibrate and test conditions. However, the attenuator does introduce certain errors which have to be taken into consideration.

Figure 6.1-65 shows a waveguide reflectometer with an oscilloscope readout. The oscilloscope has a built-in logarithmic amplifier, which gives a readout in terms of dB/cm when the calibration is done with a sliding short and the test is accomplished by using the oscilloscope's built-in attenuator. Sources of errors include the crystal detector's detector law and the oscilloscope's built-in attenuator; other errors have been discussed previously.

Error Analysis of Impedance Measurements

A summary of error factors in reflection coefficient measurements can be given as

$$\Delta\rho = A + B\rho_L + C\rho_L^2 \tag{6.1-15}$$

where $\Delta\rho$ is the error of reflection coefficient and ρ_L is the reflection coefficient being measured. A gives the first-order error and B and C the second- and third-order errors, respectively.[19] There are basically three types of reflection coefficient systems: the slotted-line technique, the reflectometer technique

[19] Ely, Paul C., Jr., "Swept-frequency Techniques," *Proc. IEEE*, Vol. 55, No. 6, June 1967.

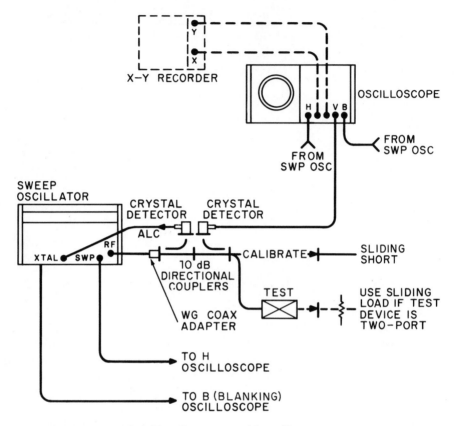

Fig. 6.1-65. Waveguide reflectometer with oscilloscope
readout.

using calibration grid technique, and the oscilloscope technique. The
following table provides information about the factors *A*, *B*, and *C* for each
of the measurement techniques in question.

Reflection Measurements of a Source

Output SWR on sweep oscillators and other microwave sources can be
checked easily with the equipment setup of Fig. 6.1-66. In the setup shown,
the sweeper is not leveled, because the basic source match is being tested.
The effectiveness of leveling to improve source match can also be checked
using the same procedure but with a leveling coupler/detector loop between
the sweeper and waveguide component shown.

The system shown utilizes the oscillator in test to supply the required
signals, thus eliminating the need for another source in the same band. With

Table 6.1-1. Summary of Error Factors in Reflection-Measuring Systems

Magnitude of System Error $= A + B\rho_L + C\rho_L^2$

	Calibration Grid Reflectometer	*Oscilloscope Reflectometer*	*Slotted Line*
$A =$	Directivity	Directivity	Probe, source, and residual reflections Slope and loss
$B =$	Directivity; attenuator accuracy	Directivity; attenuator accuracy Coupler/detector tracking; detector square-law error	Probe reflection; attenuator accuracy Detector Square-law error
$C =$	Reflectometer-source match (Γ_x)	Reflectometer-source match (Γ_x)	Probe reflection

the output terminated in the load, a calibration grid is plotted on the X-Y recorder using attenuator settings of return loss corresponding to specific source SWR values. After plotting a calibration grid, the load is replaced by a waveguide short, which reflects the incident wave back to the source. A

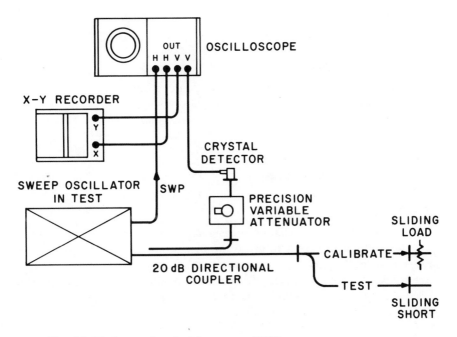

Fig. 6.1-66. System for checking source SWR on sweep oscillator without leveling.

portion of the reflected signal is reflected by the source mismatch up through the directional coupler to the detector, adding in random phase with the incident signal. When using the short, single sweeps are triggered and recorded for several small incremental changes in the short position, more than $\lambda/4$ at the lowest frequency, taking into account various phases of source reflections.

When a freely sliding short is used, only one test sweep is required while rapidly phasing the short. The envelope of the resulting plot represents source SWR as a function of frequency. Attenuator settings for the calibration grid are determined from the following equations:[20]

$$-dB_1 = 20 \log_{10} 2(1 - \rho_g) \qquad (6.1\text{-}16)$$

$$-dB_2 = 20 \log_{10} 2(1 + \rho_g) \qquad (6.1\text{-}17)$$

where ρ_g = simulated generator reflection coefficient.

EXAMPLE: A sweep oscillator with a maximum specified output SWR of 2:1 is connected for test as shown in Fig. 6.1-66. By the relationship of Eq. (6.1-14), given earlier, a 2:1 SWR is approximately equal to a ρ of 0.33. Substituting for ρ_g in Eqs. (6.1-16) and (6.1-17) and rounding off to the nearest tenth:

$$-dB_1 = 20 \log_{10} 2(1 - 0.33) = 2.5$$

$$-dB_2 = 20 \log_{10} 2(1 + 0.33) = 8.5$$

Using either Eq. (6.1-16) or Eq. (6.1-17), the attenuator setting corresponding to a source SWR of unity is found to be 6 dB. Grid lines are plotted for the three attenuator settings, and the attenuator is returned to 6 dB. The final plots are then made by using the movable short, resulting in a record similar to Fig. 6.1-67. Since the envelope of the plot remains inside the 2:1 SWR limits, the sweeper output system meets specifications.

6.1.3 COMPLEX S-PARAMETER MEASUREMENTS

Theory

Impedance and attenuation measurements were discussed in detail in previous sections. How to measure the magnitude of gain or attenuation was shown. Furthermore, the magnitude and, in some cases, phase information of impedance were discussed. The above parameters must be known to describe or analyze a network, but to achieve full information, *all* the parameters have to be known, magnitude and phase included, across the entire frequency band of interest.

[20] *Hewlett-Packard Journal*, Vol. 12, No. 4, December 1960, 5.

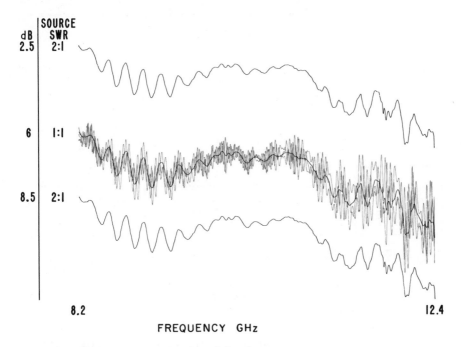

Fig. 6.1-67. Swept SWR plot of signal source.

There are network analyzer systems[21] available which enable one to characterize networks by measuring their complex small-signal parameters. Such systems at microwave frequencies measure the scattering, or s-parameters. The theory and usefulness of s-parameters at high frequencies were discussed in Chap. 3. If one wishes to use other common parameters, he can easily change over by using conversion tables provided in Table 3.3-1. But it is still worthwhile to mention that these parameters have proved to be valuable tools for design engineers because of their inherent ease of measurement, their design advantages, and the insight they provide. To understand how these measurements are performed, it is advisable to become acquainted with the *network analyzer concept* which follows naturally from network-parameter theory. If one measures scattering parameters, he is measuring (1) magnitude ratios, and (2) relative phase angles of response and applied signals at the ports of the network in question with the other ports terminated in their characteristic impedances. Figure 6.1-68 shows the basic elements of a network analyzer system designed to perform these measurements.

[21] Anderson, R. W., and O. T. Dennison, "An Advanced New Network Analyzer for Sweep-Measuring Amplitude and Phase from 0.1 to 12.4 GHz," *Microwave Journal*, Vol. 18, No. 6, February 1967.

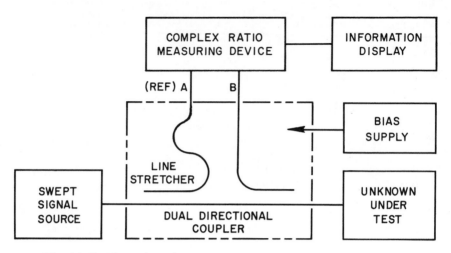

Fig. 6.1-68. Network analyzer concept.

A signal source (preferably a swept-frequency oscillator) drives the systems. The transducer instrument, which is fed by the signal source, is essentially a dual-directional coupler connected as a reflectometer (discussed in the preceding section). It converts the signal applied and the response signals from the unknown under test to a set of output signals that provide the network information. These signals are fed into the complex ratio-measuring instrument, where the magnitude and phase are measured. This setup is capable of measuring the complex reflection coefficient, s_{11}. The information display puts the data in the form that is most usable for the operator. Data can be supplied as either magnitude or phase information on meter readouts or plotted on a polar display.

Network Analyzer System

Hewlett-Packard Company has developed a network analyzer capable of making these measurements. Figure 6.1-69 shows a typical setup using the network analyzer.

It is both difficult and expensive to make accurate measurements of complex signal ratios directly at microwave frequencies. However, frequency translation is an excellent technique that allows one to use lower-frequency techniques to make these complex ratio measurements. The basic requirement of frequency translation is to use a technique in which the magnitude ratio and phase information of the signals are preserved.

The key technique that allows the microwave network analyzer to

Fig. 6.1-69. Typical test setup using new network analyzer (top center) to sweep-measure transmission of microwave filter. Magnitude and phase are measured on analyzer meter and presented as a function of frequency on oscilloscope. Magnitude and phase can also be presented in polar form on a polar display scope which plugs in, in place of phase-gain indicator and will feed external recorder.

measure complex ratio is the technique of frequency translation by sampling. The block diagram of the basic analyzer shown in Fig. 6.1-70 will help the reader to understand this technique. Sampling as used in this system is a special case of heterodyning, which translates the input signals to a lower, fixed IF frequency where normal circuitry can be used to measure amplitude and phase relationships. The principle is to exchange the local oscillator of a conventional heterodyne system with a pulse generator which generates a train of very narrow pulses. If each pulse within the train is narrow compared to a period of the applied RF signal, the sampler becomes a harmonic mixer with equal efficiency for each harmonic. Thus sampling-type mixing has the advantage that a single system can operate over an extremely wide input frequency range. In the case of the network analyzer this range is 110 MHz to 12.4 GHz.

To make the system capable of swept-frequency operation, an internal phase-lock loop keeps one channel of the two-channel network analyzer tuned to the incoming signal. Tuning of the phase-lock loop is entirely automatic. When the loop is unlocked, it automatically tunes back and forth across a portion of whatever octave-wide frequency band has been selected

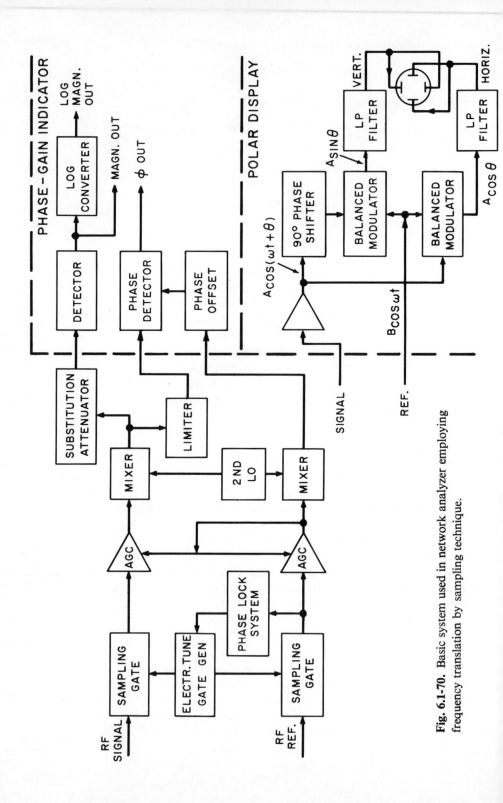

Fig. 6.1-70. Basic system used in network analyzer employing frequency translation by sampling technique.

by the user. When any harmonic of the tracking-oscillator frequency falls 20 MHz above the input frequency (i.e., when $nf_{osc} - f_{in} = 20$ MHz), the loop stops searching and locks. Search and lock-on are normally completed in about 20 μs. The loop will remain locked for sweep rates as high as 220 GHz/s (a rate corresponding to about 30 sweeps per second over the highest frequency band, 8 to 12.4 GHz).

The IF signals reconstructed from the sampler outputs are both 20-MHz signals, but, since frequency conversion is a linear process, these signals have the same relative amplitudes and phases as the microwave reference and test signals. Thus gain and phase information are preserved, and all signal processing and measurements take place at a constant frequency.

As shown in Fig. 6.1-70, the IF signals are first applied to a pair of matched AGC (automatic gain control) amplifiers. The AGC amplifiers perform two functions: they keep the signal level in the reference channel constant, and they vary the gain in the test channel so that the test signal level does not change when variations common to both channels occur. This action is equivalent to both taking a ratio and removing the effects of power variations in the signal source, of frequency response characteristics common to both channels, and of similar common-mode variations.

Before the signals are sent to the display unit, a second frequency conversion from 20 MHz to 278 kHz is performed. To obtain the desired dB and degree quantities, the *phase-gain indicator plug-in display unit* contains a linear phase detector and an analog logarithmic converter which is accurate over a 60-dB range of test signal amplitudes. Ratio (in dB) and relative phase can be read on the meter of the display unit if desired, but the plug-in also provides calibrated dc-coupled voltages proportional to gain (as a linear ratio or in dB) and phase for display on the vertical channels of an oscilloscope or X-Y recorder. If the horizontal input to the oscilloscope or recorder is a voltage proportional to frequency, the complete amplitude and phase response of the test device can be displayed.

The polar display unit converts polar quantities of magnitude and phase into a form suitable for display on a CRT. This is accomplished by using two balanced-modulator phase detectors. The phase of the test channel is shifted 90° with respect to the reference channel before being applied to the balanced modulator. The output of one modulator is proportional to $A \sin \theta$. This signal is amplified and fed to the vertical plates of the CRT. The output of the other balanced modulator is proportional to $A \cos \theta$, and this signal is applied to the horizontal plates of the CRT. Thus the polar vector can be displayed in rectangular coordinates of an oscilloscope or X-Y recorder. In other words, a Smith Chart display can be achieved on the oscilloscope face or on the X-Y recorder.

Transmission Measurements

To obtain complex transmission coefficient (s_{21} and s_{12}) measurement with the network analyzer system, a transducer device accommodating the basic requirements must be connected between the signal source and the harmonic converter-network analyzer. Such a transducer device has to provide capability to split the incoming signal to reference and test channels. Furthermore, metered electrical and mechanical line length extension capability has to be provided. The basic diagram of a transmission test transducer is given in Fig. 6.1-71.

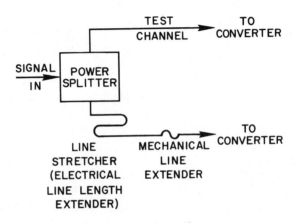

Fig. 6.1-71. Transmission test-transducer diagram.

The line stretcher allows one to make *electrical* length adjustments for balancing the phase of any device inserted in the test channel. Also, a mechanical extender of the trombone type is provided to compensate for the physical length of the device inserted into the test channel. Furthermore, standard-length air lines can also be inserted to accommodate any insertion lengths of the devices to be measured. Figure 6.1-72 shows a typical trans-

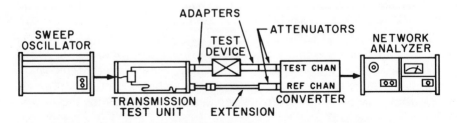

Fig. 6.1-72. Transmission measuring system.

mission-measuring setup; note the attenuators connected in both reference and test channels to reduce errors due to multiple mismatches resulting from the moderate reflection of the harmonic converter at higher frequencies.

Transmission tests can be divided into three categories: the insertion, incremental, or comparative types. Insertion tests are those in which one measures the magnitude and phase before and after the device is inserted into the test line. The difference in magnitude and phase is the transmission of interest. Incremental transmission tests are used when the test device is already inserted in the test channel, and phase shift versus such factors as frequency, temperature, or time is being measured. Comparative tests determine the relative phase and magnitude difference between the test device and another one, which could be taken as a standard, inserted in the reference channel.

The phase gain indicator's output can be connected to a dual-trace oscilloscope in which the magnitude and phase of the device's transmission coefficient can be monitored across the frequency band being swept. Usually, a network analyzer of this type, using a harmonic converter technique, is not able to accept high-power levels. In the reference channel -16 dBm is usually the maximum power which is applied. The test channel upper limit power is set to about -10 dBm. The detection range in the test channel usually extends down to the noise level which is in the order of -78 dBm. To achieve 60-dB dynamic range, the cross talk and noise levels must be greater than 60 dB below the calibrated level of the system. If one starts with -10 dBm in the test channel, the noise level will be down at least 68 dB. Cross talk from reference to test channel is usually larger than 60 dB below the reference channel level.

It is true that this dynamic range is more than adequate for many measurements, but there are applications where it is not enough, and wider dynamic ranges are needed. There are techniques in which this limitation can be extended. For instance, if one wants to measure a bandpass filter which has at least 60 dB rejection, the measurement can be done in the following manner: A 10-dB pad attenuator should be connected in the test channel and a 30-dB pad in the reference channel. Now the source power can be increased until the reference level meter reads in the operating range. This will put the test channel 20 dB above the reference channel by the 20-dB offsetting pad. Power levels for reference and test will be about -30 dBm and -10 dBm, respectively. Noise level -70 dBm and cross talk level at -90 dBm are now at levels where they will not interfere greatly with the dynamic range. Ideally the reference channel should be even farther below the test channel to decrease the effects of cross talk. Other substitution may be used to increase the measurement range of the network analyzer when measuring values of attenuation greater than 60 dB. A tabulation of possible substitutions is given in Table 6.1-2 for the equipment setup of Fig. 6.1-72.

Table 6.1-2.

Test Range* (dB)	Test Channel Atten. Value (dB)	Ref. Channel Atten. Value (dB)	Uncertainty Curve (Fig. 6.1-78)
0 to −60	10	30	A
+20 to −40	10	10	B
+40 to −20	30	10	C

* +dB = Gain in test device.
−dB = Attenuation.

Note that measurements can be displayed only over a 60-dB dynamic range. The measurement range can be extended beyond 60 dB (see Fig. 6.1-73).

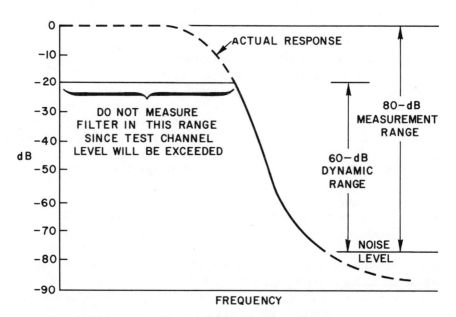

Fig. 6.1-73. RF substitution for 80-dB measurement range viewed with network analyzer 60-dB dynamic range.

Other substitution can be accomplished if the source introduced needed power and if the pads are capable of handling that power. The network analyzer in this transmission measurement setup is very capable of measuring low insertion losses or compressing the amplitude response, making high-resolution measurements. An equipment setup for making these types of tests is shown in Fig. 6.1-74.

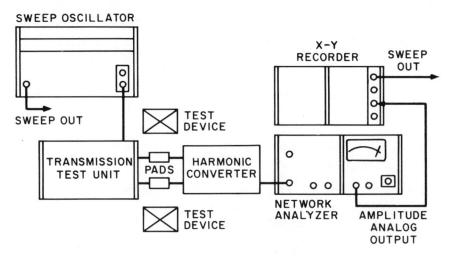

Fig. 6.1-74. Equipment setup for making high resolution measurements.

While in the calibrate configuration, the X-Y recorder should be calibrated for a full-scale deflection of ±0.5 dB with grid lines plotted at 0.1-dB increments. The grid lines can be plotted by offsetting the amplitude vernier on the network analyzer in 0.1-dB increments and plotting the residual response of the system. Then, insert a 20-dB pad in the test channel and compensate for it by increasing the test channel gain by 20 dB. The resultant high-resolution plot will show the deviation of the pad from 20 dB. Now place another 20-dB pad in the reference channel and plot the tracking between the two pads. A plot of amplitude tracking between the two 20-dB pads is shown in Fig. 6.1-75.

At this point it is worthwhile to evaluate the sources of errors involved in the network analyzer system, since they are very complex in their nature and their variables. Many of these variables can be reduced, and some can even be eliminated. The basic sources of error can be divided into the following categories:

1. Tracking between channels as a function of frequency; (This error can be eliminated by referencing the meter in CW or plotting grid lines on an X-Y recorder.)
2. The IF substitution attenuator in the network analyzer used to offset for higher resolution;
3. Noise level −78 dBm at a 10-kHz bandwidth in positively all measurements and becoming significant at large values of attenuation; (Noise can be reduced by using low-pass filters from the analog outputs or by using an X-Y recorder.)

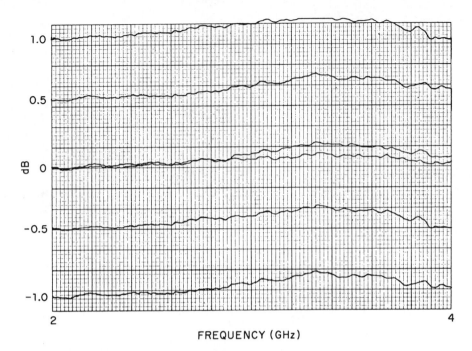

Fig. 6.1-75. Comparison test for tracking with
frequency of two 20-dB pads.

4. Cross-talk level greater than 60 dB below reference signal adding
   as a vector to the system error vector; (Pads in the reference channel
   to set the reference signal below test channel reduce this error
   vector.)
5. Mismatch uncertainty, which is present at every connection where
   the reflection coefficients are different from zero. (Mismatch adds
   as a vector error in the system; the use of good pads and connectors
   reduces this error quite significantly.)

Overall accuracy must be specified with careful consideration of all sources
of error.

Measuring Phase Response of Microwave Devices

The measurement of phase shift of a device at a CW frequency is
straightforward. It is a matter of zeroing a phase-gain meter for zero degrees
phase shift during calibration, inserting the test device, and reading the phase
shift from the meter. When making swept-frequency phase measurements,
it is necessary to consider the relative lengths of the reference and test chan-

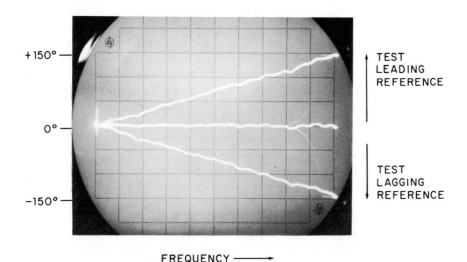

FREQUENCY ⟶

Fig. 6.1-76. Upper trace: reference channel length >
test channel length. Lower trace: reference
channel length < test channel length. Center trace: both channel lengths equal.

nels of the system. Figure 6.1-76 shows the oscilloscope patterns for calibrating position for excess length in one of the two channels. Figure 6.1-77 shows
a typical transmission test setup for measuring phase shift of a microwave
device. A line stretcher in the transmission test transducer allows the refer-

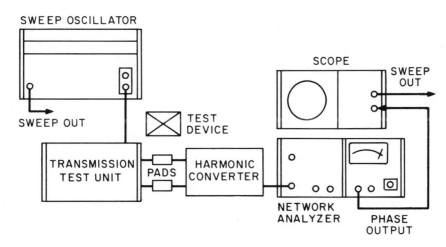

Fig. 6.1-77. Transmission test setup for measuring
phase shift of a microwave device.

ence channel to be either lengthened or shortened to calibrate the system on the swept-frequency basis. This gives a 0° phase-shift reading for all frequencies of interest ± any residual phase tracking between the two channels. The line stretcher can remove or add enough line from the test channel to compensate for the excess electrical length in the unknown device. The total electrical length of the system under test is the sum of the physical length of air line in the reference channel and the equivalent extension of the line stretcher that gives constant phase versus frequency on the oscilloscope display. The remaining phase ripple displayed on the scope is the nonlinear phase shift. Stable and accurate readings of phase and amplitude are assured by the use of rigid coaxial air lines in the reference channel. The electrical length of a device is its physical length plus any excess length due to dielectric materials, etc., within the device. The excess length of the device comes from the inherent slowing of the wave inside the device because of the materials the device is constructed from. As was seen from the use of the line stretcher, linear phase shift depicts difference in electrical length between reference and test channels.

Group delay is the time by which signal energies are delayed when traveling through a device. This parameter is an important one to control in communications systems in which any change in time delay over a frequency band of interest introduces distortion. Time delay information is necessary to determine delays from components in radar systems that might degrade overall performance of the system. Group delay is related to the phase-shift frequency ratio by the relationship

$$t_D = \frac{\Delta\phi}{\Delta f} \frac{1}{360°} \qquad (6.1\text{-}18)$$

where t_D is group delay.

An easy way to use the network analyzer to determine group delay for an electrically long device is to measure the frequency band Δf that results in a $\Delta\phi$ of 360°. Therefore, from the above equation, t_D is equal to $1/\Delta f$.

An example can be cited in which a coaxial pad's electrical length and group delay are being measured.

1. Calibrate and insert a 20-dB coaxial pad in the test channel. The resulting linear phase shift displays the effect of electrical length and group delay.
2. Electrical length

$$l_e = \frac{C}{360°} \times \frac{\Delta\phi}{\Delta f} = \frac{3 \times 10^{10}}{360} \times \frac{\Delta\phi}{\Delta f} = \frac{10^8}{1.2} \times \frac{\Delta\phi}{\Delta f} \text{ cm}$$

where ϕ is in degrees and f is in hertz.

$$l_e = 83.3 \times \frac{\Delta\phi}{\Delta f} \text{ cm}$$

where f is in megahertz. Total $l_e = 83.3 \times \dfrac{\Delta\phi}{\Delta f}$ + the reading on the transducer's mechanical extension. Group delay is the rate of change of phase with respect to frequency. Verify your result by using the line stretcher to rebalance phase shift to 0°. It should agree with $83.3 \times \dfrac{\Delta\phi}{\Delta f}$.

3. Group delay

$$t_D = \frac{\Delta\phi \text{ (radius)}}{2\pi \,\Delta f} = \frac{\Delta\phi \text{ (degrees)}}{360 \,\Delta f} = \frac{1.2}{360 \times 10^8} \times l_e = \frac{l_e}{C}$$

where $C = 3 \times 10^{10}$ cm/s.

It should be noted that total electrical length and total group delay using the transmission test unit must be calculated by considering the length of the mechanical extension of the reference channel. When measuring the electrical length of the 20-dB pad, one could use the calibrated line stretcher to obtain a level trace on the oscilloscope (pad in test channel). The reading on the line stretcher dial should then be added to the mechanical extension reading to give total electrical length. Group delay can then be calculated from this length quite accurately.

The accuracy of the network analyzer system when measuring phase shift is dependent upon the following sources of errors.

1. Tracking between channels as a function of frequency. This error can be eliminated for CW or X-Y plotted response.
2. Phase offset on the phase-gain plug-in, which allows for higher resolution.
3. Limiting circuits in the system cause some of these errors when there is a large difference between reference and test channel.
4. Cross talk adds as a vector and hence causes some phase error. Correct padding reduces this error.
5. Mismatch uncertainty, which is present at every connection where reflection coefficient magnitude and phase are different from zero. Mismatch is a vector quantity and causes phase error. The use of good pads and connectors reduces this error.
6. Noise level adds positively, 10-kHz bandwidth (in all cases noise may be reduced by filtering at a narrower bandwidth, say, 1 kHz, by correct padding, and by X-Y recording).

Overall accuracy specifications depend upon the interaction of these errors.

Transmission Test Uncertainties

Four factors determine three types of transmission test uncertainty:

Factors	*Types of Errors*
Instrument	Uncertainty introduced by the network analyzer and the harmonic frequency
Test setup	converter
	+
	Mismatch uncertainty
Test device	+
Type of test	Uncertainty introduced by the transmission test unit
	= Total system uncertainty

Figure 6.1-72 shows the recommended test setup for coaxial transmission tests, and Table 6.1-2 (p. 424) indicates values of attenuators that will minimize both the uncertainty introduced by the network analyzer and the uncertainty due to mismatch. Using the values of attenuators shown in the table will reduce the effects of signal cross talk from the reference channel to the test channel and will optimize the SWR in the test channel for the required gain/attenuation range. This will minimize the uncertainty introduced by the network analyzer. The two graphs in Fig. 6.1-78 show typical amplitude and phase uncertainty versus measured values of gain/attenuation for a given attenuator configuration. These are the amounts of uncertainty that result when IF attenuation is varied so that attenuation/gain measurements can be read from the most sensitive scale range. These curves include the uncertainty introduced by the IF attenuator.

Using the phase offset switch to allow phase readings from the most sensitive scale introduces phase uncertainty in addition to that given by the curves. This additional uncertainty is typically $\pm(0.1°) \pm (0.2°/10°$ step) up to a $1.5°$ maximum for the full $\pm180°$ range.

Incremental tests are those in which initial system calibration is made with the unknown device installed as part of the test setup. Phase and gain changes are measured from this initial reference. This type of test adds no appreciable uncertainty to the system.

Insertion tests are those in which the unknown is inserted into a calibrated setup. When making insertion tests, there is an added uncertainty if the mechanical and electrical extensions in the transmission unit are changed. This added uncertainty typically does not exceed $+0.05$ to -0.09 dB and $\pm1.7°$.

Mismatch uncertainty results from reflections between test ports during calibration and between the test and unknown ports during the measurement.

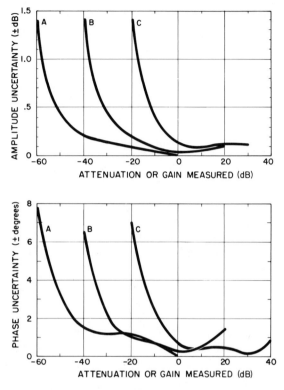

Fig. 6.1-78. Typical uncertainty versus gain or attenuation measured.

This uncertainty is minimized when attenuators are used in the test setup as shown. Typically the test port reflection coefficients are 0.07 or less. The following expressions show how to calculate the uncertainty for both calibration and test conditions.

Typically, specifications for uncertainties can be written:

AMPLITUDE MISMATCH UNCERTAINTY $= 20 \log (1 \pm |\Gamma_1\Gamma_2|)$, where Γ_1, $\Gamma_2 = 0.07$ (reflection coefficient of test ports 1 and 2). Calibration $= 20 \log (1 \pm |0.07 \cdot 0.07|) = \pm 0.04$ dB. Test $= 20 \log (1 \pm |0.07 \ \Gamma_x|)$, where $\Gamma_x =$ reflection coefficient of unknown.

PHASE MISMATCH UNCERTAINTY $= \pm\sin^{-1} |\Gamma_1\Gamma_2|$. Calibration $= \pm\sin^{-1} |0.07 \cdot 0.07| = \pm 0.3°$. Test $= \pm\sin^{-1} |0.07 \ \Gamma_x|$.

An example of how to calculate the total uncertainty in a transmission measurement is given below. Measure the incremental and insertion attenuation and phase shift of a 0- to 20-dB coaxial attenuator having 90 degrees phase shift. Assume both input and output reflection coefficients of the device are 0.1. Values shown in Table 6.1-3 are typical except for mismatch uncertainties which are worst-case resulting from typical reflections.

Table 6.1-3.

Source of Uncertainty	Uncertainty	
	Amplitude ($\pm dB$)	Phase ($\pm degrees$)
Test method A (curves)	0.1	1.2
Phase offset	—	1.0
Subtotal: instrument uncertainty for incremental test	0.1	2.2
Calibration mismatch	0.04	0.3
Input test port mismatch	0.06	0.4
Output test port mismatch	0.06	0.4
Subtotal: mismatch uncertainty	0.16	1.1
Total Incremental Test Uncertainty	± 0.26	$\pm 3.3°$
Transmission test unit	+0.05 −0.09	1.7
Subtotal: instrument uncertainty added for insertion test	+0.05 −0.09	1.7
Total Insertion Test Uncertainty	+0.31 −0.35	$\pm 5.0°$

Reflection Measurements

Measuring s_{11} or s_{22} is really an impedance-measuring procedure, which was discussed in detail in the previous section. With the help of a slotted line, the impedance of a device was measured and plotted on a Smith Chart. The reflectometer then measured the magnitude of reflection coefficient. As was mentioned in the discussion of network analyzer systems, if one uses a reflectometer setup and a phase-coherent frequency translator, instead of a pair of crystal detectors, not only the magnitude of reflection coefficient but also the information on the phase angle can be measured.

Figure 6.1-79 shows a Smith Chart, in which the phase angle of reflection coefficient is shown in the periphery of the chart in terms of $\pm$ degrees. Figure 6.1-80 shows the equipment block diagram for making such measurements.

It is necessary to calibrate this setup before measurement. Calibration involves measuring a device with a known reflection coefficient or impedance—usually a short circuit. The reflection coefficient of a short circuit

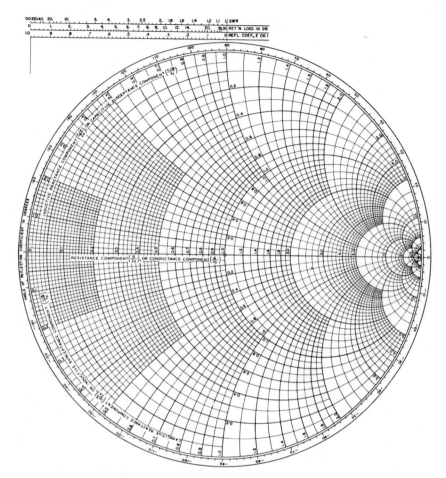

Fig. 6.1-79. Smith Chart showing the angle of reflection
coefficient on the periphery of the chart.

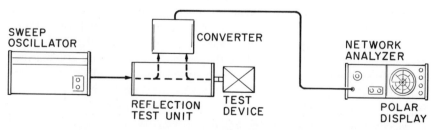

Fig. 6.1-80. Measuring reflection or impedance.

is 1 and the phase angle is 180°. So we need only measure this and adjust the controls of the instrument until the polar display shows a dot at the 0 impedance point $(0 + j0)$. The polar display is first calibrated by pressing the "beam center" button and adjusting the X-Y dc-offset control knobs to get the dot to the center of the display. Then the reflection coefficient of the short circuit can be measured. This is done by adjusting the test channel gain on the network analyzer to get the display on the screen. Generally, the display will not be a single point but will more likely resemble an arc. This means the lengths of the reference and test channels are not identical. The line stretcher must be adjusted in the reflection test unit until the display shrinks into as small a cluster of points as possible. Adjustment of the test channel gain, the amplitude vernier, and the phase vernier on the network analyzer will center this cluster over the 180° point on the display. This completes the calibration procedure of the network analyzer.

Now the reflection coefficients scale can be expanded or compressed by using the calibrated IF gain control. If frequency markers are available on the sweep oscillator, they will appear on the polar display as bright dots giving an indication of frequency. The recorder outputs make high-resolution Smith Chart plots possible with a standard X-Y recorder. Smith Chart overlays can be placed on the oscilloscope screen of the polar display. The measurement is then simply done by replacing the short circuit in the reflection test unit's measuring port with the device to be measured. The display on the screen is the polar display of reflection coefficient versus frequency. The display is related to the plane of the short circuit. If the source of reflection is not at the plane of the short circuit, a transformation must be made. This is done easily with this type of reflection test unit by simply adjusting the line stretcher to move the reference plane into the test device as far as is necessary. This will then move the display, and it will plot the impedance exactly at the point of interest where the reflection occurs. Swept impedance tests can be made with a network analyzer on both active and passive devices. Adjustable frequency markers on the sweep oscillator intensify the trace to provide a marker at any frequency when it is needed. The continuous swept display speeds testing, provides continuous data versus frequency, and allows impedance changes to be observed as parameters are being changed.

A common design problem is to minimize the mismatch of a device by a suitable compensating design. However, in many cases the plane of the mismatch is not directly accessible to the test port of impedance instrumentation, except through a length of transmission line. Normally this requires correcting the measured impedance to reflect the electrical length of the transmission line between the test port and the desired impedance test plane at each frequency. The lengthy calculation can be avoided by use of the reflection unit's line stretcher.

Figure 6.1-81 shows the impedance at the input connector of a thin-

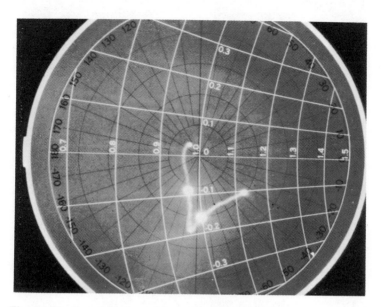

Fig. 6.1-81. Swept display of input impedance of a coaxial attenuator at the plane of the con-
nector. An expanded Smith Chart overlay
for direct readout of impedance to a full
scale $\Gamma = 0.2$ (VSWR $= 1.5$) is used. Frequency range: 10.0–12.0 GHz.

resistive-film coaxial attenuator. It is desirable to improve the mismatch by
appropriately compensating the resistive element of this attenuator. The
impedance appears primarily capacitive of the inputs of the device. The
attenuator film is actually located 3.16 cm beyond the input connector. It is
at this point that the impedance should be determined, since this is where
design adjustments will be made. The length of the line stretcher and the
incident arm of the reflection unit are now reduced, which effectively extends
the test plane of the unknown beyond the test port of the reflection unit.
The unit is calibrated to read directly in centimeters of reference plant
extension.

Figure 6.1-82 shows the impedance of the attenuator being tested; the
reference plane was extended 3.16 cm by the line stretcher. Note that only
the *phase angle* of the reflection coefficient has changed by the appropriate
value of electrical phase shift at each particular frequency. The *magnitude* of
the reflection coefficient does not change, but the display now shows the true
impedance at the actual plane of the resistive element. Thus an appropriate

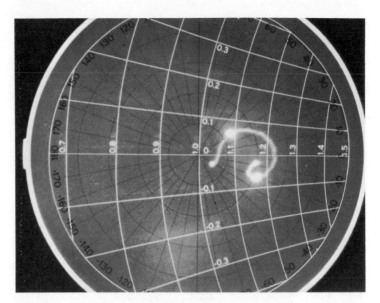

Fig. 6.1-82. The same impedance measurement of the coaxial attenuator of Fig. 6.1-81 but with the plane of measurement extended 3.16 cm into the attenuator where the resistive element is located. Impedance can be determined anywhere in the attenuator for design compensation considerations, for example, by using the line stretcher in the reflection unit to shift the plane of measurement the desired amount. Frequency range: 10.0–12.0 GHz.

compensating network can be designed to reduce the attenuator's mismatch. Impedance can be determined ultimately at any point within the line, so that compensating tuning stubs or networks can be inserted where convenient. For impedance measurements near 50 ohms, as in the last example, it is possible to expand the polar display to obtain better resolution. The scale is expanded after calibration for a reflection coefficient of 1 by using the network analyzer test channel gain control. Thus, for an expanded Smith Chart overlay corresponding to a full-scale reflection coefficient of 0.2 or 1.5-to-1 VSWR, the test channel gain is increased by $-20 \log 0.2$ or 14 dB from the calibration setting. An expanded Smith Chart overlay can be used on the polar display to provide more accurate readings near 50 ohms.

Sources of Errors

The sources of error in a reflection measurement are:

1. *Noise.* When the signal in the test channel becomes very small, the noise level in the test channel becomes significant. The noise level in the system is -78 dBm. Noise errors can be minimized by keeping the signal level in the test channel as large as possible.
2. *Cross talk.* Cross talk occurs when a signal in the reference channel induces a signal in the test channel. Cross-talk errors can be minimized in two ways: by keeping the level of the reference signal as small as possible, so that the magnitude of the signal induced in the test channel is minimized; or by keeping the level of the test channel as large as possible, so that the cross-talk signal is insignificant.
3. *Mismatch.* The system components and the device being tested are not perfectly matched. In other words, they are not all 50 ohms. As a result there will be reflections that result in mismatch uncertainty.

   It is possible to calculate mismatch uncertainties. The three errors of noise, cross talk, and mismatch can be reduced by using well-matched attenuators in front of the harmonic frequency converter. The correct values of pads will keep the signal levels in the reference and test channels at values where noise and cross-talk errors are minimized. Secondly, since the sampler has high VSWR, pads in front of this unit reduce the effective VSWR and hence reduce multiple mismatch errors.

   The table below shows the best pad values to use when the reflection coefficient is less than 1 or greater than 1 (when gain is involved). A reflection coefficient greater than 1 arises when devices with negative impedance, such as tunnel diodes, are measured.

Table 6.1-4. Attenuators in Front of the Harmonic Frequency Converter

$|\rho|$	*Reference Channel*	*Test Channel*
≤ 1	30 dB	10 dB
> 1	30 dB	20 dB

4. *Tracking errors.* The reference and test channels of the network analyzer system do not track perfectly. That is, they do not have the same amplitude and phase response as a function of frequency. Tracking errors are primarily due to coupler tracking in the reflection test unit and the tracking of the two AGC amplifiers in the network analyzer. However, tracking errors can be calibrated out by mea-

suring the reflection coefficient of a short at many different frequencies. At some frequencies the measured reflection coefficient will be $1/180°$, and there will be no error. However, due to tracking error, the measured reflection is not $1/180°$ at all frequencies. The difference between the measured value and $1/180°$ is precisely the tracking error at this frequency. Once the tracking errors have been obtained for many different frequencies, these values can be used as correction factors for future reflection measurements.

5. *Directivity error.* The directivity of the couplers in the reflection test unit is another source of error. Directivity errors can be calibrated out at single frequencies. The method of doing this will be shown later in this section.

Figure 6.1-83 shows the typical system uncertainty when tracking errors have been calibrated out. One set of graphs shows the error when directivity has been canceled out and tracking errors have been canceled.

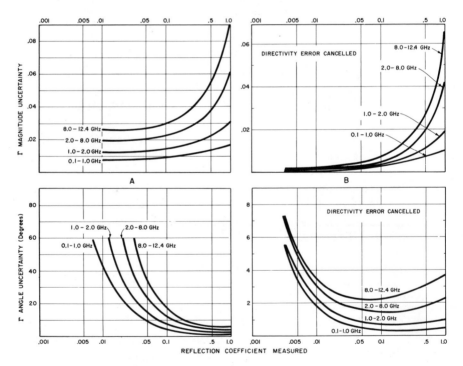

Fig. 6.1-83. Typical system uncertainty versus reflection coefficient measured.

Canceling Out Directivity Error

Directivity error can be canceled at single frequencies by using a sliding load. The procedure is to set the sweeper to a single frequency and measure the reflection coefficient of the sliding load. A small dot will be seen on the polar display, since the reflection coefficient of the sliding load is quite small. Since the reflection coefficient is so small, the test channel gain on the network analyzer should be increased. Let us increase it by 32 dB. This displaces the reflection coefficient of the sliding load from the center of the polar display. The outer circle on the polar display is now a reflection coefficient of 0.025. When the sliding load is moved, the dot on the polar display should revolve in a circle. If there were no directivity errors, the circle would rotate about the center of the polar display. However, the circle is offset from the center of the polar display by precisely the directivity error. If the horizontal and vertical controls of the polar display are now adjusted so that the circle

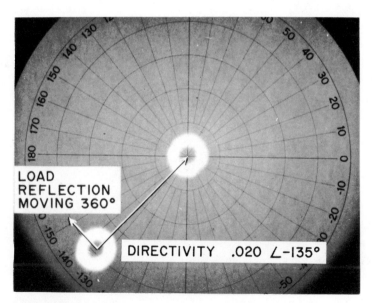

Fig. 6.1-84. Cancelling out directivity error.

rotates about the center of the chart, the directivity error at this particular frequency will be calibrated out. Figure 6.1-84 shows such a display. After calibrating, do not change any of the system controls or the line stretcher before making measurements.

Active Network Analysis

Engineers and technicians who design oscillators with tunnel diodes are vitally concerned with the impedance characteristics of this very useful device. The resistive part of the impedance presented by the tunnel diode must be negative to sustain oscillations. Negative impedance measurements can be made with a network analyzer in two ways.

Figure 6.1-85 shows a Smith Chart overlay designed for negative im-

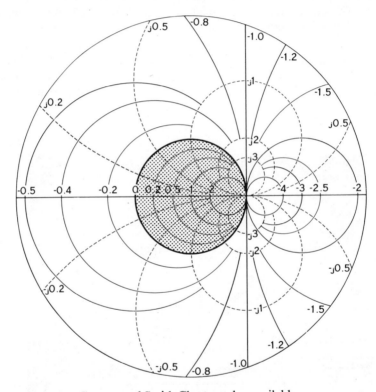

Fig. 6.1-85. Compressed Smith Chart overlay available for direct readout of negative impedance. Full scale $\Gamma = 3.16$. ($\Gamma > 1$ indicates negative real part of impedance).

pedances for a full scale of $\Gamma = 3.16$ ($\Gamma > 1$ represents negative impedance). The shaded area on this Smith Chart represents the normal or positive values of resistance. The area outside this area represents $\Gamma > 1$, which is in the negative resistance region. This overlay is used in a fashion similar to that of the expanded Smith Chart, except that now the IF test channel gain is decreased by $-20 \log 3.16$ or -10 dB from the calibration setting for a reflection coefficient of 1.0. Another technique for displaying negative impedance offers more expansion of Γ ranging from 1 to 10. Smith has shown that a plot of $1/\Gamma$ can be represented on the normal Smith Chart if the positive real axis is interpreted as having negative real values.[22] The network analyzer can be made to plot $1/\Gamma$ simply by reversing the inputs of the harmonic frequency converter so that the reflected signal, which is now greater than the incident signal for a negative resistance as applied to the reference channel by the incident signal, is connected to the test channel. With slight modification to the polar display unit, we simply reverse the voltage to the vertical deflection plate to display the correct polarity of the reflection coefficient. The range of Γ is limited to a maximum of about 10 by the reference channel dynamic range of 20 dB. Other methods can be used to expand this range, but a Γ of 10 or less is typical of most negative resistance devices.

Figure 6.1-86 is a swept impedance plot of a tunnel diode over the 110-MHz to 1.15-GHz frequency range. The frequency at which the impedance becomes positive real and thus where the tunnel diode can no longer oscillate is readily measured as 1.08 GHz. It is that point on the trace that crosses the outer perimeter of the Smith Chart.

Network Analysis in Waveguide

The wide frequency range of the network analyzer makes it comparable with the broadband width of coaxial transmission line. The use of precision 7-mm connectors reduces mismatch ambiguities at all critical points in the RF-measuring circuits and helps define reference planes accurately. However, the network analyzer is not limited to coaxial measurements. It also can be used in waveguide tests to measure s_{21} and s_{11}. The waveguide setup for measuring both transmission and reflection is shown in Fig. 6.1-87. Pads can be inserted in the reference and the test channel arms between the directional coupler and the waveguide bends to alleviate errors due to the moderate mismatch of the waveguide-coax adapters and harmonic converter.

This is arranged for waveguide impedance tests in which the power out of the sweep oscillator is divided into equal channels by a special waveguide directional coupler with two auxiliary arms. The reference channel signal feeds to an auxiliary arm of the directional coupler to a movable short. The short reflects all the signal down the line of the directional coupler and couples to the auxiliary arm of the coupler to the waveguide bend, through the waveguide coaxial adapter into the converter's reference channel. The elec-

[22] Smith, P. H., "A New Negative Resistance Smith Chart," *Microwave Journal*, June 1965.

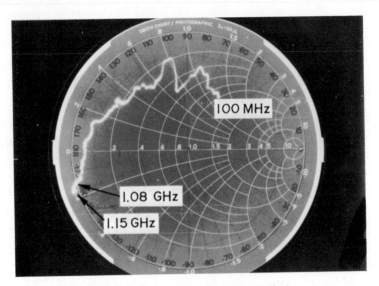

Fig. 6.1-86. Swept-impedance plot of a tunnel diode obtained with the harmonic frequency converter inputs reversed to give negative Smith Chart plot of $1/\Gamma$. Resistive coordinates are now read as negative values. Frequency range is 110 MHz–1.15 GHz. The frequency at which the resistance becomes positive is where the trace leaves the Smith Chart (Γ becomes less than 1) and, in this case, is 1.08 GHz. The tunnel diode will not sustain oscillations above that frequency.

trical length of the reference channel can be adjusted with a movable short. This adjustment provides a variable reference plane and enables calibrating to a constant phase reference for swept-frequency tests. The second arm of the power splitter goes in the same fashion through the main line of the other directional coupler, where, for calibrating purposes, a fixed short circuit is connected. This is the port where the device is connected to be measured. This then will reflect to the auxiliary arm of the directional coupler into the test channel, where it will be measured.

Small tracking variations and source mismatch of the system cause the trace to move about as frequency is swept across the waveguide band. These variations appear during rapid sweep as a large spot which introduces a

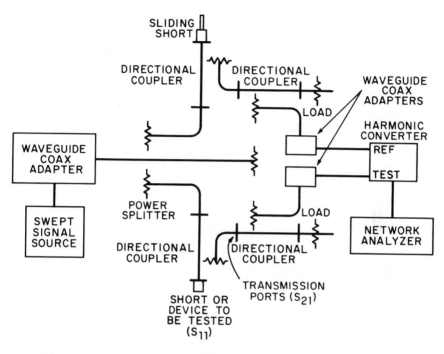

Fig. 6.1-87. Waveguide system for measuring impedance on CW and swept-frequency basis.
The sliding short acts as a line stretcher.

reflection coefficient ambiguity of about 0.02 (this is only for swept-frequency
measurements). In terms of VSWR this is only 1.04:1. Angle ambiguity at
0.2 reflection coefficient is about 6°. These ambiguities can be calibrated out
at single frequencies by manually adjusting the calibration reference exactly
to $1/180°$ at each frequency of interest. For transmission tests, the waveguide is disconnected at the auxiliary arm of the measuring port's directional
coupler, and the device to be measured is inserted there. Of course, the
reference arm has to be taken apart, and an equal length of line has to be
added to maintain mechanical line extension. This technique provides very
accurate waveguide impedance and transmission measurements with the
available high directivity waveguide directional couplers. Of course, such a
device in larger waveguide bands would be quite cumbersome. With automated, computerized test measurements, this can be neglected because
waveguide-to-coaxial adapter error and all other errors can be calibrated
out.

S-Parameter Test Sets

Transducers can be developed and used to measure complete and entire scattering parameters of two-port or multiport devices. Such a test setup is shown in Fig. 6.1-88. Port 1 of the device is driven, and port 2 is terminated

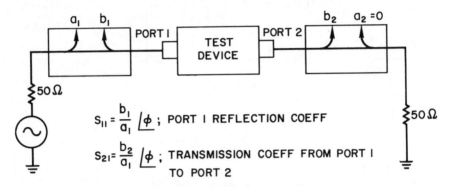

Fig. 6.1-88. Measuring s_{11} and s_{21} with s-parameter test set.

in 50 ohms. The couplers are used to separate the incident, reflected, and transmitted signals. This way the test setup is measuring s_{11} and s_{21}. Measuring s_{22} and s_{12} requires driving port 2 and terminating port 1. In other words, the signal source and the 50-ohm termination must be switched around, as shown in Fig. 6.1-89. Such a set contains, as was seen above, two dual-directional couplers and a switching system necessary to set up the circuit for measuring each parameter. The test set can be used to measure micro-wave hardware or circuit devices like transistors. Some special hardware

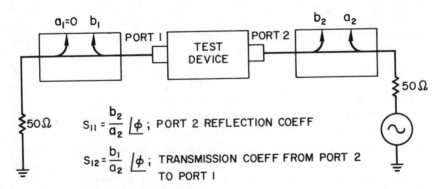

Fig. 6.1-89. Measuring s_{22} and s_{12} with s-parameter test set.

must be used with these test sets. A universal extension which consists of an air line and rotary joint is necessary to provide a flexible, effective way to connect to microwave hardware. Fixtures such as those shown in Fig. 6.1-90 enable one to measure all types of active devices, such as transistors, FET's, tunnel diodes, inductors, and capacitors. A power supply is also needed to bias these active devices. A typical measurement setup is shown in Fig. 6.1-91. Calibration of the test set has to be made right at the transistor fixture's two ports. Calibration kits provide the short circuits, 50-ohm terminations, and straight-through connections needed for this purpose. DC-biasing is introduced into the s-parameter test set at a convenient place in the test set. Measuring the input impedance of a tunnel diode is very interesting because it has a reflection coefficient greater than 1. Of course, this is due to the negative resistance of the diode when it tunnels. The formula below shows a tunnel diode resistance of -25; it will have a reflection coefficient with a magnitude of 3.

$$\rho = \left|\frac{Z_L - Z_0}{Z_L + Z_0}\right| = \left|\frac{-25 - 50}{-25 + 50}\right| = \left|\frac{-75}{25}\right| = 3 \qquad (6.1\text{-}19)$$

The Smith Chart plot of this tunnel diode is shown in Fig. 6.1-92. Since the tunnel diode has a reflection coefficient greater than 1, the compressed Smith Chart is used. The heavy inner circle on the compressed chart is the normal Smith Chart.

Computerized, Automatic Network Analyzer[23]

The use of a network analyzer to measure scattering parameters was discussed in previous sections. It was also shown that, although there are errors involved, this method of measuring scattering parameters will be satisfactory most of the time. But there are occasions when these errors are excessive—such as when making standards laboratory-type measurements in which accuracy is of primary concern. Some errors can be controlled or bypassed by using a digital computer; but, before this is shown, let us discuss the many types of errors that exist.

Usually two types of errors were determined in early error analyses. The first were the so-called *scalar errors*, in which the error term itself was a scalar value, meaning that the standard being used was calibrated to a certain accuracy. This error can be reduced by making more accurate calibrations. The other errors were the so-called *uncertainty terms*. These can be subdivided further into the so-called *vectorial errors* and the actual uncertainty terms or *random errors*. Vectorial errors are those like multiple

[23] Adam, S. F., "A New Precision Automatic Microwave Measurement System." *IEEE Trans. on Instrumentation*, Dec. 1968. Also see Hackborn, R.A., "An Automatic Network Analyzer System," *The Microwave Journal*, Vol. 11, No. 5, May 1968.

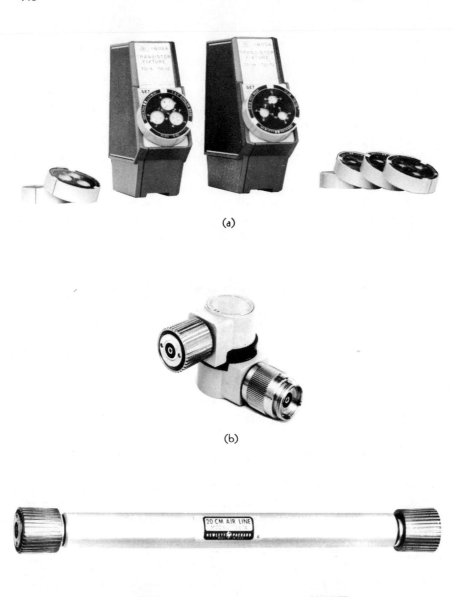

(a)

(b)

(c)

Fig. 6.1-90. Transistor fixtures: universal extension-
rotary joint.

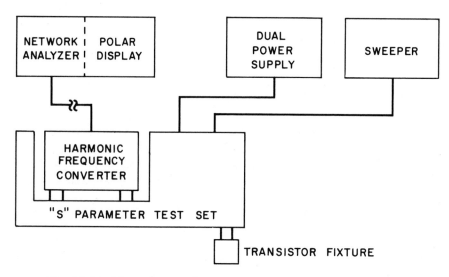

Fig. 6.1-91. Measuring transistor *s* parameters.

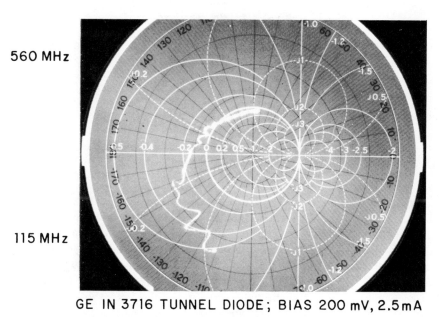

GE IN 3716 TUNNEL DIODE; BIAS 200 mV, 2.5 mA

Fig. 6.1-92. General Electric 1N3716 tunnel diode:
Bias, 200 mv, 2.5 mA.

mismatch errors which change with frequency very rapidly. These changes can be quite violent, depending upon the vectorial summation. Random errors cannot be predetermined and are different for each measurement; connector repeatability and some instrumentation repeatability are good examples. Random errors are not canceled by computerized measurement. However, if many measurements are made and the results plotted on a Gaussian curve, further reduction of them is possible. But scalar errors and vectorial errors are generally the ones which the computer-controlled network analyzer can take into very close and very accurate consideration. For example, the reflection coefficient measurement technique with the network analyzer considers different sources of errors—primarily the directivity of the directional coupler, the reflection coefficients on the measurement port, tracking errors, source mismatch errors, and the contribution of instrumentation error in the system.

The following example will explain how the computerized, automatic network analyzer operates. Imagine a perfect directional-coupler system not having any errors with a two-port network cascaded behind it characterized by s-parameters. Assume that all the errors in an actual system are concentrated in that two-port network. Figure 6.1-93 is a flow graph of that as-

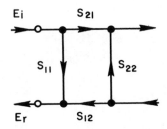

Fig. 6.1-93. Flow graph of a perfect reflection measuring system and the errors concentrated in a two-port network cascaded to that.

sumption. Superimposing the perfect system and the two-port network which includes all the errors is allowed, since the entire system with its imperfections can be seen by looking into it from either the input or the output. Consequently, characterizing the scattering parameters of the actual system could possibly characterize all the errors in terms of scattering parameters. A perfect directional coupler or reflection coefficient measuring system would only provide an incident and a reflected voltage and will measure the ratio. Let us try to analyze how each of these parameters can be measured.

What type of measurement of reflection coefficient E_r/E_i can be made

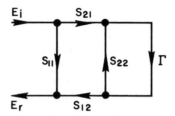

Fig. 6.1-94. Measuring the s parameters of the
two-port network.

by terminating the output of the system? First, as Fig. 6.1-94 shows, the
output is terminated in characteristic impedance where the termination Γ is
equal to 0. If Γ is equal to 0, solving the flow graph will give $E_r/E_i = s_{11}$. If
one places a short circuit on the output (which means that $\Gamma = -1$), then

$$\frac{E_r}{E_i} = s_{11} - \frac{s_{21}s_{12}}{1 + s_{22}} \tag{6.1-20}$$

will be the solution of the flow graph. Providing an open circuit on the output
where $\Gamma = +1$, the solution of the flow graph becomes

$$\frac{E_r}{E_i} = s_{11} + \frac{s_{21}s_{12}}{1 - s_{22}} \tag{6.1-21}$$

It can be clearly seen that s_{21} and s_{12} are not separated in any of the equations,
but it is enough to know their product. These equations are enough to be able
to make an accurate calibration.

Let us see how the measurement is made. The objective is to define Γ
in terms of s_{11}, s_{22}, and $s_{21}s_{12}$.

$$\Gamma \text{ meas} = \frac{E_r}{E_i} = s_{11} + \frac{s_{21}s_{12}\Gamma}{1 - s_{22}\Gamma} \tag{6.1-22}$$

shows the derivation of such a measurement. From Eqs. (6.1-20), (6.1-21),
and (6.1-22), s_{11}, s_{22}, and $s_{21}s_{12}$ are vectors that are already known. Now the
measurement on the system shows Γ meas, and Γ remains the only unknown.
From this expression, Γ can be calculated.

$$\Gamma = \frac{1}{\dfrac{s_{21}s_{12}}{\Gamma_{\text{meas}} - s_{11}} + s_{22}} \tag{6.1-23}$$

where s_{11} gives the residual error, including the directivity errors of the
couplers and the test-to-reference channel isolation; the product $s_{21}s_{12}$ gives
the tracking errors where the tracking variations of the reference and test

channels, of the frequency converter, and coupler coefficients are involved; and s_{22} involves the source match errors where the mismatch of the connector and internal mismatches down the main line of the directional couplers are involved. Of course, the mismatch at the connector is not entirely eliminated, since connector repeatability stays in the system as a random error. The true value of Γ is known.

Of course, these measurements have to be made at all the frequencies of interest, as do the calibrations. When a digital computer is used, the frequency band to be covered is divided into many fixed frequencies, and all these parameters are measured and stored in digital form in the computer's memory. When the measurement is made, the mathematical manipulations are done with the help of the computer. The final information about the reflection coefficient is in the form of vectorial values, resulting in a series of points on a Smith Chart display. Such a display is shown in Fig. 6.1-95; Fig. 6.1-95(a) shows the unrefined (raw) data and Fig. 6.1-95(b) shows the data after the computer refinement. Using the same technique for two ports, complete s-parameter measurements can be done with both input and output ports characterized by the scattering parameters. In addition, the two ports have to be connected together to make a forward and reverse transmission measurement in order actually to characterize all necessary parameters for refinement. The block diagram of such an automatic network analyzer is shown in Fig. 6.1-96.

The automatic error-correction capabilities can best be understood by considering one error associated with the measurement of RF and microwave impedance. A basic reflectometer technique is used with this system. Normally, accuracy is limited, particularly for low-reflection coefficients, by the reflectometer coupler's directivity. A termination standard is connected during the calibration, and the residual reflections are automatically measured at each test frequency, stored in computer memory, and subtracted out as a vector term in the measurement.

The printout in Fig. 6.1-97 indicates the very high effective directivity obtainable with such a technique. The minimum 70-dB return loss shown in the figure is, in effect, the directivity of the system. A 70-dB directivity is equivalent to a residual reflection coefficient error of only 0.0003. Standard coaxial coupler directivity at X-band is typically no better than 30 to 35 dB.

The error-correction program operates on many other error functions in the system, resulting in unprecedented accuracy for both transmission and reflection measurements.

Two modes of operation are possible to display output data. In one case the parameter of interest may be viewed on an analog display either on an oscilloscope, which gives quick readout, or on an X-Y plotter, for more permanent records. The analog display is convenient for quickly observing functional relationships between parameters. Significant relationships can be

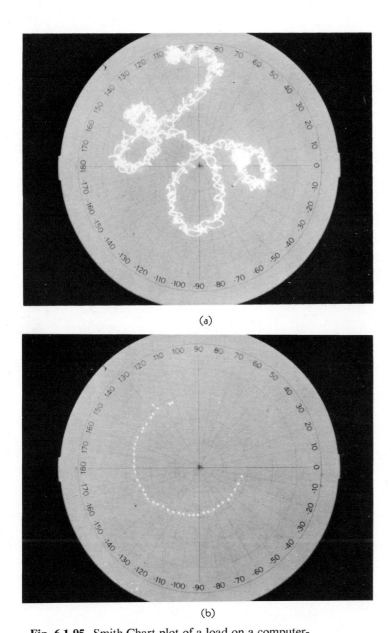

(a)

(b)

Fig. 6.1-95. Smith Chart plot of a load on a computerized network analyzer: (a) unrefined, (b)
computer corrected.

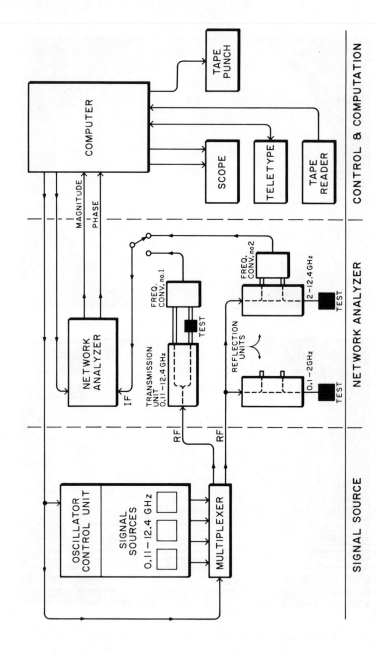

Fig. 6.1-96. Block diagram of a computerized network analyzer.

TASKS:

CONNECT DEVICE: 50 OHM REFERENCE CHECKED AGAINST ITSELF

FREQ,	VSWR	RETURN LOSS (DB)	MISMATCH LOSS (DB)
8000.	1.000	76.63	.00
8484.	1.000	76.91	.00
9012.	1.001	70.88	.00
9496.	1.001	70.18	.00
9980.	1.001	70.70	.00
10507.	1.001	69.81	.00
10991.	1.000	92.71	.00
11519.	1.001	71.63	.00
12003.	1.001	70.88	.00
12399.	1.001	70.68	.00

Fig. 6.1-97. Measurement of reflection coefficient of 50-ohm load that was also used as own calibration standard for error correction program. Corrected return loss calculation thus shows a minimum residual directivity of 70 dB for this system through 12.4 GHz.

observed very rapidly in this manner. The other mode is a digital format of the data.

An example of an analog display is shown in Fig. 6.1-98. This is a plot of the corrected phase and frequency response of a device from 2 to 4 GHz in 100-MHz increments. The display has been self-corrected for any variations in the system's phase and amplitude frequency response. This response is measured and stored in memory during the initial calibration phase, eliminating messy grease-penciling on oscilloscopes or tracking of calibration curves on X-Y recorders. All calibrations are normalized, and units of measurement can be read directly off the straight lines of the oscilloscope graticule.

A typical digital readout on the teleprinter is shown in Fig. 6.1-99. Selected test frequencies are listed in the first column, and the other columns list the calculated parameters derived from the measurement of reflection coefficient (Γ). The second column is the calculated VSWR, where $VSWR = (1 + |\Gamma|)/(1 - |\Gamma|)$. Column three is the return loss in dB where $RL = -20 \log_{10} |\Gamma|$. Column four is the mismatch loss in dB where $ML = -10 \log_{10} |1 - \Gamma^2|$. Note the high resolution of the VSWR reading in this measurement. One-part-in-a-thousand resolution is now possible with the extremely high accuracy resulting from the automatic correction program to decrease greatly directivity and vectorial errors. This system performs these typical types of calculations on data in microseconds. Even though the mathematical relationships may be complex (i.e., require vector algebra),

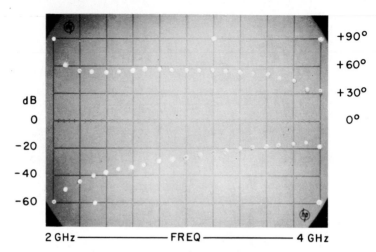

Fig. 6.1-98. Top Trace: Phase versus frequency of a device displayed in 100-MHz increments from 2 to 4 GHz. Vertical scale—30 deg/cm. Bottom Trace: Attenuation versus frequency in 100-MHz increments over same frequency range. Vertical scale—20 dB/cm. Calibrations on oscilloscope "quick look" display are self-corrected for frequency response so that units of measurement can be read directly off straight lines of graticule. Eliminates "grease penciling" or tracing of calibration curves on X-Y recorder.

measurement, error correction, calculation, and printed readout occur in a matter of minutes. A typical system for measuring s-parameters is shown in Fig. 6.1-100.

QUESTIONS

1. What are the three basic attenuation-measuring techniques?
2. What are the resistance values of a 10-dB T-pad attenuator in a 200-ohm characteristic impedance line?
3. What are the major causes of the rapid performance degradation of circuit, T, or π-type attenuators at higher microwave frequencies?

```
TASKS:

CONNECT DEVICE: HP 909 LOAD
   FREQ,        VSWR      RETURN LOSS (DB)      MISMATCH LOSS (DB)
   8000.       1.032        36.00                   .00
   8484.       1.033        35.71                   .00
   9012.       1.036        35.01                   .00
   9496.       1.039        34.27                   .00
   9980.       1.048        32.58                   .00
  10507.       1.060        30.77                   .00
  10991.       1.068        29.70                   .00
  11519.       1.075        28.86                   .01
  12003.       1.076        28.68                   .01
  12399.       1.076        28.77                   .01
```

Fig. 6.1-99. Teleprinter readout of typical calculated parameters from measured reflection coefficient of load. Note that, consistent with the very high accuracy of the 8540A, the VSWR can be read to three decimal place resolution.

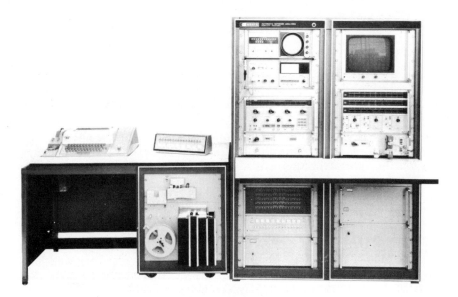

Fig. 6.1-100. The Hewlett-Packard computerized automatic network analyzer.

4. At which end of their operating frequency range do lossy-line attenuators start to deviate from nominal characteristics?

5. What ohms per square resistivity should be deposited on a distributed thin-film attenuator to maintain 50-ohm characteristic impedance if $D = 0.300$ in. and $a = 0.180$ in.?

6. What is the attenuation of a piston attenuator if the piston diameter is 0.450 in. and the piston length is 1.120 in. at 1.5 GHz?

7. What is the attenuation of a rotary vane (waveguide) attenuator if the center section card is rotated 58° out of the plane of the end section cards?

8. What is the directivity of a directional coupler in these cases?
   (a) Connected in the forward direction, the signal drops 20 dB from input of the main line to the auxiliary line output.
   (b) Connected in the reverse direction, the attenuation is 65 dB.

9. What factor determines whether an attenuator is nonreciprocal?

10. What is the phenomenon called which is employed in many isolators and circulators?

11. When is IF substitution technique used to measure attenuation?

12. What are the three types of error families in attenuation measurements?

13. How accurately can attenuation calibration be achieved in the following case? The standard variable attenuator is calibrated to 1%, uncertainties amount to 0.18 dB, RF substitution technique is employed where the source mismatch is 0.12, the standard attenuator (which is connected to the source) has a VSWR of 1.15:1 on both sides with varying phase when adjusted, the attenuator separating the standard and the unknown has a VSWR of 1.1:1 on both sides, the detector has a VSWR of 1.35:1, and the unknown attenuator has an attenuation of 20 dB and a mismatch on both sides of 0.07 (maximum possible error).

14. What is the impedance of a load if it is connected to a 50-ohm characteristic impedance line at 8 GHz, it sets up a VSWR of 1.8:1, and the minimum shift when the load is replaced with short moves 0.95 cm toward the generator?

15. The end of an antenna cable is shorted at the plane of the antenna, and a VSWR of 2.6:1 is measured at the other end of the cable. Assuming that no other reflection exists, how much attenuation does the cable have?

16. What is the VSWR of a load measured on a reflectometer if the output of the reverse directional coupler changes 18 dB while replacing a short circuit with the load under test at the measurement port?

17. What is the error of a VSWR measurement with a slotted line where a VSWR of 1.27:1 was measured? Residual errors, including slope and loss, were 1.04:1, probe reflection was 0.02, attenuator accuracy was 0.01 dB, and detector square-law error was negligible.

18. What is the error on a swept reflectometer using calibration grid technique if reflection coefficients are 0.25, the directivity of the reverse

directional coupler is 34 dB, attenuator accuracy is 1%, and the source match of the reflectometer is 0.05?

19. What are the dB limit settings when measuring source match with the phased short-circuit technique if a VSWR limit of 1.8:1 has to be marked?

20. How many frequency translations take place in the network analyzer described in the text? At what frequencies is the signal translated? At what frequency is the actual magnitude and phase measurement performed by the instrument?

21. What are the differences in making insertion, incremental, and comparative transmission measurements?

22. What is the group delay through a device if the differential phase shift of π occurs when the frequency is varied from 9.185 to 9.295 GHz?

23. If one inserts a device in the test channel of a network analyzer and reads a mechanical extension of 4.55 cm, and if sweeping from 4.000 GHz to 4.500 GHz results in a linear phase shift of 260°, what is the group delay of the device?

24. Determine with measurement method A the insertion test uncertainty of the network analyzer described in Table 6.1-3 under the following conditions: Measurements are made of the incremental and insertion attenuation and phase shift of a variable 0–10 dB attenuator having a 60° phase shift if the input and output VSWR of the attenuator is 1.2:1. Reflection coefficients of the test ports are 0.07.

25. What is the test uncertainty of the network analyzer if a reflection coefficient of 0.2 is measured in the 4- to 8-GHz frequency range? Measurement is to be done with both techniques—directivity error canceled and not canceled.

6.2 TIME DOMAIN REFLECTOMETRY[24]

Evaluation of microwave networks by measuring transmission line properties was discussed in the previous section. However, it is quite hard to measure the reflection coefficient of a port on a device by reconstructing the equivalent circuit of the device if the reflection is composed of many discontinuities at different distances. Measuring reflection coefficient versus frequency gives information about the reflection of a device in the frequency domain. The idea of a "closed-loop radar" helps one understand how time domain reflectometry (TDR) complements reflection coefficient measurement. TDR employs a step generator and an oscilloscope in a system that is best described as a closed-loop radar. A voltage step is propagated down the

[24] Oliver, B. M., "Time Domain Reflectometry," *Hewlett-Packard Journal*, Vol. 15, No. 6, February 1964.

transmission line under investigation, and the incident and reflected voltage waves are monitored by the oscilloscope at a particular point on the line. The radar sends a pulse out and listens for the echoes. If the pulse is short enough, each reflection produces a characteristic echo distinct from all others.

The pulse echo technique has been used for many years to locate faults in wideband transmission systems such as coaxial cables and long telephone lines. The time scale is such that microsecond pulses and megahertz bandwidths suffice. If microwave transmission line problems are being analyzed, closely spaced discontinuities require using nanosecond pulses and gigahertz bandwidths for high resolution. Thus pulse echo reflectometry (in other words, time domain reflectometry) as a laboratory tool has had to await the development of fast-pulse generators and oscilloscopes.

6.2.1 THE TIME DOMAIN REFLECTOMETER

Figure 6.2-1 shows the basic block diagram of a time domain reflectometer. It consists of a very-fast-rise-time pulse generator, an oscilloscope,

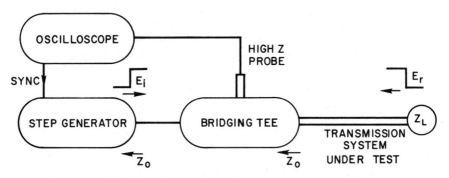

Fig. 6.2-1. Time domain reflectometer.

and a bridging tee with a high-impedance arm toward the oscilloscope that does not cause high reflections from the pulse generator to the measuring terminal in the 50Ω system. There are no complicated directional couplers to separate incident and reflected waves, since they are already separated by time. The source must be well matched to eliminate rereflections off the generator. These reflections would clutter the display if the rereflected signal were to pass again in the forward direction through the bridging tee, causing a response in the high-impedance probe of the viewing oscilloscope.

The difference in time t between two successive echoes is $2s/v$, where s

is the distance between the two successive discontinuities resulting from the echoes and v is the velocity of propagation. If $t > \tau/2$, the echoes can be resolved; in other words, they will be seen as two separate reflections. (τ stands for the system rise time.) Consequently the minimum separation providing accurate measurement is $s = V\tau/4$. Since $v = 2$ to 3×10^{10} cm/sec, and, at the present state of art, $\tau < 2 \times 10^{-11}$ sec, the minimum separation is in the order of 1 to 2 mm. If discontinuities are located closer than this, the reflected signals will produce seriously overlapping signals that complicate interpretation. Of course, this fact would also complicate frequency domain interpretation.

Discontinuities separated by more than the minimum separation determined above will produce distinctly separated echoes that write their own response separately on the oscilloscope face. Since the horizontal reflection's repetition rate is identical to the PRF of the pulse generator, it provides a time scale in the horizontal deflection of the scope. This scale can be calibrated in distance if the velocity of propagation in the transmission device under test is known. Consequently one could locate discontinuities in the system. If the device under test is open, touching the transmission line with a finger introduces added discontinuity. Moving the finger along the transmission line causes the discontinuity to move on the display, literally enabling one to put his finger on the trouble.

Small reflections can also be analyzed in the presence of large reflections at another plane on the line, since they are separated from each other on the display. This allows one to clean up the design of a device by eliminating these different discontinuities one at a time.

The magnitude of reflection coefficient due to each discontinuity can be accurately displayed, since present-day sampling oscilloscopes are able to register signals greater than 100 microvolts in magnitude and to detect reflections in the order of -80 dB (corresponding to an equivalent SWR of 1.0002:1).

The fast-rise-time signal generated should not necessarily be an impulse; it could be a step, since both waves contain the same information. The step is even simpler, and it is easier to interpret than an impulse. In TDR systems such as the Hewlett-Packard time domain reflectometers, a step generator is provided to maintain the signal.

6.2.2 TIME DOMAIN REFLECTOMETRY DISPLAYS AND THEIR INTERPRETATION

The step generator produces a positive-going incident wave that is fed into the transmission system under test. The oscilloscope's high-impedance input bridges the transmission system at its junction with the step generator. The step travels down the transmission line at the velocity of propagation of the line. If the load impedance is equal to the characteristic impedance of the line, no wave is reflected, and all that will be seen on the oscilloscope is the incident voltage step recorded as the wave passes the point on the line monitored by the oscilloscope (Fig. 6.2-2).

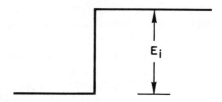

Fig. 6.2-2. Time domain display when $E_r = 0$.

If a mismatch exists at the load, part of the incident wave is reflected. The reflected voltage wave will appear on the oscilloscope display algebraically added to the incident wave (Fig. 6.2-3).

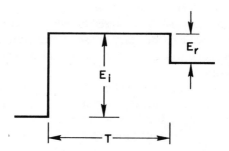

Fig. 6.2-3. Time domain display when $E_r \neq 0$.

The reflected wave is readily identified, since it is separated in time from the incident wave. This time is also valuable in determining the length of the transmission system from the monitoring point to the mismatch. Letting D denote this length,

$$D = v_p \cdot \frac{T}{2} = \frac{v_p T}{2}$$

where v_p = velocity of propagation, and
$\quad\quad T$ = transit time from monitoring point to the mismatch and back again, as measured on the oscilloscope (Fig. 6.2-3).

The velocity of propagation can be determined from an experiment on a known length of the same type of cable. For example, the time required for the incident wave to travel down and the reflected wave to travel back from an open-circuit termination at the end of a 120-cm length of RG-9A/U is 11.4 nanoseconds, giving $v_p = 21$ cm/ns $= 2.1 \times 10^{10}$ cm/s. Knowing v_p and reading T from the oscilloscope, we can determine D. The mismatch is then located down the line.

The shape of the reflected wave is also valuable, since it reveals both the nature and magnitude of the mismatch. Figure 6.2-4 shows four typical oscilloscope displays and the load impedance responsible for each.

From transmission line theory,

$$\rho = \frac{E_r}{E_i} = \frac{Z_L - Z_0}{Z_L + Z_0}$$

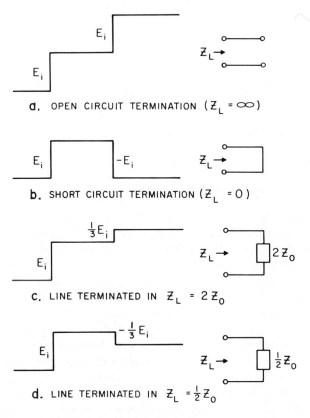

a. OPEN CIRCUIT TERMINATION ($Z_L = \infty$)

b. SHORT CIRCUIT TERMINATION ($Z_L = 0$)

c. LINE TERMINATED IN $Z_L = 2 Z_0$

d. LINE TERMINATED IN $Z_L = \frac{1}{2} Z_0$

Fig. 6.2-4. TDR displays for typical loads.

Knowledge of E_i and E_r, as measured on the oscilloscope, allows Z_L to be determined in terms of Z_0, or vice versa. In Fig. 6.2-4, for example, we may verify that the reflections are actually from the terminations specified.

$$E_r = E_i \qquad\qquad (6.2\text{-}1)$$

Therefore,

$$\frac{Z_L - Z_0}{Z_L + Z_0} = +1, \text{ which is true as } Z_L \to \infty$$

$$\therefore Z_L = \text{open circuit}$$

$$E_r = -E_i \qquad\qquad (6.2\text{-}2)$$

Therefore,

$$\frac{Z_L - Z_0}{Z_L + Z_0} = -1, \text{ which is only true (for finite } Z_0)$$

when $Z_L = 0$.

$$\therefore Z_L = \text{short circuit}$$

$$E_r = +\frac{1}{3}E_i \qquad\qquad (6.2\text{-}3)$$

Therefore,

$$\frac{Z_L - Z_0}{Z_L + Z_0} = +\frac{1}{3} \quad \text{and} \quad Z_L = 2Z_0$$

$$E_r = -\frac{1}{3}E_i \qquad\qquad (6.2\text{-}4)$$

Therefore,

$$\frac{Z_L - Z_0}{Z_L + Z_0} = -\frac{1}{3} \quad \text{and} \quad Z_L = \frac{1}{2}Z_0$$

Assuming that Z_0 is real (approximately true for high-quality commercial cable), it is seen that resistive mismatches reflect a voltage of the same shape as the driving voltage, with the magnitude and polarity of E_r determined by the relative values of Z_0 and Z_L.

If the termination is not a resistive one—if it is actually a pure capacitance or inductance—then the displays shown in Fig. 6.2-5 are obtained.

It can be seen in the capacitive termination that, at the instant the step arrives at the plane of the discontinuity, it will look like a short circuit; the dot on the display moves down exactly with the value of the original step. But the impedance increases as the capacitor charges, and, when it is fully charged, it will reach open circuit in an exponential manner. The inductive termination will act first as an open circuit, since it first resists the flow of current. Later it will let energy through in an exponential manner, since no series resistance is assumed. After a while it reaches short-circuit condition.

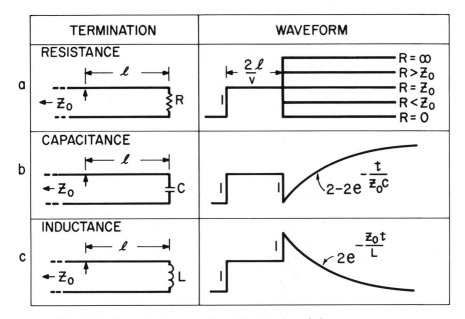

Fig. 6.2-5. The reflections produced by simple resistive
or reactive terminations.

The reflections produced by complex load impedances are also of interest. Four basic examples of these reflections are shown in Fig. 6.2-6. These waveforms can be verified by writing the expression for $\rho(s)$ in terms of specific Z_L for each example (i.e., $Z_L = R + sL$, $\dfrac{R}{1 + RCs}$, etc.), multiplying $\rho(s)$ by E_i/s, the transform of a step function of height E_i, and then transforming this product back into the time domain to find an exact expression for $E_r(t)$. This procedure is useful, but a simpler analysis is possible without resorting to Laplace transforms. The more direct analysis involves evaluating the reflected voltage at $t = 0$ and at $t = \infty$ and assuming any transition between these two values to be exponential. (For the sake of simplicity, time is chosen to be zero when the reflected wave arrives back at the monitoring point.) In the case of the series R-L combination, for example, the reflected voltage at $t = 0$ is $+E_i$. This is because the inductor will not accept a sudden change in current; it initially looks like an infinite impedance, and $\rho = +1$ at $t = 0$. Then current in L builds up exponentially, and its impedance drops toward zero. At $t = \infty$, therefore, E_r is determined only by the value of $\rho = (R - Z_0)/(R + Z_0)$ when $t = \infty$. The exponential transition of $E_r(t)$ has a time constant determined by the effective resistance seen by the inductor. Since the output impedance of the transmission line is Z_0, the inductor sees Z_0 in series with R, and $\tau = L/(R + Z_0)$.

A similar analysis is possible for the case of the parallel R-C termination. At time zero, the load appears as a short circuit, since the capacitor will not

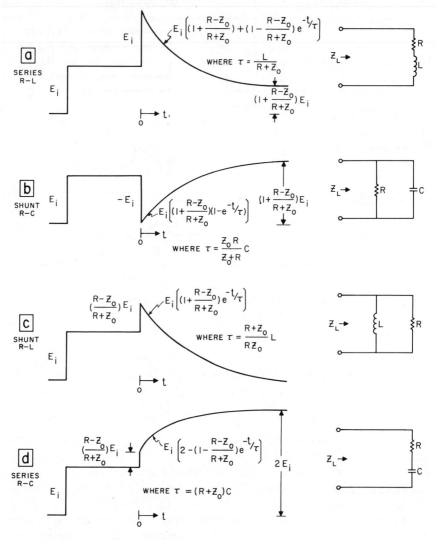

Fig. 6.2-6. Oscilloscope displays for complex loads.

accept a sudden change in voltage. Therefore, $\rho = -1$ when $t = 0$. After some time, however, voltage builds up on C, and its impedance rises. At $t = \infty$, the capacitor is effectively an open circuit:

$$Z_L = R \quad \text{and} \quad \rho = \frac{R - Z_0}{R + Z_0}$$

The resistance seen by the capacitor is Z_0 in parallel with R, and therefore the time constant of the exponential transition of $E_r(t)$ is $[Z_0 R/(Z_0 + R)]C$. The two remaining cases can be treated in exactly the same way. The results of this analysis are summarized in Fig. 6.2-6.

When one encounters a transmission line terminated in a complex impedance, determination of the element values comprising Z_L involves one of two measurements: (1) either E_r at $t = 0$ or at $t = \infty$; or (2) the time constant of the exponential transition from $E_r(0)$ to $E_r(\infty)$. Alternative (1) is a straightforward procedure from the information given in Fig. 6.2-6. Alternative (2) is most conveniently done by measuring the time to complete one-half the exponential transition from $E_r(0)$ to $E_r(\infty)$. The length of time for this to occur corresponds to 0.69τ, where τ again denotes the time constant of the exponential. Adjusting the deflection sensitivity of the oscilloscope in the TDR system so that the exponential portion of the reflected wave fills the full vertical dimension of the graticule makes this measurement very easy, as shown in Fig. 6.2-7.

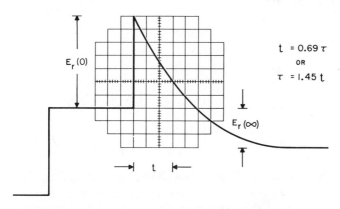

Fig. 6.2-7. Determining the time constant of a reflected
wave returning from complex Z_L.

So far, mention has been made only of the effect of a mismatched load at the end of a transmission line. Often, however, one is concerned not only with what is happening at the load but also with what is happening at intermediate positions along the line. Consider, for example, the transmission system in Fig. 6.2-8(a). The junction of the two lines (both of characteristic impedance Z_0) employs a connector of some sort. Let us assume that the connector adds a small inductor in series with the line. Analyzing this discontinuity on the line is not much different from analyzing a mismatched termination. In effect, one treats everything to the right of M in the figure as an equivalent impedance in series with the small inductor and then calls this series combination the effective load impedance for the system at the point M. Since the input impedance to the right of M is Z_0, an equivalent representation is shown in Fig. 6.2-8(b), and the pattern on the oscilloscope (Fig. 6.2-8(c)) is merely a special case of Fig. 6.2-6(a).

One of the virtues of TDR is its ability to handle situations involving more than one discontinuity. An example of this appears in Fig. 6.2-8(c). The oscilloscope display for this situation would be similar to the diagram in Fig. 6.2-8(e), drawn for the case where $Z_L > Z_0 > Z_0'$.

It is seen that the two mismatches produce reflections that can be analyzed separately. The mismatch at the junction of the two transmission lines generates a reflected wave, E_{r_1}, where

$$E_{r_1} = \rho_1 E_i = \frac{Z_0' - Z_0}{Z_0' + Z_0} E_i$$

Similarly, the mismatch at the load also creates a reflection due to its reflection coefficient $\rho_2 = (Z_L - Z_0')/(Z_L + Z_0')$. Two things must be considered, however, before the apparent reflection from Z_L, as shown on the oscilloscope, is used to determine ρ_2. First, the voltage step incident on Z_L is $(1 + \rho_1)E_i$, not merely E_i. Second, the reflection from the load is $[\rho_2(1 + \rho_1)E_i] = E_{r_L}$; but this is not equal to E_{r_2}, since a rereflection occurs at the mismatched junction of the two transmission lines. The wave that returns to the monitoring point is

$$E_{r_2} = (1 + \rho_1')E_{r_L} = (1 + \rho_1')[\rho_2(1 + \rho_1)E_i]$$

Since $\rho_1' = -\rho_1$, E_{r_2} may be rewritten as:

$$E_{r_2} = [\rho_2(1 - \rho_1^2)]E_i$$

The part of E_{r_L} reflected from the junction of Z_0' and Z_0 (i.e., $\rho_1' E_{r_L}$) is again reflected off the load and heads back to the monitoring point, only to be partially reflected at the junction of Z_0' and Z_0. This continues indefinitely, but after some time the magnitude of the reflections approaches zero. It is now seen that, although TDR is useful when observing multiple discontinuities, one must be aware of the slight complication that these discontinuities introduce when the display is analyzed. It is fortunate that most practical measuring situations involve only small mismatches (e.g., $Z_0 \approx Z_0'$) and the effect of multiple reflections is almost nil. Even in this situation, however, it is advisable to analyze and clean up a system from the generator end. The reflection from the first of any number of discontinuities is unaffected by the presence of others. Therefore, if it is remedied first and one then moves on to the second discontinuity, the complications introduced by rereflections will not exist.

In general, the source impedance of the step generator might not be equal to the characteristic impedance of the transmission line it drives. When

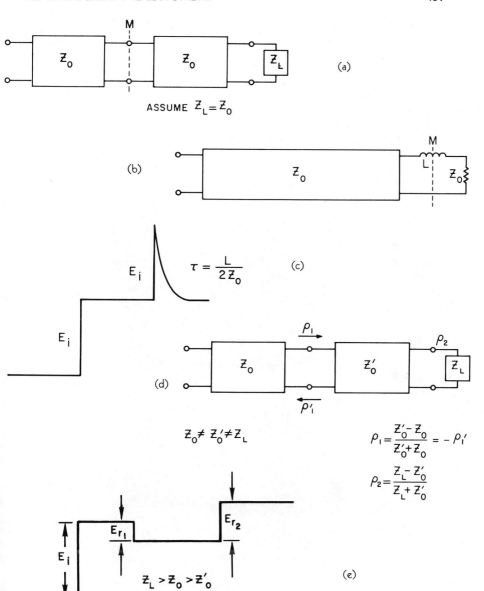

Fig. 6.2-8. (a) Transmission system, (b) equivalent representation of transmission system shown in (a), (c) TDR pattern of the circuit shown in 6.2-8(b), (d) multiple discontinuities, (e) oscilloscope display of multiple discontinuities.

this is the case, voltage waves returning from a mismatch or discontinuity in the system under test will be reflected at the generator end and will complicate the analysis of the display (Fig. 6.2-9). Therefore it is almost essential that the source impedance of the step generator match the cable it drives. Then, all reflections returning from the system under test pass the oscilloscope's monitoring point only once and are then absorbed in the source impedance of the step generator.

The photos in Fig. 6.2-9 show the oscilloscope displays of two TDR

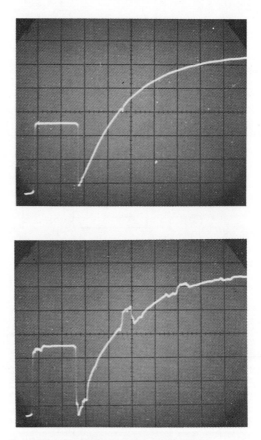

Fig. 6.2-9. Capacitive load displayed on the TDR.

systems investigating a transmission line terminated in a capacitor. In the upper photo, the source impedance of the step generator matches the characteristic impedance of the line under test ($Z_s = Z_0 = 50$ ohms). In the lower photo, however, this was not the case. There the source impedance of the step generator was altered by inserting a short length of 75-ohm cable between the step generator (with $Z_s = 50$ ohms) and the point where the sampling scope bridged the input to the line under test. The resulting mismatch caused the reflected wave returning from the capacitor to be rereflected at the source, thus launching a second incident wave down the line. This second wave sends a second reflected wave from the capacitor back to the monitoring point. The second reflected wave, in turn, launches a third incident wave down the line. This process continues indefinitely, but, unless the reflection coefficient at each end is equal to ± 1, the reflections decrease in magnitude and only the first few are noticeable.

6.2.3 AN ACTUAL TIME DOMAIN REFLECTOMETER

Figure 6.2-10 shows a block diagram of a simple TDR plug-in unit with medium-fast rise time for a general-purpose oscilloscope by Hewlett-Packard.

Figure 6.2-10 shows that the step generator output has been brought

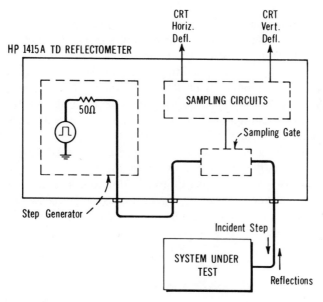

Fig. 6.2-10. The Hewlett-Packard 1415A time domain reflectometer has both a 50-ohm source impedance and a high-impedance sampling gate which bridges a feed-through section of 50-ohm transmission line.

through the front panel, making it available for even more versatile meas-urements (dealt with later in this chapter). Essentially, a sampling oscilloscope with a sampling gate composes the bridging circuit and the electronics for the oscilloscope horizontal and vertical deflection. This provides a quite high input impedance for the oscilloscope but does not interfere with the RF transmission line system by upsetting its impedance. The displays explained above assumed infinitely fast rise-time systems. If discontinuities are far apart and large enough, the scope presentation will fall very close to the ideal case shown above. But the finite rise time of the system tends to smooth out the sharp changes.

As was mentioned earlier, the *resolution* of a TDR is determined by its rise time. The above-mentioned medium-fast TDR has a rise time of $\tau <$ 10^{-10} second. Two discontinuities spaced closer than 1 cm apart would be hard to separate. The reflection observed is the sum of the reflections that would be observed if each discontinuity were isolated. It is impossible to acquire quantitative information for each discontinuity, but the composite reflection determines the effective discontinuity seen by the TDR. Since this is essentially the discontinuity that a signal with frequency components below approximately 4 GHz will see, the test is valid for this frequency range.

Treated ideally, the reflections from small inductors and small capaci-tors have very short time constants of the exponential transition from $E_r(0)$ to $E_r(\infty)$ and are only special cases of the examples of Fig. 6.2-6. Under actual measuring conditions, however, the finite bandwidth of the step generator oscilloscope system becomes a limiting factor in the display. Consider, for example, a series combination of R and L, where $R = 50\Omega = Z_0$ (of the cable feeding the R-L termination), and L is of the order of 10^{-10} henry. Ideally, the display should resemble Fig. 6.2-11(a); it actually resembles Fig. 6.2-11(b). Qualitatively, one can understand what is happening in Fig. 6.2-11(b) by realizing that the time constant of the reflected wave is so short that it decays to its final value almost before the TDR system has begun its rise. Despite this limitation, quantitative information is still available con-cerning the magnitude of the small inductor causing the reflection.

Recalling from transmission line theory that

$$\rho = \frac{E_r}{E_i} = \frac{Z_L - Z_0}{Z_L + Z_0}$$

and substituting

$$Z_L = R + j\omega L$$

then

$$\rho = \frac{(Z_0 + j\omega L) - Z_0}{(Z_0 + j\omega L) + Z_0} = \frac{j\omega L}{2Z_0 + j\omega L}$$

But, since L is small, the product ωL will be very much less than $2Z_0$ unless ω becomes very large. However, the finite rise time (i.e., limited bandwidth)

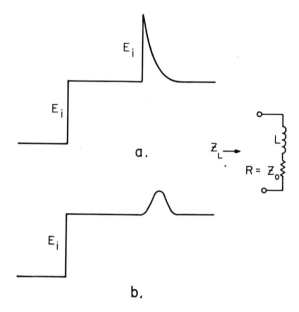

Fig. 6.2-11. Ideal versus actual displays of reflection
from a small inductor in series with
$R = Z_0$.

of the TDR system dictates that the frequency spectrum of the displayed step
will not contain frequency components beyond a certain cutoff frequency.
Therefore, ωL can be neglected with respect to $2Z_0$, and

$$\rho = \frac{E_r}{E_i} \approx \frac{j\omega L}{2Z_0} \quad \text{or} \quad E_r \approx \frac{L}{2Z_0}(j\omega E_i)$$

Continuing to talk in terms of the sinusoidal components of the displayed
waveform, if

$$E_i = E\epsilon^{j\omega t}$$

then,

$$j\omega E_i = j\omega E\epsilon^{j\omega t} = \frac{dE_i}{dt}$$

and

$$E_r \approx \frac{L}{2Z_0}\frac{dE_i}{dt}$$

Therefore the reflected wave will be a differentiated version of the incident
step with its magnitude proportional to $L/2Z_0$. Since both $E_r(t)$ and dE_i/dt
can be read from the oscilloscope display, L can be evaluated (see Figs. 6.2-12
and 6.2-13).

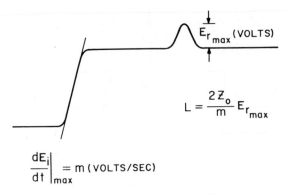

$$\frac{dE_i}{dt}\bigg|_{max} = m \text{ (VOLTS/SEC)}$$

Fig. 6.2-12. Determining the magnitude of small
inductors in series with $R = Z_0$.

The photos in Fig. 6.2-13 show an actual reflection from a small inductor in series with $R_L = Z_0$. The upper photo was taken at 50 mv/cm sensitivity and 4 ns/cm sweep speed. The lower photo is an expanded view of the reflected wave; the sensitivity is 10 mv/cm and the sweep speed is 400 ps/cm. From this photo, $E_{r_{max}}$ is seen to be 0.34 mv. Since $m \approx 3$ mv/ps and $Z_0 = 50\Omega$,

$$L \approx \frac{2(50)}{3 \text{ mv/ps}} (34 \text{ mv}) = 1.1 \times 10^{-9} \text{ H}$$

A similar analysis is possible for the case of a small capacitor in shunt with a resistor of value Z_0. The results of this analysis are summarized in Fig. 6.2-14.

In the cases where a small inductor is shunting a load resistor or a small capacitor is in series with a load resistor, there is no similar approximation to determine the value of L or C. When the exponential transition of the reflected voltage becomes faster than the rise time of the displayed step, the small shunt inductor becomes indistinguishable from a short-circuit termination and the small series capacitor becomes indistinguishable from an open circuit. As a practical matter, however, these cases are seldom encountered, unless one is actually attempting to create a short or open termination.

The prime virtue of time domain reflectometry is its ability (1) to display the transmission quality of a system for frequencies from dc to a few gigahertz, and (2) to isolate discontinuities so that they may be compensated for individually on a broadband basis. For this reason, TDR is directly applicable to measuring cable parameters: Z_0 (either its absolute value or uniformity with distance), loss (either series or shunt), and length. TDR is also useful for broadband reflection coefficient measurements of terminations, for broadband evaluations of individual components such as con-

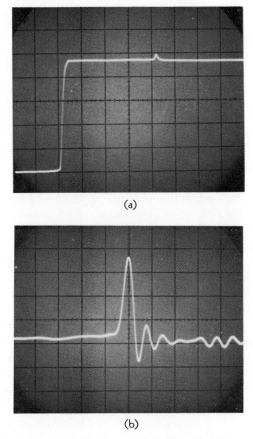

(a)

(b)

Fig. 6.2-13. Actual photographs of TDR displays of inductive reactance (a) total display, (b) enlarged display.

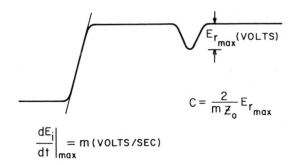

$E_{r_{max}}$ (VOLTS)

$$C = \frac{2}{m\,Z_0} E_{r_{max}}$$

$$\left. \frac{dE_i}{dt} \right|_{max} = m\,(\text{VOLTS/SEC})$$

Fig. 6.2-14. Determining the magnitude of small capacitors in shunt with $R = Z_0$.

nectors and tees, and for determination of the broadband transmission quality of entire systems.

6.2.4 APPLICATIONS OF TIME
DOMAIN REFLECTOMETRY

With the advent of subnanosecond (picosecond) pulse generators and oscilloscopes with equivalent bandwidths, the application of the pulse reflection technique to high-frequency transmission systems has become practical. Distance resolution has shrunk from hundreds of yards to fractions of an inch, and the new generation of sampling oscilloscopes permits accurate measurement of reflections only a thousandth of a volt in amplitude.

The time domain reflectometer is thus a type of closed-loop, one-dimensional radar system in which the transmitted signal is a very fast step function, and the reflected signals are monitored on the oscilloscope screen. The faster the step, the greater the distance resolution, since distance is related to time in this technique. The amplitude of the reflected step is directly related to impedance, so any slight deviation from the 50-ohm output of the time domain reflectometer can be recognized and measured easily. Thus cables, connectors, baluns, strip lines, tapered sections, and a host of other broadband devices can be analyzed with TDR.

Faults occur in even the best high-frequency transmission systems and can cause substantial losses of power or can severely distort the transmitted signal. Dielectrics may deteriorate and change, water may leak into cables or connectors, contacts may corrode, conductors may open or short, the cable may be cut or damaged, or a clamp may be fastened too tightly. The time domain reflectometer treats such occurences as discontinuities, or unexplained abrupt changes in the otherwise constant characteristic impedance of a transmission system. Since time is readily convertible into distance, the exact location of the discontinuity can be found.

Aside from discontinuities, the transmission line itself has a number of relevant properties. It has a characteristic impedance, which may or may not change with frequency and may vary over its length. It has a certain dielectric, or velocity of propagation. It has a certain attenuation per unit length, which does vary with frequency. The time domain reflectometer will provide quantitative as well as qualitative information on any transmission cable impedance, loss, rise time, electrical length, and discontinuities in a single measurement.

As in any measurement technique, there are limitations imposed both by the state of present-day technology and by the technique itself. Since the time domain reflectometer relates time to distance, the finite rise time of the incident or reflected pulse limits the possible resolution of distance. It also limits the system bandwidth—the frequency range over which the measurements are valid. Reflections generated in waveguide systems, unlike those in coaxial systems, travel at various propagation velocities, depending on the mode of propagation. Analysis of waveguide reflections is therefore very complex and is further compounded by the inherent low-frequency cutoff that makes it a narrow-band measurement.

Standing-wave ratio (SWR) measurements provide an immediate overall indication of a transmission line's performance, whereas time domain reflectometry (TDR) measurements isolate the line's characteristics in time, i.e., location. Multiple reflections due to a number of discontinuities or numerous impedance changes complicate the TDR measurement but are nonetheless easily evaluated.

The TDR may be used conveniently to measure transmission line propagation characteristics, such as attenuation, phase constant, and characteristic impedance. However, these constants actually will vary along the line, and one section may have quite different characteristics from another. TDR isolates such variations in time (distance), providing an impedance profile along the length of the line. Therefore an entire cable may be tested completely by a single measurement.

An ideal cable (Fig. 6.2-15) should appear as a resistive load, with no

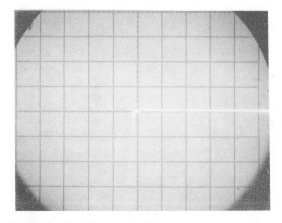

Fig. 6.2-15. Ideal cable (50-ohm load).

reflections occurring, except at the beginning and end of the cable. In practice, reflections will occur along the line. This indicates changes in characteristic impedance (Fig. 6.2-16). They may appear as steps or pedestals, which generally indicate that a section of different impedance cable has been spliced into the line (Fig. 6.2-17). Or, the reflections might appear as small bumps, which indicate a fault or discontinuity (Fig. 6.2-18). The profile

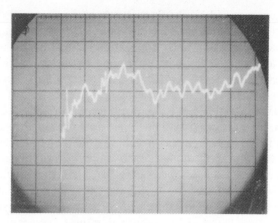

Fig. 6.2-16. Actual cable.

might show a slowly rising or falling characteristic, which indicates a series or shunt loss in the cable.

If the loss over the length of the line is less than about 0.25 dB and the total variation of impedance along the line is less than about ±10 ohms, the impedance profile is valid. The trace displayed on the screen is then an accurate presentation of the impedance at all points along the line, within the accuracy limits of the system.

The CRT is calibrated in reflection coefficient, which is related to impedance by the formula

$$\rho = \frac{Z - 50}{Z + 50} \qquad \text{or} \qquad Z = 50 \left(\frac{1 + \rho}{1 - \rho} \right)$$

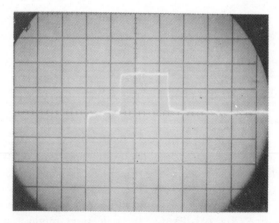

Fig. 6.2-17. Section of 95-ohm cable spliced in length
of 50-ohm line.

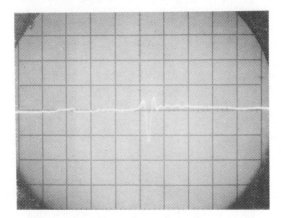

Fig. 6.2-18. Discontinuity caused by BNC "T"
connector in line.

Since 50Ω characteristic impedance is used most widely in the electronics
industry, most TDR systems are referenced and built in 50Ω systems. If
a cable with a different characteristic impedance is to be measured, corrections
have to be made in the interpretation of the display.

A typical application of the impedance profile is checking the impedance
specification of a cable. In Fig. 6.2-19, a 50-ohm cable has been specified
to be 50 ohms ±1% (±0.5 ohms), so a trace in the shaded area represents
a "no go" condition.

Although a swept-frequency reflectometer measurement could provide a
similar final figure, it cannot determine the location of a discontinuity. The
TDR display may well show the problem to be a poorly mated connector and
not a cable problem at all.

TDR's are usually calibrated in the distance/time scale. This enables
one to locate the discontinuity; in the case of the cable, the location of the
connectors can easily be located. If long cables are tested, applying a tempo-
rary pressure to the outside of the cable will show up on the display and
enable the operator to identify the physical location on the cable, locating
points of interest viewed on the TDR display.

When accuracies beyond the capability of the time scale of a TDR are
needed, such as for multiple-antenna phasing, relative measurement techniques should be used. Two cables may be parallel and the distance between
the end-cable reflections noted. One line may then be shortened or lengthened
until the two reflections are superimposed, and the lines will be properly
phased. It is good practice to terminate such lines with a small discontinuity,
such as a T or an improperly mated connector. It is easier to line up two
small discontinuities than two open or short circuits.

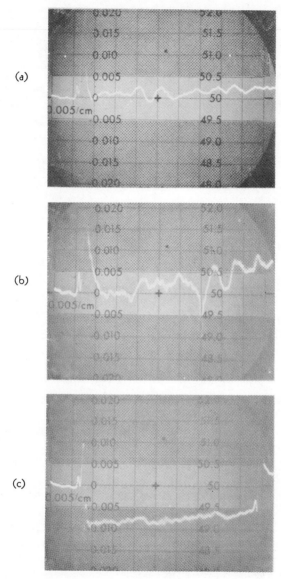

Fig. 6.2-19. The first centimeter of each trace is an air line calibrated at 50.1 ohms = The use of a known impedance immediately preceding the test table avoids the effect of drift or pulse base line shift by providing a constant reference. The two positive spikes are reflections from connectors at the beginning and end of the test cable. (a) "Go." (b) "No Go." Kink in cables causes 1% reflection. (c) "No Go." Nominal Z_0 less than 49.5 ohms.

If *lossy lines* are connected into the system, the rise time will become degraded because loss increases going down the line. Of course, the ability of the TDR to resolve small discontinuities or changes of impedance will also be impaired after the lossy section. To make reasonable measurements, for instance on a cable, the loss has to be determined.

The low-frequency loss in the distortionless 50-ohm line (Fig. 6.2-20)

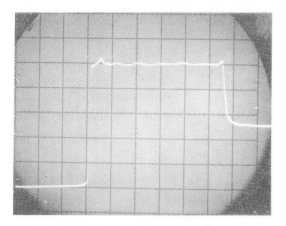

Fig. 6.2-20. Short at end of cable measures dc loss.

can be measured readily by shorting the end of the cable and noting the amplitude of the reflection. If the reflection coefficient is 0.1, for example, the total loss (down and back) is 20 dB, and thus the one-way loss is 10 dB. Normally, the low-frequency loss is quite low for distortionless lines, and a low reading such as 0.1 probably indicates a 10-dB pad somewhere in the system.

High-frequency losses (Fig. 6.2-21) are shown by a degradation of the incident pulse. The reflected pulse rise time will be

$$T_{rr} = \sqrt{2T_{rc}^2 + T_{rl}^2}$$

where T_{rl} is the measured incident pulse rise time, T_{rc} is the rise time of the cable, and T_{rr} is the measured rise time of the reflected pulse. If the pulse shapes are all identical, then the cable rise time is

$$T_{rc} = \sqrt{\frac{T_{rr}^2 - T_{rl}^2}{2}}$$

or

$$T_{rc} \cong 0.707 \, T_{rr} \quad \text{if} \quad T_{rr} \gg T_{ri}$$

The most common case involves a series loss that exceeds the shunt loss, causing a very long line to approximate an open circuit at dc. The TDR

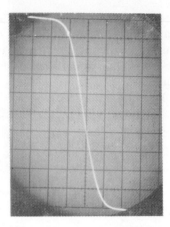

Fig. 6.2-21. Fall time of reflected pulse indicates HF
loss.

characteristic is a slowly rising one (Fig. 6.2-22) and approximates a series
capacitance termination. If the line is near 50 ohms, the dc attenuation may
again be measured by reading the short-circuit reflected pulse height and the
bandwidth by the reflected rise time.

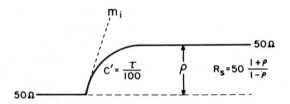

Fig. 6.2-22. Displayed waveform of lossy line.

The actual value of the series cable resistance is a function of the conductor skin depth and therefore varies with frequency. It would be erroneous,
then, to assume that the equivalent lumped circuit (Fig. 6.2-23) is valid at all
frequencies. However, for very broadband usage or pulse work, the equiva-

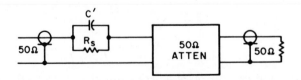

Fig. 6.2-23. Approximation of equivalent circuit.

lent circuit is a good approximation of the cable's performance, so long as its length is not changed. Note that the trace shown in Fig. 6.2-22 gives no indication of the actual attenuation in the line. It is valid only if the shunt conductance of the cable is known to be zero. Attenuation must still be measured by placing a known mismatch (such as a short) at the end of the line and measuring its amplitude and rise time.

Measurement of impedances at the end of the line may also be made easily by first placing a short at the end. The TDR unit may then be set so that the reflection from the short reads ρ of -1. Then the end of the line effectively becomes the calibrated 50-ohm sampling point, and impedances may be measured in the normal manner.

The relative value of series resistance for cables of the same Z_0 may be compared by measuring the initial slope of the rising waveform. This method provides a quick check for excessive series loss if the other end of the cable cannot be reached. A similar display will be obtained if the shunt losses of a cable exceed the series loss, except that the trace will have a falling, rather than rising, characteristic.

Up to this point we have been dealing with 50-ohm systems, since that is the source impedance of time domain reflectometers. When any other impedance system is used, the mismatch at the sampler output affects both the outgoing pulse and the incoming reflections. More generally, each discontinuity will affect the discontinuities which follow it.

Figure 6.2-24 clearly illustrates this phenomenon. The jump from a 50-ohm to a 93-ohm cable is evident and follows TDR rules, but the step from the 93-ohm cable to the open circuit does not. Instead of jumping to $+1$ (the reflection coefficient for an open circuit), the trace actually exceeds that

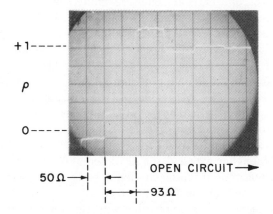

Fig. 6.2-24. Multiple reflections caused by two discontinuities.

value. Finally, at the extreme right-hand edge of the trace the rereflections between the two discontinuities diminish, and the level settles down to the true value.

Even if three or more major discontinuities exist, the display of the first two discontinuities is not affected by the following ones. Both reflections may be measured and represented by an impedance. However, the third discontinuity may produce a reflection which is mixed in with the rereflection caused by the first two discontinuities. Because of this ambiguity, we can say that quantitative measurements should be restricted to the first two discontinuities.

One may, of course, calibrate the oscilloscope for direct reading of reflection coefficients based on a system other than 50 ohms. The simplest way to calibrate is to short- or open-circuit a short length of the cable and adjust the reflection coefficient vernier until a reading of 1.0 is obtained from the display of the reflection from the open or short. The measurement must be taken immediately after the open or short, because the first rereflection will cause an error. It is also important to remember that any measurement after the first major discontinuity is likely to be invalid. The effects of small discontinuities can be largely overcome by using a tapered section, which provides a gradual transfer from 50 ohms to the new Z_0.

As with any measurement method, there are good and bad techniques for time domain reflectometry. Measurements will be more accurate if reflections in the early part of the system are kept low. It is generally a good idea to use a length of air line or very good cable between the time domain reflectometer and the test circuit. This is done so that reflections between the sampler and the pulse generator will have died down to an unnoticeable level before reflections from the test circuit are measured.

Most oscilloscope measurements are limited by visual accuracy, so differential measurement techniques are employed: TDR is no exception. A 93-ohm impedance can be measured to within ± 0.1 ohm if compared to a known 90-ohm load but only to ± 2 ohms if compared to a 50-ohm standard.

When measuring impedances near 50 ohms, a calibrated air line placed immediately before the test cable provides the best reference, since a rigid line has a more constant Z_0 over its length than a flexible cable does.

The most accurate method by far is the "standard mismatch" technique used by the National Bureau of Standards in Boulder, Colorado. A long length of calibrated air line is used between the TDR and the unknown, and a length of calibrated line with a Z_0 slightly different from 50 ohms is inserted in the line. The reflection from this mismatch serves to calibrate the system, and only a relative measurement need be made.

To achieve a resolution compatible with this high accuracy, an X-Y recorder is used to record the final results. The gain of the X-Y recorder may be increased to the point where system noise is the limiting factor. This noise will be far less than that seen on the CRT, since the limited bandwidth of the X-Y recorder will average out most of the noise.

A swept-frequency reflectometer measurement (using a sweep oscillator) and a single-frequency slotted-line SWR measurement remain the only accurate methods for determining the characteristics of a transmission system at any given frequency of a band of frequencies. Although the user can arrive at a single frequency characteristic by using TDR measurements, a Smith Chart, and the formula for multiple reflections, the measurement capability of TDR will suffice for only "low Q" systems. To attempt accurate analysis of highly resonant systems with TDR is like trying to read a pocket slide rule to six significant digits. However, to determine the cause of an unwanted resonant condition, the TDR will provide a quick and ready answer.

When a single, resistive discontinuity, such as a mismatched load, is encountered, conversion may be made directly to SWR by the equation

$$\text{SWR} = \frac{1 + \rho}{1 - \rho}$$

When reactive discontinuities or multiple discontinuities are encountered, however, the Smith Chart is the best technique. Reactive discontinuities are first determined in units of capacitance or inductance and then converted to complex impedances. Similarly, distances are converted to wavelengths for Smith Chart use. The following hypothetical example should illustrate the application. A 50-ohm line runs to a 300-ohm antenna with a 50:300 ohm balun used to match the antenna to the line (Fig. 6.2-25). The TDR display

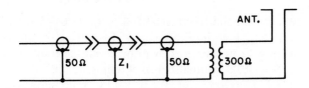

Fig. 6.2-25. Equivalent circuit of antenna system.

(Fig. 6.2-26) shows, however, that a section of cable which is not 50 ohms has been inserted from A to B and that there is an inductive discontinuity at C and D, probably caused by the balun. Finally, at E, the antenna does not match the 300-ohm line. Frequency is 180 MHz.

The first step is to determine the antenna impedance. The base line is known to be 300 ohms, so the measured ρ of 0.09 indicates an impedance of

$$Z_E = 300\,\frac{1.09}{0.91} = 360\Omega$$

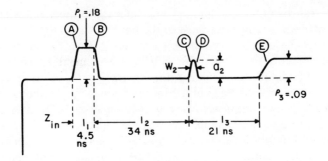

Fig. 6.2-26. TDR display of antenna system.

Actually, the impedance transfer from 50 to 300 ohms is irrelevant in this case, so the antenna impedance can be related to 50 ohms instead. Then,

$$Z_E = 50 \frac{1.09}{0.91} = 1.2(50) = 60\Omega$$

Plotting a point at $1.2 + j0$ (point E) on the Smith Chart (Fig. 6.2-27), a clockwise arc is generated equal to l_3, which is 21.5 ns or 3.87 wavelengths. The resulting normalized impedance at point D is then $0.97 + j0.18$. The normalized inductance of the balun is

$$L_{(50)} \approx 2a_2w_2 = 2 \times 0.07 \times 0.3 \times 10^{-9}$$

$$= 0.042 \times 10^{-9}$$

$$= +j0.048 \text{ at } 180 \text{ MHz}$$

The total impedance at point C is then $0.97 + j0.18 + j0.048 = 0.97 + j0.23$, and this point is plotted on the Smith Chart. From here the arc corresponding to l_2 (34 ns or 6.12 wavelengths) is drawn, and the resulting impedance at point B is $1.27 + j0.02$, normalized to 50 ohms.

The cable (AB) has a measured ρ of $+0.18$, or an impedance of

$$Z_1 = 50 \frac{1 + 0.18}{1 - 0.18} = 71.9\Omega$$

so the Smith Chart must now be normalized to this new impedance. The last point is now

$$\frac{50}{71.9}(1.27 + j0.02) = 0.88 + j0.014$$

and this point is plotted. Then the arc representing the length of the cable l_1 (4.5 ns, 0.81λ) is made, and the impedance at the beginning of the cable

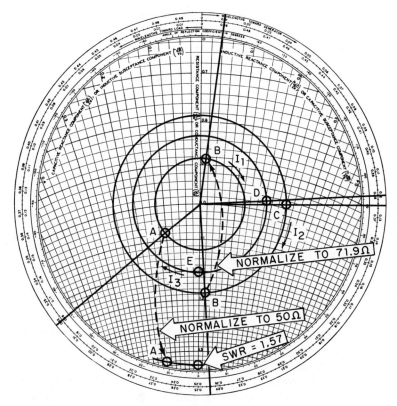

Fig. 6.2-27. Smith Chart for antenna system.

is $1.08 - j0.1$, normalized to 71.9 ohms. Returning the normal figure to 50 ohms makes this impedance

$$\frac{71.9}{50}(1.08 - j0.1) = 1.55 - j0.14$$

and this impedance will create an SWR of 1.57.

With a VSWR of 1.57 at the source, mismatch loss can be calculated to be approximately 5% if the generator has a 50-ohm source impedance. This type of analysis may seem laborious compared to a simple slotted-line measurement, but the fact that a complete Smith Chart has been generated means a wealth of useful information is available:

1. If the section of 72-ohm cable were replaced by a 50-ohm section, the VSWR would be reduced to 1.27.

2. If, in addition, the balun is improved to eliminate its inductive discontinuity, the VSWR would be further reduced to 1.2.
3. If instead of improving the balun, it is moved $\frac{1}{4}$ wavelength in either direction, its inductance will help offset the antenna mismatch and the VSWR can be reduced to 1.13.
4. If additional turns are added to the balun to better match the antenna impedance to 50 ohms, only the inductance of the balun (VSWR = 1.055) will contribute to the overall standing-wave ratio.
5. If a single-frequency SWR measurement is made and found to be significantly different from the Smith Chart calculation, the signal generator or transmitter mismatch to the 50-ohm system is probably faulty.

Successful application of the time domain reflectometer requires a thorough knowledge of its advantages and limitations. Cable testing is but one of many applications, others being connector and antenna testing, stripline and switch design, impedance measurement, and the testing of practically any broadband passive circuit in the VHF-and-up range.

A time domain reflectometer can prove to be a very effective design tool for TEM wave transmission line design. Because complicated transmission devices intended to work in the TEM mode of propagation are not necessarily restricted to coaxial form, they can have many sources of discontinuities that are quite cumbersome to analyze. Even if one chooses to build a preliminary scale model of the device in order to make measurements that test the feasibility of the theory at higher frequencies, he could use only *frequency* domain measurement techniques, due to the bandwidth limitations of *time* domain reflectometers. Of course, if many sources of reflections have to be optimized at widely different levels, it becomes very cumbersome to do in the frequency domain.

Time domain reflectometry can complement frequency domain techniques by making a *mechanically scaled-up* version of the device to separate closely spaced reflections as they will be resolvable on TDR. Furthermore, by linear scaling of the device, the characteristic impedance will stay the same; but the reactances will scale linearly, achieving the effect of equivalent bandwidth increase. For example, if a TDR system has an equivalent bandwidth of 3.8 GHz, scaling up the device to be tested to its five-times scale enables one to see reactances and resolution equivalent to those of a 19-GHz TDR. Further conveniences are that the increased size allows adjustments to be made without involving very small parts and tight tolerances. Discontinuities can be corrected on the scaled-up model; when all discontinuities are optimized, it is only a matter of scaling down the dimensions and testing the newly built scaled-down-to-size device with frequency domain techniques for possible further refinements. (H. C. Poulter proved the validity of the scaling factor.[25]) Precision coaxial connectors were designed using scaled-up TDR techniques. Strip-line structures and slab-line devices offer excellent applications for using this technique.

C. G. Sontheimer reported how TDR was used in the design of balanced mixers.[26] Figure 6.2-28 shows the schematic of the balanced mixer

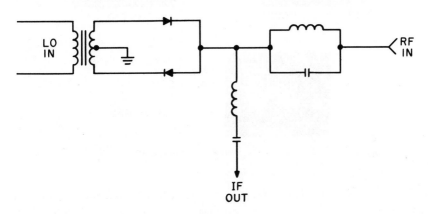

Fig. 6.2-28. Schematics of a balanced mixer.

under consideration. The diodes are ideally open-circuited when back-biased; when conducting, they provide a good short circuit to ground. Both circuits are tuned to be resonant at the IF. The series circuit prevents signal power from being dissipated in the IF load.

The parallel circuit blocks transmission of IF energy to the signal-source impedance. Except for diode resistance and capacity, the principal factors limiting the performance of a practical broadband mixer are leakage inductance in the transformer secondary, which adds to the resistances of the diodes in the conducting state, and reflections generated in the signal input path by the two tuned circuits.

Figure 6.2-29 illustrates a use of TDR in the evaluation of balanced-mixer designs. Each waveform shows the reflections from the signal port of a mixer with the diodes alternately back- and forward-biased. The two upper waveforms were taken with a commercial mixer covering the range of 10 MHz to 1 GHz. In one case, 1N 831 diodes were used; in another case, hot-carrier diodes were used. The greater on-off impedance variation obtained with hot-carrier diodes is evident.

[25] Poulter, H. C., "Mechanical Scaling Enhances Time Domain Reflectometry Use," paper presented at WESCON 1965 (Hewlett-Packard Application Note 75).

[26] Sontheimer, C. G., "Some Uses of Time Domain Reflectometry (TDR) in the Design of Broadband UHF Components," Anzac Electronics, Inc., Norwalk, Conn. (Hewlett-Packard Application Note 75).

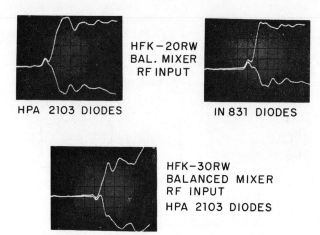

HFK−20RW
BAL. MIXER
RF INPUT

HPA 2103 DIODES   IN 831 DIODES

HFK−30RW
BALANCED MIXER
RF INPUT
HPA 2103 DIODES

Fig. 6.2-29. Reflections from signal of balanced mixer.

The waveform also shows an inductive reflection immediately preceding the diodes. This observation caused design effort to be concentrated on the signal input circuits, with the result shown in the lower waveform. In this mixer the inductive discontinuity has been greatly reduced, and low conversion loss is obtained over the range from a few megahertz to 2.4 GHz.

TDR equipment has also proved useful in efforts to improve LO-to-signal isolation. This is necessary for local oscillator noise reduction and low noise figure. Transmission, rather than reflection, is used in this case, as shown in Fig. 6.2-30. The purpose of the 10 dB pad is to reduce the diode

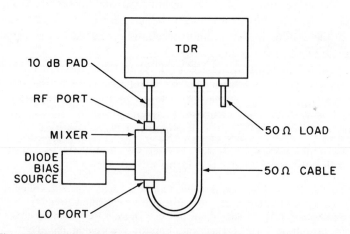

TDR

10 dB PAD

RF PORT

MIXER   50 Ω LOAD

DIODE
BIAS
SOURCE   50 Ω CABLE

LO PORT

Fig. 6.2-30. Block diagram of isolation test setup.

current caused by the incident pulse to an order of magnitude below the diode-biasing current.

Typical waveforms are shown in Fig. 6.2-31. The upper trace was made

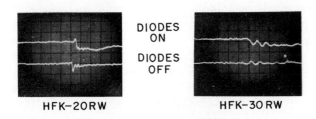

e =.02/cm THROUGH 10dB PAD

Fig. 6.2-31. Transmission from local oscillator to
signal port in balanced mixer.

with the diodes back-biased and the lower trace with the diodes conducting. Very nearly identical traces were obtained with different sets of diodes, indicating that the feedthrough is due to capacitive and inductive unbalance in the transformer driving the diodes. The left-hand picture was obtained with the 10 to 1,000 MHz mixer referred to earlier. Using such pictures as a guide, work was concentrated on the transformer. The right-hand picture in Fig. 6.2-31 shows the results achieved in a new mixer with an upper-frequency limit of 2,400 MHz. Conventional measurements show isolation improved by 10 to 15 dB.

QUESTIONS

1. What is the minimum separation of a TDR system if the system rise time is 50 ms in an air-dielectric transmission line?
2. How far from the test port is a discontinuity in a transmission line filled with a dielectric material having $\epsilon_r = 2.54$ if the scope display shows a distance of 2.66 ns from the step to the discontinuity?
3. How far will the TDR display of a resistive termination shift up or down in terms of E_i in a 50-ohm characteristic impedance system if the termination is 41 ohms?
4. What is the value of a parallel capacitance on a transmission line if the TDR display shows a deviation of $e_{r_{max}} = 28$ mv? System rise time information is $m = 3$ mv/ps in a 50-ohm characteristic impedance system.
5. What is the rise time of a cable if the reflected pulse rise time is 108 ps and the incident pulse rise time is 28 ps?

6.3 NOISE FIGURE MEASUREMENTS[27]

The weakest signal that can be detected in microwave communications is usually determined by the amount of noise added by the receiving system. Consequently any decrease in the amount of noise generated in the receiving system will produce an increase in the output signal-to-noise ratio equivalent to a corresponding increase in received signal. From a standpoint of performance, increasing the signal-to-noise ratio by reducing the amount of noise in the receiver is more economical than increasing the received signal level by raising the power of the transmitter. For example, a decrease of 5 dB in the receiver noise is equivalent to a 3:1 increase in the transmitter power. The ultimate sensitivity of a detection system is set by the noise presented to the system with the signal. In addition, many detection systems have additional noise in the detection and amplification process. Since the noise contribution of the detection system is usually the larger of these two, the level of input noise is generally beyond control. The approach is to study, measure, and attempt to minimize the noise contribution of the detection system. A microwave input termination has a certain amount of available noise power which it can deliver to a matched system.

To establish definitions, it is wise first to consider lower frequencies, where it is possible to work with lumped constants. A resistance R at temperature T generates a voltage across its open-circuited terminals resulting from the random motion of free electrons thermally agitated. This noise voltage e_n is infinitely broadbanded and can be defined by the equation

$$\frac{e_n^2}{B} = 4kTR \ (\text{volts})^2/\text{unit frequency bandwidth} \qquad (6.3\text{-}1)$$

where k = Boltzmann's constant (1.374×10^{-23} joule/°Kelvin),
$\quad\ \ T$ = absolute temperature (°Kelvin),
$\quad\ \ R$ = resistance (or resistive component of impedance), and
$\quad\ \ B$ = bandwidth.

Generally, noise considerations will be concerned with a finite bandwidth, and the more familiar notation of the equation might be used:

$$e_n^2 = 4kTRB \qquad (6.3\text{-}2)$$

If the resistance R is connected to a matched load resistance ($R_L = R$), maximum transfer of the noise power will occur. Noise power P_N dissipated in the load resistance R_L due to the noise voltage generated in the original resistance R will have the value

$$P_n = kTB \qquad (6.3\text{-}3)$$

Equation (6.3-3) defines the available noise power from the original resistance.

[27] "Noise Figure Primer," Hewlett-Packard Application Note 57.

The actual noise power dissipated in the load resistance could be affected by any loss in the connecting leads, a less than perfect match to the original resistance, and the noise power generated within the load resistance itself. In systems operating at frequencies where voltages and resistances cannot be clearly defined, Eq. (6.3-3) becomes the usable expression containing terms that can be measured. In certain low-noise applications, a deliberate and carefully determined mismatch is created between the input terminations and the detection device. This technique couples less than the available termination noise power (kTB) into the detection device. However, this system consideration is beyond the scope of this study. In a microwave receiver, the input termination is antenna-coupled to the atmosphere; in an IF strip, the input termination is generally a mixer of some sort. In either case the termination has an available noise power given by $P_n = kTB$. If there were a perfect amplifier or receiver which added no noise in the amplification process, and if it were perfectly matched to its input termination, its output noise power would be $kTBG$, where G is the power gain of the system. The figure of merit for an actual microwave receiver or IF amplifier is the ratio of actual output noise power where $T = 290°$ to the theoretical minimum. Noise figure F is this figure of merit referred to room temperature ($T = 290°K$). Thus

$$F = \frac{N_1}{kT_0BG} \qquad (6.3\text{-}4)$$

where N_1 is the measured noise power output to the system when $T_0 = 290°K$. Of course, the perfect amplifier would have to have a noise figure of 1 (0 dB). Of the total noise power output of a system ($N_1 = kT_0BGF$), it is known that a specific portion is the result of amplified input noise (kT_0BG). The amount of noise power added by the receiver (N_r) is the difference, or

$$N_r = N_1 - kT_0BG = (F - 1)kT_0BG \qquad (6.3\text{-}5)$$

Signal Generator–Power Meter Method

Noise figure ($F = N_1/kT_0BG$) can be calculated from measurements taken by a stable, well-attenuated signal generator and suitable power meter. We can measure T, which is the temperature of the input termination in °K; N_1 can be measured with a power meter at the system output.

To determine the gain bandwidth product BG accurately, it is necessary to plot the response of the system and integrate the curve graphically—a time-consuming process. Approximations can be made on the basis of the 3-dB bandwidth, which shortens the process. After measuring the 3-dB points, there are graphs that may be used to obtain an approximation of the effective noise bandwidth. Gain, of course, may be easily measured with a proper signal generator and power meter.

From these data the noise figure may be calculated. Often, however, this is only a first step, because the primary objective is to minimize the noise figure by repositioning components, by substituting crystals or tubes, or by tuning filter networks. Clearly, the many-step process represented by the signal generator–power meter method has limited value in such a situation, since a new measurement must be made at each readjustment.

The Noise Source as a Broadband Signal Generator

If a known level of broadband noise can be introduced at the input of a device under test, a differential power measurement at the output would indicate a gain bandwidth product of the device.

A gas-discharge noise source operates as an input termination at a very high temperature and has an available noise power much higher than the normal termination. The effective thermal agitation of an argon tube noise source, for example, represents an equivalent temperature of approximately 10,000°K, compared with room temperature of 290°K.

Available excess power from the fired noise source can be expressed in the same terms as the input termination:

$$P_{ns} = k(T_2 - T_0)B$$

where T_2 is the equivalent absolute temperature of the noise source. Since a measurement of the device output, both with and without the additional noise power input, will give an indication of gain-bandwidth product, it is possible to compute the noise figure with no further measurements, because all independent variables of Eq. (6.3-4) are known.

Using the system of Fig. 6.3-1, consisting of an input termination, an

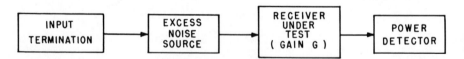

Fig. 6.3-1. A typical noise figure measuring system.

excess-noise source, a receiver under test, and an output power detector, it is possible to measure N_1 with the excess-noise source "cold" and to measure N_2 with the excess-noise source fired. N_1 is graphically illustrated in Fig. 6.3-2(a) and consists of the amplified input termination noise plus the noise generated within the receiver. N_2, illustrated in Fig. 6.3-2(b), consists of N_1 plus the amplified excess-noise power viewed at the receiver output.

Taking the ratio of these measured powers, we have:

$$\frac{N_2}{N_1} = \frac{(\text{input termination}) \times G + (\text{receiver}) + \text{excess noise} \times G}{(\text{input termination}) \times G + (\text{receiver})}$$

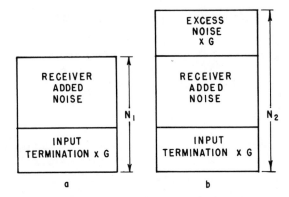

Fig. 6.3-2. Representation of total noise power output
   of the system in Fig. 6.3-1 when (a) the excess noise source is "cold" (N_1) and (b) the
   excess noise source is "fired" (N_2).

Substituting from previous equations and assuming measurement conditions
of $T = T_0 = 290°K$,

$$\frac{N_2}{N_1} = \frac{kT_0BG + (F - 1)kT_0BG + k(T_2 - T_0)BG}{kT_0BG + (F - 1)kT_0BG}$$

$$= \frac{FT_0 + (T_2 - T_0)}{FT_0}$$

and, finally,

$$F = \frac{(T_2 - T_0)}{T_0} \times \frac{1}{\left(\dfrac{N_2}{N_1} - 1\right)}$$

Converting to logarithmic notation,

$$F_{dB} = 10 \log \frac{(T_2 - T_0)}{T_0} - 10 \log \left(\frac{N_2}{N_1} - 1\right) \qquad (6.3\text{-}6)$$

In Eq. (6.3-6), the ratio $(T_2 - T_0)/T_0$ is a measure of the relative excess-noise
power available from a noise source and is specified by the manufacturer.
In the case of argon gas tubes, this ratio is 33:1; 10 log $(T_2 - T_0)/T_0$ is 15.2
dB. When using such a tube, Eq. (6.3-6) simplifies to

$$F_{dB} = 15.2 - 10 \log \left(\frac{N_2}{N_1} - 1\right)$$

Noise Figure Measurement with an Excess-Noise Source

Equation (6.3-6) opens the door to several measurement techniques
utilizing the excess-noise source. We shall consider the "twice-power" and

"Y-factor" manual techniques and an automatic approach to noise figure measurements.

TWICE-POWER METHOD OF MANUAL NOISE FIGURE MEASUREMENT. In actually measuring the "N_1" and "N_2" of Eq. (6.3-6), if N_2 was set to be twice N_1, then Eq. (6.3-6) would reduce to

$$F_{dB} = 10 \log \frac{(T_2 - T_0)}{T_0} - 10 \log (1) = 10 \log \frac{(T_2 - T_0)}{T_0}$$

With the proper equipment, the condition of $N_2 = 2N_1$ can be established by varying the relative excess-noise power of the noise source. With the equipment of Fig. 6.3-3, the procedure would be as follows:

1. Set a convenient reference on the power detector with the excess-noise source "cold" and the 3-dB pad out. This is N_1.
2. Insert the 3-dB pad and fire the excess-noise source.
3. Vary the rotary vane attenuator until the original power detector reference point is reached. This creates a condition of $N_2 = 2N_1$.

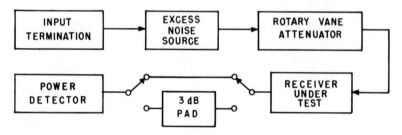

Fig. 6.3-3. The "twice-power" method of manual
noise figure measurement.

Figure 6.3-4 illustrates this condition, in which the output noise power contributed by the excess-noise source exactly equals the sum of the amplified input termination noise plus the receiver noise contribution. Since this excess-noise ratio was adjusted with the attenuator equal to input termination noise plus receiver noise (thereby causing N_2 to equal $2N_1$), it can be seen from Eq. (6.3-6) that the attenuated excess-noise ratio is equal to the noise figure of the receiver. In the case of an argon source, it can be read as 15.2 dB minus the attenuator setting (in dB).

Although the attenuator reduces the amount of excess noise injected into the system, it has no effect on input termination noise power if the termination and attenuator are at the same temperature. This is because, regardless of the amount of attenuation, when the excess-noise source is cold, the receiver input is still looking at a matched input at temperature T.

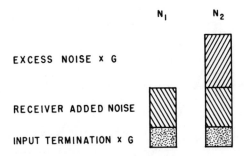

Fig. 6.3-4. Representation of total noise power output for the "twice-power" method of manual noise figure measurement.

Y-FACTOR METHOD OF NOISE FIGURE MEASUREMENT. A method closely resembling the "twice-power" method involves the determination of the numerical ratio N_2/N_1 (which is called Y-factor) and the calculation of noise figure by substitution in Eq. (6.3-6).

In practice, the Y-factor method generally makes use of an IF attenuator with a power indicator set to a convenient reference. The IF attenuator change in going from source OFF to source ON then yields the Y-factor, which is then entered in the equation. Graphs are also available for specific values of relative excess noise, with coordinates calibrated in "Y-factor" and noise figure (see Fig. 6.3-5).

AUTOMATIC NOISE FIGURE MEASUREMENTS. Although manual measurements yield valid results, they represent a tedious process and are often not easily accomplished by unskilled personnel. In addition, since there is no continuous indication of noise figure, the work of reducing the noise figure is considerably slowed by the necessity of new measurement after each circuit change.

To fill the need for a direct-reading, continuously indicating noise figure meter, at least three automatic systems have been devised. All depend on the periodic insertion of known excess noise into the system. This results in a pulse train of two pulse levels, N_2 and N_1. The pulse train typically is amplified in an IF strip and is then separated into two distinct levels by selective gating. These levels, together with the amount of excess-noise insertion, contain the information needed to indicate directly the noise figure on a meter face.

The three automatic systems differ in their method of indication. One approach uses a special ratio-resolving meter movement that responds to the ratio of the two signal levels in a manner similar to a wattmeter movement. However, such a meter movement is quite expensive, and in general it has not achieved wide acceptance in this country.

Another method uses AGC in its IF amplifier to hold the value

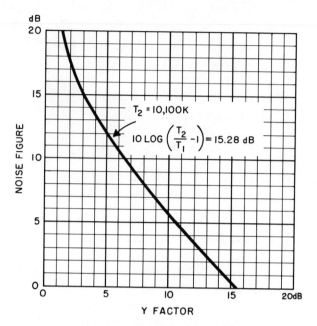

Fig. 6.3-5. A "Y-factor" chart for determining noise figure with a 15.28-dB excess noise source.

$(N_2 + N_1)/2$ constant and measures $N_2 - N_1$. This contains the necessary information to measure the noise figure and is especially useful in very-high-noise-figure cases, where N_1 is close in value to N_2. In such cases the difference of the levels is more definitive than their ratio.

A third approach actually measures the ratio N_2/N_1 and displays this ratio on a meter face calibrated by the equation

$$F_{dB} = 15.2 - 10 \log \left(\frac{N_2}{N_1} - 1 \right)$$

Such an instrument is shown in simplified block-diagram form in Fig. 6.3-6.

In operation, the gating source pulses the noise source at a rate of 500 hertz; N_1 and N_2 pulses arrive at the IF amplifier. Noise sources have a finite noise buildup time, so the IF amplifier is gated to pass only the final amplitudes of N_1 and N_2 to the square-law detector. The detected N_2 pulse is switched to an AGC integrator, where a voltage for gain control of the IF amplifier is derived. The time constant of this circuit is made long enough to control the IF amplifier gain during the N_1 pulse. Measurement of the N_1 pulse is in effect a measurement of the pulse ratio. The N_1 pulse is measured by switching it to the meter integrator and meter.

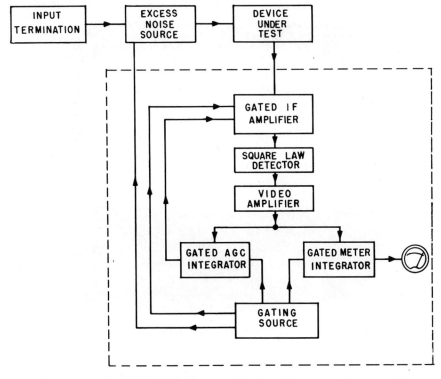

Fig. 6.3-6. Simplified block diagram of automatic
noise figure measurement system.

Convenient internal calibration of the meter is accomplished by ar-
tificially creating readings of "$+\infty$" and "$-\infty$." By pulsing the noise source
during both the N_2 and N_1 time periods, we obtain a condition of $N_2 = N_1$.
In the equation $F_{dB} = 15.2 - 10 \log (N_2/N_1 - 1)$, this condition results in
a noise figure of $+\infty$. The artificial condition of $F = -\infty$ would correspond
to an "N_1" value of "0." This can be created by gating "off" the IF amplifier
during the "N_1" time period. If the metering circuit is designed to be a linear
indicator of the power of "N_1" (square-law detector) and the meter minimum
position is calibrated as $-\infty$ and the full-scale deflection as $+\infty$, all other
points on the meter face can be calculated by the equation $F_{dB} = 15.2 -
10 \log (N_2/N_1 - 1)$. For example, an "$N_1/N_2$" ratio of $\frac{1}{2}$ would bring about
a midscale reading. From the equation this midscale reading is calculated
to be 15.2. The balance of the scale is calibrated in a similar fashion.

Figure 6.3-7 shows an actual meter face from an automatic noise figure
meter. In addition to calibration for 15.2-dB excess-noise sources, it is cali-

Fig. 6.3-7. Meter face from an automatic noise figure
meter.

brated for 5.2-dB temperature-limited diode sources. The linear current scale
is used in adjusting noise source excitation current.

Networks in Cascade

The effects of the noise contribution of networks in cascade can be seen
in Fig. 6.3-8. This illustration shows the input termination and three networks, each with gain G and noise figure F. The power graphs assume that
each network is active with gain greater than 1. However, the analysis will
be equally valid for passive networks with gain less than 1.

The input termination supplies a noise power kTB which is amplified
by the three networks and appears at the output as $kTBG_1G_2G_3$. The noise
contribution of network 1, by Eq. (6.3-5), is $(F_1 - 1)kTBG_1$. When further
amplified by networks 2 and 3, the noise power appears at the output as
$(F_1 - 1)kTG_1G_2G_3$. Similarly, the output noise contributed by network 2 is
$(F_2 - 1)kTBG_2G_3$; and from network 3, $(F_3 - 1)kTBG_3$. The system noise
figure (F_s) is the ratio of actual noise power output to noise power output
contributed by the input termination.

$$F_s = \frac{\text{Total noise output}}{kTBG_1G_2G_3}$$

$$= \frac{kTBG_1G_2G_3 + (F_1 - 1)kTBG_1G_2G_3}{kTBG_1G_2G_3} + \frac{(F_2 - 1)kTBG_2G_3 + (F_3 - 1)kTBG_3}{kTBG_1G_2G_3}$$

$$= F_1 + \frac{(F_2 - 1)}{G_1} + \frac{(F_3 - 1)}{G_1G_2}$$

The general equation for the noise figure of networks in cascade is thus

$$F_s = F_1 + \frac{(F_2 - 1)}{G_1} + \frac{(F_3 - 1)}{G_1G_2} + \cdots + \frac{(F_n - 1)}{G_1G_3 \ldots G_{(n-1)}}$$

It can be seen that the overall noise figure of any cascaded amplifying system depends primarily on that of the first stage. The effects of subsequent stages are reduced by the gain up to that point. On the other hand, the use of a

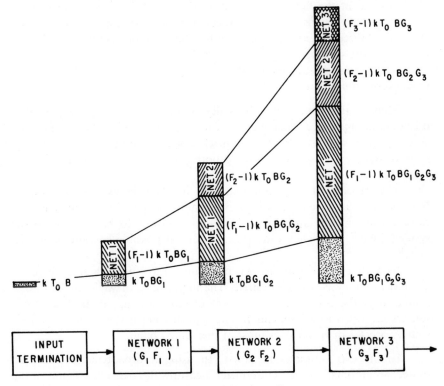

Fig. 6.3-8. The effects of noise contribution of
networks in cascade.

passive stage with gain less than 1 increases the importance of the subsequent stage's noise figure.

Accuracy Considerations

Up to this point we have deliberately ignored several possible sources
of error. Actual measurement techniques must consider the possibility of
system errors caused by mismatch, temperature, and image and spurious
responses. The instrument accuracy of the noise source and noise figure meter
should also be considered.

TEMPERATURE. In the derivation of Eq. (6.3-6),

$$F_{dB} = 10 \log \frac{(T_2 - T_0)}{T_0} - 10 \log \left(\frac{N_2}{N_1} - 1 \right)$$

the ambient temperature was assumed to be 290°; hence T_0 canceled out of
all terms except the figure "$10 \log (T_2 - T_0)/T_0$." T_2 is the equivalent fired
temperature of the noise source. In specifying the relative excess-noise power
of a noise source, the manufacturer knows the value of T_2 and rates the tube
in terms of the standard temperature, 290°K (62.6°F). For example, a variation of 20° from the assumed 290° would cause an error of about 0.3 dB
in measured noise figure.

MISMATCH. Noise power obeys all power transfer laws. However, since
it is random in phase, mismatches cause ambiguous errors rather than known
amounts of power loss. In the automatic noise figure meter measuring the
ratio of N_2/N_1, a mismatch affecting both pulses does not affect accuracy,
since the ratio remains unchanged.

The critical matching situation, then, involves the excess-noise source.
Noise sources are rated in *available* excess-noise power; thus mismatches will
cause an ambiguous amount of excess-noise power to be coupled to the
system. Figure 6.3-9 shows the effect of several possible conditions of mismatch.

Note that the possible error is maximum at low-noise figures where the
greatest accuracy is usually desired. The importance of well-matched noise
sources over the entire frequency range of interest is apparent. Waveguide
sources are rated as 1.2 maximum and typically are better than 1.1. Diode
sources are 1.3 or better over most of their rated band.

IMAGE AND SPURIOUS RESPONSE. In using a broadband excess-noise
source, an automatic noise figure meter measures the true noise figure of the
total passband of the device under test. If, in its operation, the device does
not utilize the full passband for signal information (as would be the case of a
radar receiver with an image response), its operating noise figure will be

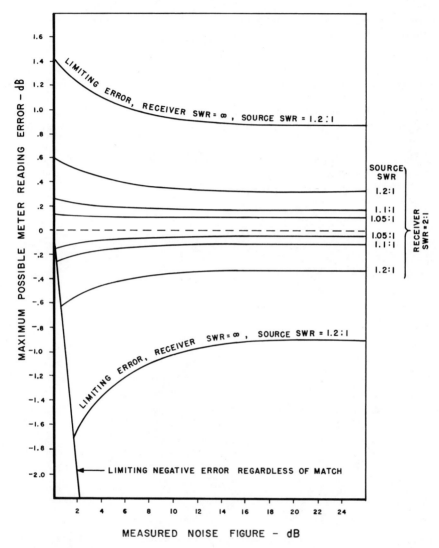

Fig. 6.3-9. Typical error effects for several possible conditions of mismatch between noise source and receiver.

higher than the measured noise figure. This apparent noise figure can be calculated by the equation

$$\text{Operating } F_{\text{dB}} = F_{\text{dB}} \text{ (reading)} + 10 \log \left(B_t / B_u \right)$$

where B_t is the total passband of the device and B_u is the operational passband. This equation is a convenient simplification that assumes constant gain in the device under test.

QUESTIONS

1. What is the noise figure in dB of an amplifier if the measured noise power output at room temperature is 0.05 mw and the equivalent gain-bandwidth product is 5.2×10^{15}?
2. Using an argon gas tube noise source, determine the noise figure in dB for the following cases: (a) 1.8 mw power output with noise source on; (b) 0.08 mw power output with noise source off.

INDEX

A

Adam, Stephen F., 262, 356, 403, 445
Admittance and impedance transmission line relationship, 25–34
Ammonia maser, 201
Amplifier, traveling-wave tube, 118–24
Amplitude-modulated signals, Fourier spectra, 276–78
Anderson, R. W., 417
Atoms, simulated, in lasers, 170–72
Attenuation, 351–53
    attenuators, 353–71
    in coaxial transmission lines, 39–41
    sources, 18
    in waveguide transmission lines, 64–67
Attenuation measurements, 99–101, 351–53, 371–81
    audio frequency substitution technique, 372
    calibration method, 371–72
    dc substitution technique, 372
    errors, 374–81
    IF substitution technique, 372–74
    RF substitution techniques, 371–72
Attenuators, 353–71
    coaxial, 353–61
    cutoff, 358
    distributed thin-film, 356–57
    fixed-waveguide, 363
    flap, 361
    lossy-line, 355
    nonreciprocal, 367–71
    pi circuits, 354–55
    piston, 358–59
    rotary-vane, 362–63
    side-vane, 361–62
    step, 357–58, 360–61
    "T" circuits, 353–54
    variable coaxial, construction, 357–60
    variable waveguide, construction, 366–67
    for waveguide transmission lines, 361–67
Atwater, H. A., 71
Audio frequency substitution technique of attenuation measurement, 372, 379–81
Automatic leveling system, 142–45

B

Backward-wave oscillator, 125–34
    amplification mechanism, 128–34
Badger, Anthony S., 262

Barretters, 215
Beads and stubs, in coaxial transmission lines, 45–49
Beatty, R. W., 232, 352
"Black box," 86–87
Bolometers, 215
Bolometer mounts, 219–24
    broadband coaxial thermistor mount, 221–22
    tunable waveguide bolometer mounts, 223–24
    waveguide bolometer mounts, 222–24
Bolometric power measurements, 214–29
    accuracy, 229–44
    bolometer mounts, 219–24
    dc-to-microwave substitution error, 241
    dual-element mount error, 242–43
    instrumentation error, 242
    mismatch errors, 231–40
    power meter bridge, 215–19
    RF losses, 240
    temperature-compensated power meters, 224–27
    temperature-compensated thermistor mounts, 227–29
    thermoelectric effect error, 241–42
Branch elimination, 95

C

Calorimeter, dry, 247
    flow, 244–47
Calorimetric power measurements, 244–47
    dry calorimeters, 247
    flow calorimeters, 244–47
    microcalorimeter design, 247–48
Calorimetric power meters, 244–47
    accuracy, 246
    head efficiency, 246
    instrumentation error, 247
    mismatch error, 246
Carlson, Rod, 327
Carter, P. S., Jr., 162
Cavity Q measurements, 271
Cesium beam standards, 201, 202–3
Coaxial attenuators, 353–61
    variable, construction, 357–60
Coaxial transmission line resonators, in frequency measurement, 261–63
Coaxial transmission lines, 35–51
    attenuation, 39–41

505